Communications
in Computer and Information Science 431

T0236359

Andrzej Kwiecień   Piotr Gaj
Piotr Stera (Eds.)

Computer Networks

21st International Conference, CN 2014
Brunów, Poland, June 23-27, 2014
Proceedings

 Springer

Volume Editors

Andrzej Kwiecień
Piotr Gaj
Piotr Stera

Silesian University of Technology
Institute of Informatics
ul. Akademicka 16
44-100 Gliwice, Poland

E-mail: {andrzej.kwiecien, piotr.gaj, piotr.stera}@polsl.pl

ISSN 1865-0929   e-ISSN 1865-0937
ISBN 978-3-319-07940-0   e-ISBN 978-3-319-07941-7
DOI 10.1007/978-3-319-07941-7
Springer Cham Heidelberg New York Dordrecht London

Library of Congress Control Number: 2014940788

Typesetting: Camera-ready by author, data conversion by Scientific Publishing Services, Chennai, India

Printed on acid-free paper

Springer is part of Springer Science+Business Media (www.springer.com)

Preface

The ongoing and intense development of computer science influences the fast growth of computer networks. Computer networks, as well as the entire area of computer science, are the subjects of regular changes resulting from general technological advancement and the development of new IT technologies. Novel methods, together with tools for designing and modeling computer networks, are regularly extended. The essential issue is the fact that the scope of their application is increased thanks to the results of new research and new application proposals that regularly appear. Solutions of this kind were not even taken into consideration in the past decades. These recent applications stimulate the progress of scientific research. This book is the result of these considerations described. It includes contributions that refer to a broad spectrum of contemporary issues and problems.

21st International Science Conference
Computer Networks

The book collates the work of scientists from numerous notable research centers. It was created as a collection of papers presented on the Computer Networks (CN) 2014 International Conference in Brunów Palace. The venue was a small village called Brunów located near Lwówek Śląski (Poland). The event took place during June 23–27, 2014. The conference, which has been organized regularly since 1994 by the Institute of Informatics from the Faculty of Automatic Control, Electronics, and Computer Science of Silesian University of Technology in Gliwice, is the oldest event related to computer networks in Poland, and one of the first events of this kind. Moreover, it attained international status six years ago. The current edition was the 21st edition in a row and the seventh international one. This year, like last year, the conference took place under the auspices of IEEE (technical co-sponsor). Moreover, the iNEER (the International Network for Engineering Education and Research) became a conference partner. As in previous years, the review of each of the papers was double-blind, performed by at least two reviewers. The contents of this book are organized thematically, including the following domains:

− Computer Networks
− Teleinformatics and Communications
− New Technologies
− Queueing Theory
− Innovative Applications
− Networked and IT-Related Aspects of e-Business

Each domain includes very stimulating studies that should be of interest to a wide readership.

To sum up, on behalf of the Program Committee we would like to express our appreciation to all the authors for sharing their research results and for their assistance in creating the proceedings, which in our judgment are a reliable reference on computer networks. We also want to thank the members of the international Program Committee for their participation in the reviewing process.

April 2014   Andrzej Kwiecień
                                                                    Piotr Gaj

Organization

CN 2014 was organized by the Institute of Informatics from the Faculty of Automatic Control, Electronics and Computer Science, Silesian University of Technology (SUT) and supported by the Committee of Informatics of the Polish Academy of Sciences (PAN), Section of Computer Network and Distributed Systems in technical cooperation with the IEEE and consulting support of iNEER organization.

Executive Committee

All members of the Executing Committee are from the Silesian University of Technology, Poland.

Honorary Member:	Halina Węgrzyn
Organizing Chair:	Piotr Gaj
Technical Volume Editor:	Piotr Stera
Technical Support:	Aleksander Cisek
Technical Support:	Jacek Stój
Office:	Małgorzata Gładysz
Web Support:	Piotr Kuźniacki
PAN Coordinator:	Tadeusz Czachórski
IEEE PS Coordinator:	Jacek Izydorczyk
iNEER Coordinator:	Win Aung

Program Committee

Program Chair

Andrzej Kwiecień — Silesian University of Technology, Poland

Honorary Members

Win Aung	iNEER, USA
Klaus Bender	TU München, Germany
Adam Czornik	Silesian University of Technology, Poland
Andrzej Karbownik	Silesian University of Technology, Poland
Bogdan M. Wilamowski	Auburn University, USA

Technical Program Committee

Anoosh Abdy	Realm Information Technologies, USA
Iosif Androulidakis	University of Ioannina, Greece
Tülin Atmaca	Institut National de Télécommunication, France
Rajiv Bagai	Wichita State University, USA
Zbigniew Banaszak	Warsaw University of Technology, Poland
Leszek Borzemski	Wrocław University of Technology, Poland
Markus Bregulla	University of Applied Sciences Ingolstadt, Germany
Tadeusz Czachórski	Silesian University of Technology, Poland
Andrzej Duda	INP Grenoble, France
Alexander N. Dudin	Belarusian State University, Belarus
Max Felser	Bern University of Applied Sciences, Switzerland
Holger Flatt	Fraunhofer IOSB-INA, Germany
Jean-Michel Fourneau	Versailles University, France
Rosario G. Garroppo	University of Pisa, Italy
Natalia Gaviria	Universidad de Antioquia, Colombia
Erol Gelenbe	Imperial College, UK
Roman Gielerak	University of Zielona Góra, Poland
Adam Grzech	Wrocław University of Technology, Poland
Edward Hrynkiewicz	Silesian University of Technology, Poland
Zbigniew Huzar	Wrocław University of Technology, Poland
Jacek Izydorczyk	Silesian University of Technology, Poland
Jürgen Jasperneite	Ostwestfalen-Lippe University of Applied Sciences, Germany
Jerzy Klamka	IITiS Polish Academy of Sciences, Gliwice, Poland
Demetres D. Kouvatsos	University of Bradford, UK
Stanisław Kozielski	Silesian University of Technology, Poland
Henryk Krawczyk	Gdańsk University of Technology, Poland
Wolfgang Mahnke	ABB, Germany
Francesco Malandrino	Politecnico di Torino, Italy
Kevin M. McNeil	BAE Systems, USA
Michele Pagano	University of Pisa, Italy
Nihal Pekergin	Université de Paris, France
Piotr Pikiewicz	College of Business in Dąbrowa Górnicza, Poland
Jacek Piskorowski	West Pomeranian University of Technology, Poland
Bolesław Pochopień	Silesian University of Technology, Poland
Oksana Pomorova	Khmelnitsky National University, Ukraine
Silvana Rodrigues	Integrated Device Technology, Canada
Vladimir Rykov	Russian State Oil and Gas University, Russia

Akash Singh	IBM Corp, USA
Mirosław Skrzewski	Silesian University of Technology, Poland
Maciej Stasiak	Poznań University of Technology, Poland
Kerry-Lynn Thomson	Nelson Mandela Metropolitan University, South Africa
Oleg Tikhonenko	Częstochowa University of Technology, Poland
Leszek Trybus	Rzeszów University of Technology, Poland
Arnaud Tisserand	IRISA, France
Bane Vasic	University of Arizona, USA
Miroslaw Voznak	VSB-Technical University of Ostrava, Czech Republic
Sylwester Warecki	Peregrine Semiconductor Inc., USA
Tadeusz Wieczorek	Silesian University of Technology, Poland
Józef Woźniak	Gdańsk University of Technology, Poland
Hao Yu	Auburn University, USA
Grzegorz Zaręba	University of Arizona, USA

Additional Reviewers

Iosif Androulidakis	Jürgen Jasperneite	Mirosław Skrzewski
Zbigniew Banaszak	Jerzy Klamka	Maciej Stasiak
Rajiv Bagai	Stanisław Kozielski	Kerry-Lynn Thomson
Leszek Borzemski	Henryk Krawczyk	Oleg Tikhonenko
Tadeusz Czachórski	Andrzej Kwiecień	Leszek Trybus
Jean-Michel Fourneau	Wolfgang Mahnke	Arnaud Tisserand
Rosario G. Garroppo	Francesco Malandrino	Bane Vasic
Natalia Gaviria	Michele Pagano	Miroslaw Voznak
Erol Gelenbe	Piotr Pikiewicz	Tadeusz Wieczorek
Roman Gielerak	Jacek Piskorowski	Józef Woźniak
Adam Grzech	Oksana Pomorova	Hao Yu
Edward Hrynkiewicz	Silvana Rodrigues	Grzegorz Zaręba
Zbigniew Huzar	Vladimir Rykov	
Jacek Izydorczyk	Akash Singh	

Sponsoring Institutions

Organizer: Institute of Informatics, Faculty of Automatic Control, Electronics and Computer Science, Silesian University of Technology
Coorganizer: Committee of Informatics of the Polish Academy of Sciences, Section of Computer Network and Distributed Systems
Technical cosponsor: IEEE Poland Section

Technical Partner

Conference partner: iNEER.

Table of Contents

Efficient Packet Selection for Deflection Routing

Loubna Echabbi[1], Jean-Michel Fourneau[2], and Franck Quessette[2]

[1] INPT, Rabat, Morocco
[2] PRiSM, Université de Versailles-Saint-Quentin,
CNRS UMR 8144, Versailles, France

Abstract. Deflection routing is the usual algorithm proposed for all-optical packet networks. We study the selection part of this algorithm: how to choose the packets which must be misdirected when resources are not sufficient. Using graph and algorithmic arguments, we prove some optimal and fast selection techniques which can be easily implemented.

Keywords: deflection routing, optical packet networks.

1 Introduction

Due to the high bandwidth they could offer, all optical networks have received considerable attention during the last years (see for instance [1] and references therein). Optical Packet Switching is the most promising technology because it can manage flows of packets with a granularity smaller than a wavelength. One of the major drawback of this technology is the lack of large buffers which can be used to store packets waiting for a free link during the routing. Therefore routing algorithms are quite different of the algorithms designed for store and forward networks based on electronic buffers. Several packet routing strategies without intermediate storage of data packets have been designed in the literature [2,3] and deflection routing [4] is clearly the simplest solution proposed so far.

In shortest-path Deflection Routing, switches attempt to forward packets along the shortest hop path to their destinations. Each link can send a finite number of packets per time-slot (the link capacity). No packets are queued. At each slot, incoming packets have to be sent immediately to their next switch along the path. If the number of packets requesting a link is larger than the link capacity, then there is some contention. Only some of the packets will receive the link they ask for and the other ones have to be misdirected or deflected. Thus, deflected packets have to travel through longer paths to their destination. These routing algorithms are known to clearly avoid deadlocks but livelocks could occur (packets move but never reach their destination). Thus the average end to end delay is not sufficient to analyze the performance. The main question is the number of packets which are heavily deflected and which cannot arrive on time at destination. Simulations show that typically more than 1 % of the packets may suffer from an extremely large number of deflections [5]. Note that this may also trigger a domino effect: due to old packets in the network, new packets are not allowed to enter or experience a larger number of deflections. Deflection

A. Kwiecień, P. Gaj, and P. Stera (Eds.): CN 2014, CCIS 431, pp. 1–13, 2014.

routing algorithms do not explain how to choose the packets to be misdirected, or even the number of such packets. The selection and the routing decision are local choices without global knowledge and must be made in every node and at each slot. We consider that the packets in a switch are synchronous. Let d be the output degree of a switch (i.e. the number of output channels). Let v the number of links and f the number of wavelengths multiplexed on the link. To model the routing problem, we consider a bipartite directed graph $G = (V_1, V_2, E)$. Nodes of V_1 represent the incoming packets while nodes of V_2 represent outputs. The edges represent the directions that a packet may use to follow a shortest path to its destination. Nodes of V_1 may have an output degree between 1 and d because several links can be the starting step of a shortest path. Such a graph will be denoted as a routing configuration (see the left-upper part of Fig. 1). The nodes of V_2 have an arbitrary degree but every node i of V_2 has a capacity $\lambda(i)$. This is the maximal number of packets which may use the output link represented by node i. The selection algorithm has to optimize the number of packets which are sent in a direction they request. If $\lambda(i) = 1$ for all i, the selection of the packets which join their wished output is a matching, and the optimal choice is a maximal matching. Remember that a matching of a bipartite graph is a regular subgraph of degree 1. A polynomial algorithm based on alternating paths to find a maximal matching in a bipartite graph is known for long. Although the complexity of the maximal matching algorithm is polynomial, it does not fulfill the time requirements of the optical switches (a few hundreds of nanoseconds).

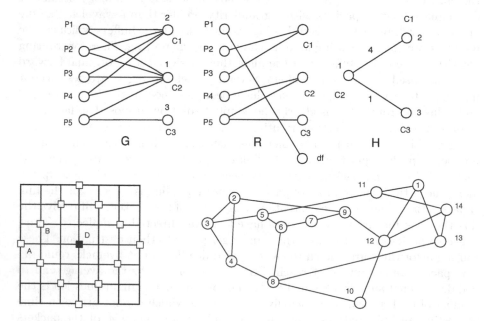

Fig. 1. Routing configuration G with 5 packets and 3 links and capacity vector (upper-left), resulting routing affectation (upper-middle), associated graph H (upper-right), 2D-grid and nodes at distance 3 of D (lower-left), NSFNET (lower-right)

Several heuristics or greedy algorithms have been proposed for assigning packets to links. Recently a fast algorithm was proposed by Quessette and Mneimneh [6,7]. But the assumptions of the algorithm about the topology are not satisfied by some core networks.

In the following, we show that the fast algorithm approach developed on [7] does not give a good answer for some networks topology such as NSFNET. Then we derive some properties for a locally optimal algorithms and we show that their complexity is linear in the number of links. An algorithm is locally optimal if is minimizes the number of deflected packets at each time step and for each switch independently. It is not a globally optimal algorithm for the complete network and the whole time of operation. We also prove that these properties are sufficient to find the optimal solution for NSFNET. We first want to stress how important is the choice of the packets to be deflected. We report some simulation results for the distribution of the delay when we have an optimal number of deflection or a trivial selection algorithm based on random choice. In these simulations, we have considered a grid topology as suggested by the ROM project [8] to model a core network. We also assume that the link capacity is only 1. The main idea behind this assumption is that the network will be based on a multi-wavelength packet format such as in the HOTARU project [9]. Note that this model is also useful for the analysis of networks without wavelength conversion. The size of the grid is 11×11. This is quite large for a core network which usually has between 15 and 30 nodes. In a grid, a packet may have one or two good directions which are the first step of a shortest path. The greedy algorithm proceeds as follows: a packet is chosen randomly among the packets present in the switch and it receives one of the link it requires if it is free. Otherwise the packet is deflected. Then one iterates with the routing process of the next packet.

We report in the following figures the number of deflections experienced by packets rather than the end to end delay to obtain results independent of the physical distance between nodes. For every distribution we have depicted the head and the tail in two separate figures because of the order of magnitude of the probabilities. In the left part of Fig. 2, we have reported the head of the distribution for a locally optimal algorithm and for the greedy algorithm. The tails of the distributions are given in the right part of Fig. 2 using a logarithmic scale. In both figures the traffic matrix is uniform. In the representation of the tails of the distribution, we have used intervals of length 5 rather than a single value. We have obtained similar results for the distribution when the traffic is not uniform. In these experiments, the arrivals are Poisson and the sources are uniform but the destinations are more concentrated on a 5×5 set of nodes at the center of the grid. Again the system is observed at heavy load. Clearly in both examples the optimal algorithm is much more efficient than the greedy one. This is also clear when we compare the average end to end delay. We have computed it for the same traffic matrices. At heavy load, in the hot spot model, the average is 14.6 hops with the greedy while it is only 7.8 hops with the optimal algorithm implemented. We obtain similar variation of results for the other cases as well. The greedy selection of packet to be deflected is not efficient. Of course its complexity is linear in the number of nodes. But many packets are heavily deflected and they are logically lost because they arrive too late.

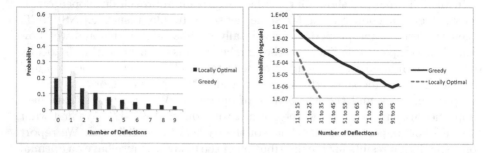

Fig. 2. Head (left) and tail (right) of the distribution of the number of deflection for an uniform traffic at heavy load in an 11×11 grid

The technical part of the paper is organized as follows: Section 2 is devoted to the presentation of MQ approach. In Section 3, we present some rules to obtain an optimal solution. And we conclude with the rules to route in a NSFNET. We only consider one single example here due to lack of space.

2   The Linear Algorithm by Mneimneh and Quessette

Authors in [7] have proposed a locally optimal algorithm for non capacitated networks. The algorithm requires that at each node in the network, there exists at most two outgoing links towards any given destination using only shortest paths. The routing configuration is denoted as a BG_2. The algorithm also requires a more complex condition denoted as "adjacent BG_2" which implies some restrictions on the links that a node may require.

Definition 1. *A BG_2 is a bipartite graph (V_1, V_2, E) such that $d(i) \leq 2$ for all $i \in V_1$. Furthermore, an adjacent BG_2 is a BG_2 graph such that for every j in V_2, j can reach at most two other vertices in V_2 by paths of length 2.*

The first condition implies that the packet set could be partitioned into two types sets: S_i and T_{ij}. Packets in S_i are those requiring only the outgoing i and T_{ij} packets are those requiring either i or j to route to their destination. The second condition implies that each outgoing link i could correspond to only two sets of T type. We can refer to them by $T_{i\cdot}$ and $T_{\cdot i}$. The algorithm suggests to serve packets with respect to the capacity constraint in the following priority order: Serve S type packets first, then serve packets from T_{ij} on outgoing link i (referred to as left advance). Finally, considering remaining packets, serve packets from T_{ij} on outgoing link j (referred to as right advance). The served packets, saturated outgoing links and corresponding incident edges are removed accordingly. The complexity of the algorithm is proved to be linear (one affectation left-advance and one right-advance are always sufficient) and the algorithm is proved to be locally optimal.

Clearly, 2D grids and some other graphs provide routing configurations which are adjacent BG_2. Let us consider for instance a 2D grid associated to a geographical representation of the directions which will be denoted as "North", "East", "South", and "West". First, packets have one or two links used by shortest paths. For instance in the lower left Fig. 1, node A has only one good direction (denoted as "East") to reach D by a shortest path. Similarly B has two favorable output links (South and East) to reach node D. Sets of directions are S_{North}, S_{South}, S_{East} and S_{West} for the first type and $T_{\text{South,West}}$, $T_{\text{South,East}}$, $T_{\text{North,East}}$ and $T_{\text{North,West}}$. Clearly due to the grid topology, a set with directions North and South is useless and no packets will require both directions as first step of a shortest path. We have a similar argument for couple (West, East). Thus the grid provides adjacent BG-2 routing configurations and the algorithm proposed in [7] can be applied.

Let us now turn to a core network topology which is less regular than the grid and which is typically used to study all optical networks the NSFNET topology (see lower right part of Fig. 1). It contains 14 nodes and 22 links. The first condition (i.e. BG_2) is not satisfied for the NSFNET topology since some packets could require more than two outgoing links: for example packets arriving to node 1 that are destinated to node 14 (also from 2 to 6, from 9 to 8, ...). We note by the way that the second condition is not satisfied either. Indeed, if we consider node 8, outgoing link $(8,6)$ is adjacent to $(8,4)$ and $(8,10)$ since they are all on shortest paths towards node 9. It is also adjacent to $(8,13)$ since they are on shortest paths towards node 11. The optimal algorithm by Mneimneh and Quessette cannot be applied to realistic core topologies.

3   How to Obtain an Optimal Number of Deflection

We now study several theoretical aspects of the selection algorithm to improve the performance. In the following, $d(j)$ will denote the degree of node j. Let x be an arbitrary node, $s(x)$ will denote an arbitrary successor or predecessor of x, without confusion as the graph is not directed. Let $\mathcal{S}(x)$ be the set of predecessors of x. If G is a graph and x a node of G, $G - \{x\}$ will denote the graph where node x and edges incident to x have been removed. Similarly, if x and y are nodes, $G + \{(x,y)\}$ will represent the graph where the edge (x,y) has been added. Finally, e_x is a null vector except component indexed by x which is equal to 1, $a^+ = \max(0,a)$ and $|S|$ is the cardinal of an arbitrary set S. We prove some results for the optimal number of deflections. These properties allow to design some simple rules to optimize the delay. The input of the routing algorithm is a routing configuration G and a capacity vector λ (see the upper left part of Fig. 1). The output is a routing affectation graph R which described where the packets are sent. The output graph must satisfy the capacity constraints. An extra node, called df, is added into the routing affectation to represent the packets which are deflected.

Definition 2. *Let G be an arbitrary routing configuration and λ be an arbitrary capacity vector, we denote as $dfln(G, \lambda)$ the minimal number of deflected packets for configuration G and capacity λ. Let $G = (V_1, V_2, E)$ be an arbitrary routing configuration and λ be an arbitrary capacity vector. The output of the selection and routing algorithm is called a routing affectation graph R (see again the upper center part of Fig. 1). It is defined as a graph $R = (V_1, V_2 \bigcup \{df\}, F)$ such that:*

1. *the degree in R of an arbitrary node y in V_2 is smaller than its capacity $\lambda(y)$,*
2. *for each node x in V_1, its degree in R is 1,*
3. *for all arc $(x, y) \in F$ we have $(x, y) \in E$ or $y = df$.*

We also define a locally optimal routing (or optimal routing in the following) as a routing which realizes a minimal number of deflections.

Definition 3 (Union). *Let $G1 = (V1, V2, E)$ and $G2 = (U1, U2, F)$ be two routing configurations. We define the union (denoted as $\cup$) as follows: $G = (W1, W2, H) = G1 \cup G2$ if and only if $V2 = U2 = W2$ and $V1$ and $U1$ are a proper partition of $W1$, and E and F are a proper partition of H.*

When the nodes of V_1 have degree at most 2, we consider a more convenient graph representation.

Definition 4. *Let $G = (V_1, V_2, E)$ be a connected routing configuration such that every node of V_1 has degree two, we build a labelled graph H such as:*

- *the nodes of H are those of V_2, and the label of a node u in H is $\lambda(u)$,*
- *u, v is an edge of H if and only if there exist a node x is V_1 such that $(x, u) \in E$ and $(x, v) \in E$,*
- *the label of an edge (u, v) in H is the number of nodes x in V_1 such that $(x, u) \in E$ and $(x, v) \in E$. This label is denoted as $D(u, v)$.*

Note that several optimal routing affectation graphs may exist with the same deflection number. The algorithms we propose will find one of them. First, we introduce some simple technical lemmas whose proofs are omitted:

Lemma 1 (Connectivity and decomposition). *Let G be a configuration which is composed of two connected components G_1 and G_2. Let λ_1 and λ_2 be the corresponding decomposition for λ, we have: $dfln(G, \lambda) = dfln(G_1, \lambda_1) + dfln(G_2, \lambda_2)$.*

Lemma 2 (Exceeded Capacity). *Let j be an arbitrary node of V_2, if $d(j) \le \lambda(j)$, we can decrease $\lambda(j)$ to $d(j)$ while the optimal number of deflections is kept constant: $dfln(G, \lambda) = dfln(G, \min(\lambda, d))$, where the $\min$ operator is applied componentwise.*

Lemma 3 (Monotonicity). *Let x be an arbitrary node of V_2, we have: $dfln(G, \lambda + e_x) \le dfln(G, \lambda)$. Consider $G(V_1, V_2, E)$ and $G'(V_1, V_2, E')$ two instances of our problem such that $E \subseteq E'$. By induction, we also have $dfln(G', \lambda) \le dfln(G, \lambda)$.*

Let us now prove some results about the effect of a routing decision for a single packet or an increase in the capacity on the number of optimal deflections.

Lemma 4 (Bound). *We know that $dfln(G,\lambda) \leq dfln(G,\lambda-e_y) \leq 1+dfln(G,\lambda)$ for an arbitrary node y of V_2. Similarly, let x be an arbitrary node of V_1, for all y in $\mathcal{S}(x)$, we have $dfln(G,\lambda) \leq dfln(G - \{x\},\lambda - e_y) \leq 1 + dfln(G - \{x\},\lambda)$.*

Proof. The first inequality is simply Lemma 3. To prove the last part of the relation, we consider an optimal solution for configuration G. Let $\lambda(y)$ be the capacity of node y. We have two cases: if node y is saturated (i.e. the number of accepted incoming edges is equal to the capacity). Thus decreasing $\lambda(y)$ by one create a new deflection for a packet which was already accepted in the previous solution. If node y is not saturated, we can decrease $\lambda(y)$ by one without increasing the number of deflections (this is a simple consequence of Lemma 2).

Thus, we have a solution with an upper bound of $1+dfln(G,\lambda)$ deflections. Of course the best solution for configuration G and capacity $\lambda - e_y$ will be smaller than this upper bound. Thus, $dfln(G,\lambda - e_y) \leq 1 + dfln(G,\lambda)$. The second relation comes from the fact that, conditioning on the choice of route for packet x, we have:

$$dfln(G,\lambda) = \min\left(1 + dfln(G - \{x\},\lambda), \min_{y \in \mathcal{S}(x)}(dfln(G - \{x\},\lambda - e_y))\right) .$$

The first term of the r.h.s. describes the effect of a deflection for packet x, while the others terms are associated to routing packet x on output y. Using the first part of the Lemma 4 concludes easily the proof. □

Lemma 5. *Let $G = (V_1, V_2, E)$ be a connected routing configuration such that every node of V_1 has degree 2. Let x be an arbitrary node of V_1 and u and v be the successors of x in V_2. Let $W_{u,v}$ the set of nodes y such that $(y,u) \in E$ and $(y,v) \in E$. Assume capacity vector λ. Without loss of generality we assume that $\lambda(u) \geq \lambda(v)$. If $|W_{u,v}| > \lambda(v)$ then there exists an optimal routing affectation R such that packet x is routed on output u (proof omitted).*

Lemma 6. *Let $G = (V_1, V_2, E)$ be a connected routing configuration such that all the nodes of V_1 has degree 2. Let us build the associated graph H and assume that there exist two nodes (u,v) connected in H such that we have $D(u,v) > \min(\lambda(u), \lambda(v))$. Assume without loss of generality that $\lambda(u) \geq \lambda(v)$. Let $W_{u,v}$ be the subset of V_1 which contains the nodes connected to u and v in G, and let x be an arbitrary node of $W_{u,v}$, then we have: $dfln(G,\lambda) = dfln(G-\{x\},\lambda-e_u)$.*

Proof. We have three choices for the routing of x:

- deflection. We have 1 deflection (i.e. packet x) and the optimal number of deflection for the remaining graph is $dfln(G - \{x\},\lambda)$,
- routing x to u. The deflection number is $dfln(G - \{x\},\lambda - e_u)$,
- routing x to v. The deflection number is $dfln(G - \{x\},\lambda - e_v)$.

Therefore:

$$dfln(G, \lambda) = \min\left(1 + dfln(G - \{x\}, \lambda), dfln(G - \{x\}, \lambda - e_u), dfln(G - \{x\}, \lambda - e_v)\right).$$

Due to Lemma 4, we have:

$$dfln(G, \lambda) = \min\left(dfln(G - \{x\}, \lambda - e_u), dfln(G - \{x\}, \lambda - e_v)\right).$$

As $D(u, v) > \lambda(v)$ and as all the node of V_1 has degree 2 the assumptions of Lemma 5 are satisfied. Thus there exist at least one optimal solution such that the route of x begins with u. Therefore: $dfln(G, \lambda) = dfln(G - \{x\}, \lambda - e_u)$. □

Definition 5 (Broken Wheel). *Let $G = (V_1, V_2, E)$ be a connected routing configuration, G is a broken wheel (see left part of Fig. 3) iff V_2 has m nodes numbered from 1 to m, V_1 is partitioned into $m + 1$ non empty sub-sets denoted as S_0 to S_m and each node in S_0 is connected to nodes 1 to $m - 1$ in V_2, each node in S_i is connected to node i and $i + 1$ in V_2, for $1 \leq i \leq m - 1$, and each node in S_m is connected to node 1 and m in V_2.*

We can now derive several properties of the optimal solutions and some rules.

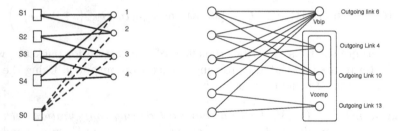

Fig. 3. *G* is a broken wheel (left), Routing configuration at node 8 on NSFNET and the decomposition of set V_2 into V_{bip} and V_{comp}

Lemma 7 (Degree 1 node of V_1). *Let x be a node V_1 such that $d(x) = 1$ and $\lambda(s(x)) > 0$, we have: $dfln(G, \lambda) = dfln(G - \{x\}, \lambda - e_{s(x)})$.*

Proof. We have two cases: we can accept $(x, s(x))$ or deflect node x.

- If we deflect node x, we add one deflection and then we have to consider a modified graph $G - \{x\}$ and the same capacity vector λ. Thus, after this choice, the optimal number of deflections will be equal to $1 + dfln(G - \{x\}, \lambda)$.
- If we decide to route x to $s(x)$, we do not increase the deflection number at this step. The graph becomes $G - \{x\}$ and the capacity vector is now $\lambda - e_{s(x)}$. Thus the optimal number of deflections is equal to $dfln(G - \{x\}, \lambda - e_{s(x)})$.

According to lemma 4, the last value is not larger than the first one. Thus, it is better to accept $(x, s(x))$ and the following rule is proved. □

Rule 1. *if a packet has only one possible direction, one must accept its request to obtain the minimal deflection number.*

Lemma 8 (Node of V_2 without conflict, proof omitted for the sake of conciseness). *Let y be a node of V_2 such that $d(y) \leq \lambda(y)$, we have: $dfln(G, \lambda) = dfln(G - \{s(y)\}, \lambda - e_y)$, where $s(y)$ be an arbitrary predecessor of y.*

Rule 2. *Assume that the configuration graph has a non saturated node in V_2, accept all the request which requires that link.*

These two properties state that we can iterate on accepting packets with one possible destination and packets without conflict. After each accepted request, accepted nodes of V_1 are removed and capacity is decreased. Nodes of V_2 with null capacity are removed and edges incident to removed nodes are also removed. Thus nodes which satisfy Lemmas 7 or 8 may appear. However this is not always sufficient to complete the selection process. We now explore other cases.

Lemma 9 (Ending with two nodes). *Let G be a configuration such that V_2 has only two nodes y and z and all nodes of V_1 are connected to y and z (i.e. $G = K_{|V_1|,2}$). We have: $dfln(G, \lambda) = (|V_1| - \lambda(y) - \lambda(z))^+$. Note that in general we have an inequality as the deflection number is larger than the difference between the number of packets and the number of resources.*

Rule 3. *Assume that the routing configuration graph is a complete bipartite graph (i.e. usually denoted as $K_{n,m}$), it is sufficient to make any random assignment of the n packets on the m links, consistent with the capacity.*

Rule 4. *Assume that the routing configuration $G = K_{n,m} \cup G1$ with capacity λ, then it is optimal to proceed as follows:*

1. *compute the optimal routing for $G1$ with capacity vector λ,*
2. *compute the optimal routing for $K_{n,m}$ with the remaining capacity λ_1,*

and we have: $dfln(G, \lambda) = dfln(G1, \lambda) + dfln(K_{n,m}, \lambda_1)$.

Proof. Such a routing is a feasible solution of the problem. Therefore we have $dfln(G, \lambda) \leq dfln(G1, \lambda) + dfln(K_{n,m}, \lambda_1)$. We now show that it is not possible to improve the solution. As $K_{n,m}$ is a complete bipartite graph the deflection number is $dfln(K_{n,m}, \lambda_1)$ is $(n - \sum_i \lambda_1(i))^+$. We consider two cases:

– if $dfln(K_{n,m}, \lambda_1) = 0$, then it is not possible to improve the solution as the deflections for packets of $G1$ have been obtained without competition with the packets of $K_{n,m}$ and with the initial capacity vector λ,
– if $dfln(K_{n,m}, \lambda_1) > 0$, the optimal solution for the routing of $dfln(G1, \lambda)$ use the same number of resources. Therefore the remaining number of resources (i.e. $\sum_i \lambda_1(i)$) is constant and the number of deflections in the optimal solution for $K_{n,m}$ is not dependent of the optimal solution for $G1$. Moreover if we give one more resource to $K_{n,m}$ after provoking one more deflection

in $G1$, one may decrease at most by one the number of deflections in $K_{n,m}$. Therefore the total number of deflection is, at the best, kept constant. Thus the solution provided by the rule is optimal.   □

Lemma 10 (Generalized Saturation). *Let G be an arbitrary routing configuration with capacity vector λ. Assume that there exists a node h in $V2$ such that for all x, z in $\mathcal{S}(h)$, $\mathcal{S}(x) = \mathcal{S}(z)$. Then it is optimal to begin with the saturation of node h with an arbitrary subset R of $\mathcal{S}(h)$ with size equal to $\min(d(h), \lambda(h))$. Let $G1$ be $G - R$ and $\lambda_1 = \lambda$ except that $\lambda_1(h) = (\lambda(h) - d(h))^+$, we have: $dfln(G, \lambda) = dfln(G1, \lambda_1)$.*

Proof. We begin with the saturation of h and we end with an optimal solution for the remaining configuration and we obtain a feasible solution, thus: $dfln(G, \lambda) \leq dfln(G1, \lambda_1)$. We have routed the maximum number of packets through h and the remaining capacity in h cannot be used by the packets which are not in $\mathcal{S}(h)$. Furthermore the assumption on $\mathcal{S}(x)$ for all x in $\mathcal{S}(h)$ implies that all the packets x in $\mathcal{S}(h)$ are in some sense equivalent and we can choose an arbitrary subset R of packets to be routed. The important feature is to take the maximal possible size without deflections: (i.e. $\min(d(h), \lambda(h))$) to get optimality.   □

We now have to take care of node 8 in NSFNET which exhibits a different configuration. we have to consider a partition of $V_2 = (V_{\mathrm{bip}}, V_{\mathrm{comp}})$ where V_{bip} is the set of nodes that are connected to all packets in V_1 and V_{comp} contains the remaining ones (see right part of Fig. 3). We assume that we have already removed non saturated outgoing links by Rule 2. We can suppose then that all nodes in V_{comp} are saturated. We also assume that for each two nodes x and y in V_{comp} either x and y have no packets in common (i.e. $\mathcal{S}(x) \cap \mathcal{S}(y) = \emptyset$) or they are connected to the same packets (i.e. $\mathcal{S}(x) = \mathcal{S}(y)$).

Lemma 11 (Node 8 NSFNET). *It is sufficient to serve packets linked to an element in V_{comp} in a first step and then make a random assignment on remaining packets from V_1 consistent with the capacity constraint.*

Proof. We can partition the subgraph obtained by considering only V_{comp} into different connected components where each component is a complete bipartite graph. For each connected component we have two cases: If the sum of capacities of outgoing links in the connected component is enough to satisfy the corresponding packets. Then no deflection is necessary and we can remove the packets from V_1 and corresponding outgoing links as in Lemma 8. Let V_1' be the remaining packets and V_2' the remaining outgoing links. Otherwise there is not enough capacity. We have a saturated component and the number of packets that can be accepted is the capacity sum. As the different saturated components have no packets in common, we can conclude that the overall packets that could be accepted is obtained by summing all capacities in saturated components. Let V_{SAT} be the set of outgoing links from V_{comp} that are in a saturated component. The maximum number of packets that could be accepted on V_{SAT} is $\sum_{x \in V_{\mathrm{SAT}}} \lambda(x)$.

If we first serve packets linked to an element in V_{SAT}, the number of accepted packets is $\sum_{x \in V_{\mathrm{SAT}}} \lambda(x)$. If we make then a random assignment on remaining packets from V_1' consistent with the capacity, then according to Lemma 9 the minimum number of deflections is $(V_1' - \sum_{x \in V_{\mathrm{SAT}}} \lambda(x) - \sum_{y \in V_{\mathrm{bip}}} \lambda(y))^+$. As this value is exactly equal to the minimum number of deflections if we had a complete bipartite graph between V_1' and V_2' which is a lower bound, we conclude that we are at the optimum. While considering V_1, we do not increase the number of deflections since all additional packets are accepted as stated in the first case.   □

The idea is to accept packets which have the most choices on the non shared outgoing link in order to give other packets the maximum chances to be accepted. However, we have to identify V_{comp} in order to apply Lemma 11 under the given assumptions. We argue that it is sufficient to serve packets with a higher priority on outgoing link with minimum degree. Indeed, nodes in V_{comp} have a degree less than the ones in V_{bip}.

Rule 5. *If there is no more degree 1 packets and no more unsaturated outgoing links, accept as much as possible packet on a minimum degree outgoing link.*

Lemma 12. *Let G be a broken wheel configuration with m nodes in V_2. It is optimal to saturate node m to optimize the number of deflections.*

Proof. We have two cases. If $d(m) \leq \lambda(m)$. The saturation of node m does not imply any deflection, and node m is not requested by other packets. Saturation of m is optimal.

And if $d(m) > \lambda(m)$, we consider $G1$ the sub-graph of G obtained after deletion of the nodes of S_0. The nodes of $V(G1)$ have degree 2 and we can build the associated graph $H1$. Clearly $H1$ is a cycle. We can apply Mneimneh's method to obtain an optimal selection of packets. Now consider this selection. If node m is saturated, the result holds. If node m is not saturated, some packets which have requested m and 1 or m and $m-1$ have not been routed to node m. And there exists such a packet as node m is not saturated and $d(m) > \lambda(m)$. Therefore we re-route it to node m until node m is saturated. This selection is still optimal as we do not increase then number of deflections.

Thus, after this first step, node m is removed from the graph and the remaining graph is now the union of a $K_{n,m-1}$ and a configuration graph $G1$. Thus we can use Rule 4 to complete the computation of the deflected packets.   □

Theorem 1. *All these rules have a linear complexity in the number of packets.*

4   Conclusions: Application to NSFNET

We now turn back to NSFNET, a topology already presented in the literature which are supposed to be realistic enough. We first build the routing table for NSFNET based on the enumeration of all the shortest paths. After applying

Rule 1 and Rule 2, we remove the unnecessary outgoing links (i.e. the saturated or not required ones), the packets which have been routed during this phase and the packets that have to be deflected because there are no more possible outgoing links. We end up in one of these three situations:

- The problem is already solved: this is the case for nodes 3 and 5.
- We end up with two outgoing links required by all remaining packets (the configuration is a $K_{n,2}$): nodes 4, 7, 10, and 11. We conclude with Rule 3.
- We end up with two connected components that are both $K_{n,2}$: node 12. Some packets require outgoing links toward nodes 1 and 14, while the others must route through nodes 9 and 10. According to Lemma 1 we can solve separately both sub-problems. Applying Rule 3 on both is sufficient.
- We need first to remove some outgoing links using Lemma 10, then we end up with a $K_{n,2}$ (nodes 1, 2, 9 ,13 and 14). For example for node 1, we first remove outgoing link 12 then we end up with a $K_{n,2}$ using outgoing links 11 and 13.
- Node 8 (see right part of Fig. 3): applying Lemma 11 and Rule 5 is sufficient.
- Node 6 : We end up with a broken wheel; thus we apply lemma 13 to know how to route the packets.

The rules we have obtained have been successfully tested on many topologies found in the literature. The selection of packets to be deflected is a very important features to derive an efficient implementation of a deflection algorithms for all optical core networks. We must take into account the complexity of the algorithm (only linear algorithms are possible when we deal with high speed routers) and efficient selection to avoid that too many packets experience very important end to end delays.

Acknowledgment. This work is partially supported by a grant between CNRS and CNRST.

References

1. Mukherjee, B.: Optical WDM Networks. Springer (2006)
2. Schuster, A.: Optical Interconnections and Parallel Processing: The Interface. In: Bounds and Analysis Techniques for Greedy Hot-potato Routing, pp. 284–354. Kluwer Academic Publishers (1997)
3. Barth, D., Berthomé, P., Chiaroni, D., Fourneau, J.M., Laforest, C., Vial, S.: Mixing convergence and deflection strategies for packet routing in all-optical networks. J. Opt. Commun. Netw. 1(3), 222–234 (2009)
4. Baran, P.: On distributed communication networks. IEEE Trans. on Communications Systems CS-12, 1–9 (1964)
5. Barth, D., Berthomé, P., Borrero, A., Fourneau, J.M., Laforest, C., Quessette, F., Vial, S.: Performance comparisons of Eulerian routing and deflection routing in a 2d-mesh all optical network. In: ESM, Prague, pp. 887–891 (2001)
6. Mneimneh, S., Quessette, F., Damm, G., Verchère, F.: Minimum deflection routing in bufferless networks (2004),
   http://www.freepatentsonline.com/20040022240.html

7. Mneimneh, S., Quessette, F.: Linear complexity algorithms for maximum advance deflection routing in some networks. In: IEEE Workshop on High Performance Switching and Routing, Poland, pp. 127–133 (2006)
8. Gravey, P., et al.: Multiservice optical network: Main concepts and first achievements of the ROM program. Journal of Ligthwave Technology 19, 23–31 (2001)
9. Watabe, K., et al.: 80Gb/s multi-wavelength optical packet switching using PLZT switch. In: Tomkos, I., Neri, F., Solé Pareta, J., Masip Bruin, X., Sánchez Lopez, S. (eds.) ONDM 2007. LNCS, vol. 4534, pp. 11–20. Springer, Heidelberg (2007)

Waterfall: Rapid Identification of IP Flows Using Cascade Classification

Paweł Foremski[1], Christian Callegari[2], and Michele Pagano[2]

[1] The Institute of Theoretical and Applied Informatics
of the Polish Academy of Sciences,
Bałtycka 5, 44-100 Gliwice, Poland
pjf@iitis.pl
[2] Department of Information Engineering, University of Pisa,
Via Caruso 16, I-56122, Italy
{c.callegari,m.pagano}@iet.unipi.it

Abstract. In the last years network traffic classification has attracted much research effort, given that it represents the foundation of many Internet functionalities such as Quality of Service (QoS) enforcement, monitoring, and security. Nonetheless, the proposed works are not able to satisfactorily solve the problem, usually being suitable for only addressing a given portion of the whole network traffic and thus none of them can be considered an ultimate solution for network classification.

In this paper, we address network traffic classification by proposing a new architecture – named Waterfall architecture – that, by combining several classification algorithms together according to a cascade principle, is able to correctly classify the whole mixture of network traffic.

Through extensive experimental tests run over real traffic datasets, we have demonstrated the effectiveness of the proposal.

Keywords: network management, traffic classification, machine learning, multi-classification, classifier selection, cascade classification.

1 Introduction

Internet traffic *classification* – or *identification* – is the act of matching IP packets to the applications that generated them [1]. For example, given the packets generated by the Skype program, a traffic classifier would group the packets in traffic *flows* and assign a *label* of *Skype* to them [2]. Traffic classification is important for network management, e.g. for Quality of Service (QoS), routing, and network diagnostics.

The field of network traffic classification needs a method for integrating results of various research activities. Many new papers describe methods that in principle propose a set of traffic features optimized for a set of network protocols [1–7]. Researchers promote their methods for classifying network traffic, which are usually quite effective, but none of them is able to exploit all observable phenomena in the Internet traffic and identify all kinds of protocols.

A. Kwiecień, P. Gaj, and P. Stera (Eds.): CN 2014, CCIS 431, pp. 14–23, 2014.

The question arises: could we integrate these approaches into one system, so that we move forward, building on the achievements of our colleagues? How would this improve classification systems, in terms of accuracy, functionality, completeness, and speed? Answering these questions can open new perspectives. A robust method for combining classifiers can promote research that is more focused on new phenomena in the Internet, rather than addressing the same old issues. We need a way to complement and develop our existing methods further.

This paper proposes a new, modular architecture for traffic identification systems: the Waterfall architecture. In a nutshell, it connects several classification modules in chain and query them sequentially, as long as none of them replies with a positive answer – i.e. the first module that identifies a flow wins. Typically, each module is a specialized and very accurate classifier that targets a subset of network protocols, i.e. supports the *rejection option* (the "Unknown" class) [8]. The modules are ordered from the most reliable and specific to the most general and CPU-intensive. Waterfall follows the scheme of *cascade classification*, which is a type of *classifier selection* approach in the field of *multi-classification* [9].

The proposed architecture solves the integration problem. Each module can exploit different traffic features and address different kinds of network protocols, for example traditional client-server traffic, Peer-to-Peer (P2P), or tunneled traffic. The system can be iteratively extended and updated as new network protocols emerge or new functionality requirements arise. Surprisingly, adding more classifiers can significantly reduce the total computation time (assuming proper ordering of the modules), which is the main advantage of Waterfall over popular *classifier fusion* approaches, e.g. Behavior Knowledge Space (BKS) [9, 10].

This paper presents a novel method with the following contributions:

1. It is the first application of cascade classification to the field of traffic classification (to the best of the knowledge of the authors). It represents an alternative to the BKS method (see Sect. 2 and 3).
2. Waterfall lets for integration of independent algorithms and for iterative development of traffic identification systems, in a way similar to the *divide and conquer* algorithm design paradigm (see Sect. 3 and 4).
3. It has an open source implementation in Python that shows excellent performance on real traffic and classifies flows in under 10 seconds of their lifetime (see Sect. 4 and Experiment 1 in Sect. 5).
4. Practical operation shows reduction in computation time with the increase in the number of modules, and that majority of traffic can be successfully classified using simple methods (see Experiments 1 and 2 in Sect. 5).
5. Proposes a new avenue for the future directions in the field of traffic classification (see Sect. 6, which concludes the paper).

2   Background

A naïve approach to the integration problem would be to survey recent papers for traffic features and apply them as long feature vectors classified with a decent machine learning algorithm. Even with adequate techniques employed, this could quickly lead us to the *curse of dimensionality* [8]: an exponential growth in the demand for training data as the feature space dimensionality increases. Besides, network flows differ in the set of available features, e.g. only a part of Internet flows evoke DNS queries [3]. Some features need more packets to be computed, e.g. port number is available after 1 packet, whereas payload statistics need 80 packets in [6]. This means that different tools are needed for different protocols: some flows can be classified immediately using simple methods, while others need more sophisticated analysis. Finally, from the software engineering point of view, a big, monolithic system could be hard to develop and maintain.

Instead, researchers adopt multi-classification – in particular the BKS combination method that fuses outputs of many classifiers into one final decision. In principle, the idea behind BKS is to ask all classifiers for their answers on a particular problem $\mathbf{x}$ and then query a look-up table $\mathbf{T}$ for the the final decision. The table $\mathbf{T}$ is constructed during training of the system, by observing the behavior of classifiers on a labeled dataset. For example, if an ensemble of 3 classifiers replies (A, B, A) for a sample with a ground-truth label of B, then the cell in $\mathbf{T}$ under index (A, B, A) is B (see [9], pp. 128). This powerful technique can increase the performance of traffic classification systems – as shown by Dainotti et al. in [10] – but comparing to Waterfall, it inherently requires *all* modules to be run on each traffic flow, with the drawback that the more modules are used, the more processing power is required.

In this paper, the idea of cascade classification is employed, which is also a multi-classifier, but so far it was not applied to traffic classification. Interestingly, L. Kuncheva in her book on multi-classification writes "Cascade classifiers seem to be relatively neglected although they could be of primary importance for real-life applications." (in [9], pp. 106). This paper picks up this thought.

3 The Waterfall Architecture

The Waterfall idea is presented in Fig. 1. The input to the system is an IP flow in form of a feature vector $\mathbf{x}$, which contains all the features required by all the modules, but a particular module will usually use only a subset of $\mathbf{x}$.

The system sequentially evaluates *selection criteria* that decide which *classification modules* to use for the problem $\mathbf{x}$. If a particular criterion is fulfilled, the associated module is run. If it succeeds, the algorithm finishes. Otherwise, or if the criterion was not satisfied, the process advances to the next step. When there are no more modules to try, the flow gets rejected and is labeled as "Unknown". More precisely,

$$Dec_i(\mathbf{x}) = \begin{cases} Class_i(\mathbf{x}) & Crit_i(\mathbf{x}) \text{ satisfied} \wedge Class_i(\mathbf{x}) \text{ successful} \\ Dec_{i+1}(\mathbf{x}) & \text{otherwise} \end{cases}, \quad (1)$$

$$Dec_{n+1}(\mathbf{x}) = Reject , \quad (2)$$

where Dec_i is the decision taken at step $i = \{1, 2, \ldots, n\}$, n is the number of modules, $Class_i(\mathbf{x})$ is the protocol identified by the module i, and $Crit_i(\mathbf{x})$ is the associated criterion.

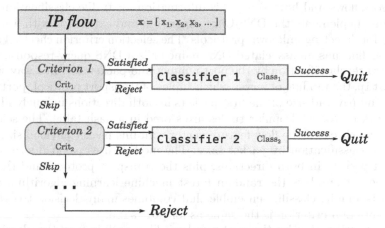

Fig. 1. The Waterfall architecture. A flow enters the system and is sequentially examined by the modules. In case of no successful classification, it is rejected.

The selection criteria are designed to skip ineligible classifiers quickly. For example, in order to implement a module that classifies traffic by analyzing the payload size of the first 5 packets in a flow, the criterion could check if at least 5 data packets were sent in each direction. If this condition is true, a machine learning algorithm could be run to identify the network protocol. Probably, a huge amount of IP flows would be skipped, saving computing resources and avoiding classification with an inadequate method. On the other hand, if a flow satisfies this criterion, it would be classified with a method that does not need to support corner cases. The selection criteria are optional, i.e. if a module does not have an associated criterion, it is always run.

4   Practical Implementation

A system using the Waterfall architecture was created and is available as open source[1]. It was implemented in the C and Python languages in two parts: *Flowcalc*, which prepares the flow feature vectors in form of ARFF files, and *Mutrics*[2], which classifies the flows. The *Mutrics* classifier has several modules, described below.

1. `dstip`: classification by destination IP address. During training, the module observes which remote destinations uniquely identify network protocols. If such particular IP address is popular enough, it is used as a rule for quick protocol identification by single lookup in a hash table.
2. `dnsclass`: classification by DNS domain name of the remote host. In [3], the authors described how to obtain the textual host names associated with

---

[1] See http://mutrics.iitis.pl/
[2] The name comes from "Multilevel Traffic Classification".

network flows and how to use this information for traffic classification. This module implements the DNS-Class algorithm and extends it with a mechanism for detecting unknown protocols. The selection criterion checks if a particular flow has an associated DNS name or is a DNS query-response flow.

3. `portsize`: classification by the port number and packet size. In a way similar to `dstip`, the module observes which tuples of transport protocol, port number, and payload size of the first packets in both directions uniquely identify network protocols. Popular tuples are stored in a hash table. The selection criterion checks if the flow feature vector contains packet payload sizes.

4. `npkts`: classification by packet sizes. The module uses payload sizes of the first 4 packets in both directions, plus the transport protocol and the port number. It employs the random forest machine learning algorithm, which itself is a multi-classifier ensemble that combines many decision trees [9, 8]. The selection criterion is the same as for `portsize`.

5. `port`: classification by the port number. The module uses the classic pair of transport protocol and port number to find the pairs that uniquely and reliably identify network protocols, similarly to `dstip` and `portsize`. Classification requires a single lookup in a hash table.

6. `stats`: classification by flow statistics. The module uses the same machine learning algorithm as `npkts`. As features, it uses the following statistics of packet sizes and inter-arrival times in both directions: the minimum, the maximum, the average, and the standard deviation – i.e. a total of 16 statistics.

The *Flowcalc* part of the system, responsible for computing the feature vectors, was limited to only consider the first 10 seconds of traffic in each flow. This simulates a real-time scenario in which the network protocol must be identified in given time limit. All experiments presented in the next section were run with such constraints to demonstrate real-time traffic classification.

5   Experiments

In this section, the results of two experiments are presented: 1) classification performance on real network traffic, and 2) effect of adding new modules. First, the methodology is described below.

Four traffic datasets were used for experimental validation: a) *Asnet1*, collected at a Polish ISP company serving <500 residential customers, b) *Asnet2*, collected at the same network, c) *IITiS1*, collected from the network of the IITiS institute serving <50 academic users, and d) *Unibs*, collected from the campus network of the University of Brescia serving 20 workstations[3]. The *Asnet1* and *Asnet2* datasets were collected at the same gateway router, but with a time gap of 8 months. The *Asnet1* and *IITiS1* datasets were collected at different networks, but at the same time. The *Unibs* dataset was collected a few years earlier than the other datasets, contains no packet payload, and has the IP addresses anonimized. Details are presented in Table 1e.

---

[3] Downloaded from `http://www.ing.unibs.it/ntw/tools/traces/`

Table 1. Experiment 1: classification performance and the datasets. Metrics for validation on 4 real traffic datasets: a) *Asnet1*, b) *Asnet2*, c) *IITiS1*, and d) *Unibs*. Part e) presents details on the datasets used in the paper.

a)

Protocol	Flows	A	B	C	D	E	F	G	H	I	J	K	L	M	%TP	%FP
A = BitTorrent	799,234	99.9%							<0.1%				<0.1%		99.9%	<0.1%
B = DNS	723,478		100%												100%	0%
C = eMule	75,555			99.9%									<0.1%		99.9%	<0.1%
D = Jabber	1,388				100%										100%	0%
E = Kademlia	148,824					100%									100%	<0.1%
F = Kaspersky	17,479						100%								100%	0%
G = Mail	7,982							100%							100%	0%
H = NTP	3,506								100%						100%	0%
I = Skype	55,070	<0.1%		<0.1%		<0.1%				99.9%					99.9%	<0.1%
J = SSH	3,620										100%				100%	0%
K = Steam	50,963											100%			100%	0%
L = STUN	6,828												100%		100%	0%
M = WWW	1,695,282													100%	100%	<0.1%
	3,680,200													Avg.	99.9%	<0.1%

b)

Protocol	Flows	A	B	C	D	E	F	G	H	I	J	K	L	M	%TP	%FP
A = BitTorrent	4,788,510	99.9%		<0.1%					<0.1%	<0.1%			<0.1%	<0.1%	99.9%	<0.1%
B = DNS	1,810,622		100%												100%	0%
C = eMule	172,676	<0.1%		99.9%					<0.1%						99.9%	<0.1%
D = Jabber	3,110				100%										100%	0%
E = Kademlia	308,926	<0.1%				99.9%			<0.1%				<0.1%		99.9%	<0.1%
F = Kaspersky	46,475						96%						4%		96%	<0.1%
G = Mail	42,055							100%							100%	0%
H = NTP	10,940								100%						100%	0%
I = Skype	112,748	0.1%				0.5%				99.4%					99.4%	<0.1%
J = SSH	6,223										100%				100%	<0.1%
K = Steam	78,167	<0.1%										99.9%			99.9%	0%
L = STUN	12,009	<0.1%											99.9%		99.9%	<0.1%
M = WWW	3,828,118						<0.1%							99.9%	99.9%	<0.1%
	11,220,579													Avg.	99.6%	<0.1%

c)

Protocol	Flows	A	B	C	D	E	F	G	H	%TP	%FP
A = BitTorrent	8,811	100%								100%	0%
B = DNS	1,018,193		100%							100%	0%
C = Jabber	137			100%						100%	0%
D = Mail	108,296				100%					100%	0%
E = NTP	4,827					100%				100%	0%
F = Skype	21						100%			100%	0%
G = SSH	17,764							100%		100%	0%
H = WWW	624,511								100%	100%	0%
	1,782,560								Avg.	100%	0%

d)

Protocol	Flows	A	B	C	D	E	%TP	%FP
A = BitTorrent	2,928	99.8%	0.1%		<0.1%		99.8%	<0.1%
B = eMule	5,600		99.9%		<0.1%		99.9%	<0.1%
C = Mail	1,971			99.3%	0.7%		99.3%	<0.1%
D = Skype	1,484	0.1%			99.8%	0.1%	99.8%	0%
E = WWW	18,798			<0.1%		99.9%	99.9%	0.2%
	30,781					Avg.	99.7%	0.1%

e)

Dataset	Start	Duration	Src. IP	Dst. IP	Packets	Bytes	Avg. Util	Avg. Flows (/5 min.)	Payload
Asnet1	2012-05-26 17:40	216 h	1,828 K	1,530 K	2,525 M	1,633 G	18.0 Mbps	7.7 K	92 B
Asnet2	2013-01-24 16:26	168 h	2,503 K	2,846 K	2,766 M	1,812 G	25.7 Mbps	12.0 K	84 B
Iitis1	2012-05-26 11:19	220 h	32 K	46 K	150 M	95 G	1.0 Mbps	753.7	180 B
Unibs	2009-09-30 11:45	58 h	27	1 K	33 M	26 G	0.9 Mbps	111.7	0 B

For establishing the ground-truth labels on the datasets a)–c), Deep Packet Inspection (DPI) was employed. DPI is not perfect – as shown by Dusi et al. [11] – but it is the most popular method used in the literature, and often the only practically available. The *libprotoident* v. 2.0.7 was used as the DPI software (reported to offer very good accuracy in [12]). The *Unibs* dataset already contained ground-truth and was not suitable for DPI because of no payload data.

Finally, the datasets were sanitized by dropping flows that had no data transmitted in both directions, e.g. incomplete TCP sessions and empty UDP flows. The datasets contain different subsets of network protocols (see Table 1).

For measuring the classification accuracy for a given protocol p, the popular $\%TP_p$ and $\%FP_p$ metrics were employed:

$$\%TP_p = \frac{|TP_p|}{|F_p|} \cdot 100\,\% \ , \quad \%FP_p = \frac{|FP_p|}{|F_p'|} \cdot 100\,\% \ , \tag{3}$$

where TP_p is the set of true positives for protocol p, F_p is the set of all testing flows for protocol p that were classified, FP_p is the set of false positives for p, and F_p' is the set of all testing flows for all protocols except p that were classified. For evaluating the overall accuracy, the average values of these metrics were used – the $\%TP$ and $\%FP$ metrics – which were complemented with the $\%Unk$ metric that measures the amount of rejected flows:

$$\%Unk = \frac{|U|}{|F|} \cdot 100\,\% \ , \tag{4}$$

where U is the set of rejected flows, and F is the set of all testing flows.

For dividing the data into training and testing parts, a 60%/40% split was used on *Asnet1* and on *Unibs*, i.e. 60% of their flows were randomly selected for training, and 40% for testing. The classifier trained on *Asnet1* was validated on the rest of *Asnet1* (to evaluate the "classical" classification performance), and on the whole *Asnet2* and *IITiS1* datasets (so as to demonstrate stability in time and space). The classifier trained on *Unibs* was validated only on the rest of the *Unibs* dataset: this tested operation on a trace without packet payloads.

In the Experiment 1, the system was evaluated for classification performance on the datasets a)-d), with 5 modules enabled: `dstip`, `dnsclass`, `portsize`, `npkts`, and `port`. For the *Unibs* dataset, `stats` was used instead of `dnsclass`, because this dataset had no DNS payload packets.

The results in form of confusion matrices and performance metrics are presented in Table 1, parts a)–d). For all datasets, the $\%TP$ and $\%FP$ metrics were close to 100% and 0% respectively, which indicates high classification performance of the system. The classifier successfully identified all protocols, including: *BitTorrent, Skype, Kademlia, SSH, STUN, WWW*, and more. For the *IITiS1* dataset, the system made no errors in classifying over 1.5 million flows; for other datasets, the number of errors was well below 0.1%. The $\%Unk$ metric was 0.1%, 0.4%, 1.1%, and 1.4%, for *Asnet1, Asnet2, IITiS1*, and *Unibs*, respectively – i.e. almost all flows were classified.

Figure 2 shows traffic progress through the system: the figure presents the percentage of IP flows at the input of successive modules. An IP flow leaves the system as soon as it gets classified, so the figure visualizes how many flows get through the end of the waterfall. It is apparent that the amount of traffic that a particular module can classify depends on the dataset. For all datasets, more than half of the flows were classified using simple methods – namely the `dstip`, `dnsclass`, and `portsize` modules – without the need to run the `npkts` module, which employs a sophisticated machine learning algorithm.

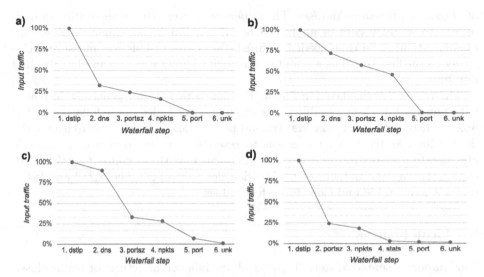

Fig. 2. Experiment 1: amount of traffic passing through successive waterfall steps, for datasets: a) *Asnet1*, b) *Asnet2*, c) *IITiS1*, and d) *Unibs*. The classifier works with 5 modules enabled; "dns" means `dnsclass` and "portsz" means `portsize`.

In the Experiment 2, the effect of increasing the number of modules was studied. The system started in configuration with only one module present: the `npkts` module, which is CPU intensive. In each iteration, one new module was added – usually at the front of the waterfall – and the whole system was given the task

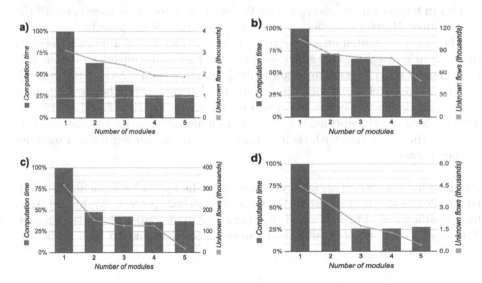

Fig. 3. Experiment 2: effect of adding modules on computation time (bars) and on the number of unknown flows (lines): a) *Asnet1*, b) *Asnet2*, c) *IITiS1*, and d) *Unibs*

of classifying the same dataset. The experiments were run in separation of each other, on a single core of an Intel Core i7 machine[4] (i.e. single-threaded implementation). The time needed to finish the computations was measured relatively to the first run. The amount of unclassified flows was measured in absolute numbers.

The results are displayed in Fig. 3. The computation time generally decreased with new modules being enabled. The modules were added in the following order: `npkts`, `dnsclass`, `portsize`, `dstip`, and `port` – that is, from most to least CPU demanding. Motivation for this was to resemble a scenario in which a generic classification algorithm is iteratively augmented with specialized methods. In each iteration, the number of unclassified flows dropped. The $\%TP$ and $\%FP$ metrics were stable and close to perfect values.

6   Conclusions

This paper presented Waterfall – a novel, modular architecture for traffic classification that lets for integration of many algorithms and exhibits a decrease in the total computation time with new modules being added. A practical, open source implementation of the system – the *Mutrics* classifier – showed very good performance results on 4 real traffic datasets, in a scenario that resembles real-time traffic classification. The paper experimentally proved that the majority of IP flows can be immediately classified using simple methods, which exploit traffic features like destination IP address, DNS domain name, and packet size.

The paper concludes with a positive answer to the question whether many independent traffic classification algorithms could be integrated, giving good results in terms of many metrics. The Waterfall architecture – together with the open source *Mutrics* classifier – are a basis for future developments.

The paper presented an alternative to the classifier fusion approach, recently introduced in the field of traffic classification. The main conclusion is that cascade classification is less computationally demanding than e.g. the BKS method. The future research could focus on a more detailed comparison between classifier fusion and classifier selection in context of traffic classification. Another interesting problems are the optimal selection of training instances for the classification modules (in accordance with their criteria), and the proper sequence of modules in the cascade.

Acknowledgments. This work was funded by the Polish National Science Centre, under research grant nr 2011/01/N/ST6/07202: project "Multilevel Traffic Classification", http://mutrics.iitis.pl/. This paper closes the project. The *Asnet1* and *Asnet2* traffic traces were collected thanks to ASN Sp. z o.o., http://www.asn.pl/.

---

[4] Intel Core i7-930 2.80 GHz, 8 GB RAM, 128 GB SSD.

References

1. Foremski, P.: On different ways to classify Internet traffic: a short review of selected publications. Theoretical and Applied Informatics 25(2) (2013)
2. Adami, D., Callegari, C., Giordano, S., Pagano, M., Pepe, T.: Skype-Hunter: A real-time system for the detection and classification of Skype traffic. International Journal of Communication Systems 25(3), 386–403 (2012)
3. Foremski, P., Callegari, C., Pagano, M.: DNS-Class: Immediate classification of IP flows using DNS. International Journal of Network Management (accepted, 2014)
4. Fiadino, P., Bär, A., Casas, P.: HTTPTag: A Flexible On-line HTTP Classification System for Operational 3G Networks. In: International Conference on Computer Communications, INFOCOM 2013. IEEE (2013)
5. Bermolen, P., Mellia, M., Meo, M., Rossi, D., Valenti, S.: Abacus: Accurate behavioral classification of P2P-TV traffic. Computer Networks 55(6), 1394–1411 (2011)
6. Finamore, A., Mellia, M., Meo, M., Rossi, D.: KISS: Stochastic packet inspection classifier for udp traffic. IEEE/ACM Transactions on Networking 18(5), 1505–1515 (2010)
7. Dusi, M., Crotti, M., Gringoli, F., Salgarelli, L.: Tunnel hunter: Detecting application-layer tunnels with statistical fingerprinting. Computer Networks 53(1), 81–97 (2009)
8. Duda, R.O., Hart, P.E., Stork, D.G.: Pattern classification. John Wiley & Sons (2012)
9. Kuncheva, L.I.: Combining Pattern Classifiers: Methods and Algorithms. Wiley (2004)
10. Dainotti, A., Pescapé, A., Sansone, C.: Early classification of network traffic through multi-classification. Traffic Monitoring and Analysis, 122–135 (2011)
11. Dusi, M., Gringoli, F., Salgarelli, L.: Quantifying the accuracy of the ground truth associated with Internet traffic traces. Computer Networks 55(5), 1158–1167 (2011)
12. Bujlow, T., Carela-Espanol, V.: Comparison of Deep Packet Inspection (DPI) Tools for Traffic Classification. Technical report, Polytechnic University of Catalonia (2013)

Modeling Packet Traffic with the Use of Superpositions of Two-State MMPPs

Joanna Domańska[2], Adam Domański[1], and Tadeusz Czachórski[2]

[1] Institute of Informatics,
Silesian Technical University,
Akademicka 16, 44-100 Gliwice, Poland
adamd@polsl.pl
[2] Institute of Theoretical and Applied Informatics,
Polish Academy of Sciences,
Baltycka 5, 44-100 Gliwice, Poland
{joanna,tadek}@iitis.gliwice.pl

Abstract. The aim of this paper is to use the superposition of two-state Markov Modulated Poisson Processes to replicate the statistical nature of internet traffic over several time scales. This paper characterizes of network traffic using Bellcore data and LAN traces collected in IITiS PAN. The fitting procedure for matching second-order self-similar properties of real data traces to that of two-state MMPP's has also been described.

Keywords: two-State MMPPs, Markov Modulated Poisson Processes.

1 Introduction

The growing variety of IP networks services and applications has been resulting in an increase of requirement to make reliable measurement of packet flows and to describe them though appropriate traffic models.

Traditionally, the traffic intensity, has been regarded as a stochastic process, was represented in queueing models by short term dependencies [1]. However, the analysis of measurements shows that the traffic establishes also long-terms dependencies and has a self-similar character. The problem of self-similarity has been described in Sect. 2. This feature of a traffic was observed on various protocol layers and in different network structures [2–7].

Several models have been introduced for the purposes of modeling self-similar processes in the network traffic area. Pioneering work in [6] motivated other researchers [8] to model network traffic using fractional Brownian Motion. On-Off source model [9] provides an opportunity for different model based on Stable Levy Motion [10] depending on order of consideration of limits. Other models of traffic sources use chaotic maps [11], α-stable distribution [12], fractional Autoregressive Integrated Moving Average (fARIMA) [13] and fractional Levy Motion [14] for modeling of network traffic. All above mentioned traffic models are based on non-Markovian approach. The advantage of these models is that

A. Kwiecień, P. Gaj, and P. Stera (Eds.): CN 2014, CCIS 431, pp. 24–36, 2014.

they give a good description of the traffic with the use of few parameters. Their drawbacks consist in the fact that they do not allow the use of traditional and well known queueing models and modeling techniques for computer networks performance analysis.

Some researchers use Markov based models to generate a self-similar traffic over a finite number of time scales [15–20]. This approach makes possible the adaptation of traditional Markovian queueing models to evaluate network performance. In our research we have made a fitting of a superposition of two state Markov Modulated Poisson Process (MMPP) proposed in [21] to real traffic data. We have used Ethernet traffic data of Bellcore Morristown Research and Engineering facility which is available for research purpose [22] and data captured in our Institute LAN [23].

The rest of this article is organized as follows. Section 2 briefly describes the mathematics underlying the theory of self-similarity. Section 3 describes the MMPP traffic model used in our study. Section 4 explains the fitting procedure for matching second-order self-similar properties of observed data traces. Obtained results are presented in Sect. 5. Conclusions about the work are drawn in Sect. 6.

2   Self-similarity of Internet Traffic

The term self-similar was introduced by Mandelbrot [24] for explaining water level pattern of river Nile observed by H. Hurst. This term was also known as *Hurst Effect*. The degree of self-similarity is expressed by *Hurst parameter*, denoted by H.

A real valued stochastic process:

$$X = \{X(t)\}_{t \in R}$$

is self-similar with $H > 0$, if for any $a > 0$,

$$\{X(at)\}_{t \in R} \stackrel{d}{=} \{a^H X(t)\}_{t \in R}$$

where $\stackrel{d}{=}$ denotes equality in finite dimensional distribution sense [25]. Above definition evidently shows that self-similar processes are scale invariant.

Mathematically, the difference between short-range dependent processes and long-range ones (self-similar) is as follows [26]:

For a short-range dependent process:
- $\sum_{r=0}^{\infty} \text{Cov}(X_t, X_{t+\tau})$ is convergent,
- spectrum at $\omega = 0$ is finite,
- for large m, $\text{Var}\left(X_k^{(m)}\right)$ is asymptotically of the form $\text{Var}\left(X\right)/m$,
- the aggregated process $X_k^{(m)}$ tends to the second order pure noise as $m \to \infty$;

For a long-range dependent process:
- $\sum_{r=0}^{\infty} \text{Cov}(X_t, X_{t+\tau})$ is divergent,

- spectrum at $\omega = 0$ is singular,
- for large m, $\mathrm{Var}\,(X_k^{(m)})$ is asymptotically of the form $\mathrm{Var}\,(X)m^{-\beta}$,
- the aggregated process $X_k^{(m)}$ does not tend to the second order pure noise as $m \to \infty$,

where the spectrum of the process is the Fourier transformation of the auto-correlation function and the aggregated process $X_k^{(m)}$ is the average of X_t on the interval m:

$$X_k^{(m)} = \frac{1}{m}(X_{km-m+1} + \ldots + X_{km}) \qquad k \geq 1 \ .$$

Estimation of Hurst parameter is the most frequently used method to check if a process is self-similar: for non-self-simlar processes $H = 0.5$; for $0.5 < H < 1$ process is self-similar; the closer H is to 1, the greater the degree of persistence of long-range dependence. Hurst parameter H can be estimated by various methods. One of the method is called *Aggregated Variance* [27, 25], where aggregated sequence is created (as discussed above). This method is based on the analysis of variance-time plot. The variation of aggregated self-similar process is equal to:

$$Var(X_k^{(m)}) = Var(X)m^{-\beta}$$

so the log-log plot of $\frac{Var(X_k^{(m)})}{Var(X)}$ versus m is a line with slope β.

Another technique of estimation is called $R-SPlot$, which is based on *Central Limit Theorem*. Both the above techniques are time-domain based; in frequency domain one can estimate H using *Periodogram* in log-log scale.

As mentioned earlier, many empirical and theoretical studies have shown the self-similar characteristics of the network traffic. These features have a great impact on a network performance. They enlarge mean queue lengths at buffers and increase packet loss probability, reducing this way the quality of services provided by a network [28]. That is why it is necessary to take into account this feature when you want to create a realistic model of traffic sources [16].

3   MMPP Model of Packet Traffic Source

Markov chains and Markov-modulated processes (MMP) are well-known modeling techniques which are successful in wide variety of fields. These models are often motivated by the idea of capturing the long-range dependence (LRD) which is seen in real internet traffic and replicating the the Hurst parameter H which characterizes LRD [17].

Two-state Markov Modulated Poisson Process (MMPP) is also known as the Switched Poisson Process (SPP). The superposition of MMPP's is also an MMPP which is a special case of Markovian Arrival Process (MAP).

A MAP is defined by two square matrices $\mathbf{D_0}$ and $\mathbf{D_1}$ such that $\mathbf{Q} = \mathbf{D_0} + \mathbf{D_1}$ is an irreducible infinitesimal generator for the continuous-time Markov chain (CTMC) underlying the process, and $D_0(i,j)$ (respectively $D_1(i,j)$) is the rate of hidden (respectively observable) transitions from state i to state j [29]. Two-state MAP is a Markovian arrival process with square matrices as follows:

$$\mathbf{D_0} = \begin{bmatrix} -\sigma_1 & \lambda_{1,2} \\ \lambda_{2,1} & -\sigma_2 \end{bmatrix}$$

$$\mathbf{D_1} = \begin{bmatrix} \mu_{1,1} & \mu_{1,2} \\ \mu_{2,1} & \mu_{2,2} \end{bmatrix}$$

where $\lambda_{i,j} \geq 0$, $\mu_{i,j} \geq 0$, for all i, j. The diagonal elements of matrix $\mathbf{D_0}$ are $\sigma_1 = \lambda_{1,2} + \mu_{1,1} + \mu_{1,2} > 0$ and $\sigma_2 = \lambda_{2,1} + \mu_{2,2} + \mu_{2,1} > 0$ such that underlying continuous-time Markov chain Matrix $\mathbf{Q}$ has no absorbing states.

Following the model proposed in [21], a LRD process (used in our study) can be modeled as the superposition of d two-state MMPPs. The i-th MMPP ($1 \leq i \leq d$) can be parameterized by two square matrices:

$$\mathbf{D_0^i} = \begin{bmatrix} -(c_{1i} + \lambda_{1i}) & c_{1i} \\ c_{2i} & -(c_{2i} + \lambda_{2i}) \end{bmatrix}$$

$$\mathbf{D_1^i} = \begin{bmatrix} \lambda_{1i} & 0 \\ 0 & \lambda_{2i} \end{bmatrix} .$$

The element c_{1i} is the transition rate from state 1 to 2 of the i-th MMPP and c_{2i} is the rate out of state 2 to 1. λ_{1i} and λ_{2i} are the traffic rate when the i-th MMPP is in state 1 and 2 respectively. The sum of $\mathbf{D_0}^i$ and $\mathbf{D_1}^i$ is an irreducible infinitesimal generator $\mathbf{Q}^i$ with the stationary probability vector:

$$\overrightarrow{\pi}_i = \left(\frac{c_{2i}}{c_{1i} + c_{2i}}, \frac{c_{1i}}{c_{1i} + c_{2i}} \right)$$

The superposition of these two-state MMPPs is a new MMPP with 2^d states and its parameter matrices, $\mathbf{D_0}$ and $\mathbf{D_1}$, can be computed using the Kronecker sum of those of the d two-state MMPPs [30]:

$$(\mathbf{D_0}, \mathbf{D_1}) = \left(\oplus_{i=1}^{d} \mathbf{D_0}^i, \oplus_{i=1}^{d} \mathbf{D_1}^i \right) .$$

Let N_t^i be a number of arrivals from the i-th MMPP in time slot $(0, t]$. The variance time for this MMPP can be expressed as:

$$Var\{N_t^i\} = (\lambda_i^* + 2k_{1i})t - \frac{2k_{1i}}{k_{2i}}(1 - e^{-k_{2i}t})$$

where:

$$\lambda_i^* = \frac{c_{2i}\lambda_{1i} + c_{1i}\lambda_{2i}}{c_{1i} + c_{2i}}$$

$$k_{1i} = (\lambda_{1i} - \lambda_{2i})^2 \frac{c_{1i}c_{2i}}{(c_{1i} + c_{2i})^3}$$

$$k_{2i} = c_{1i} + c_{2i}$$

The second-order properties are determined by this three entities: λ_i^*, k_{1i} and k_{2i}. The covariance function of the number of arrivals in two time slots of size Δt is expressed by [21]:

$$\gamma_i(k) = \frac{(\lambda_{1i}-\lambda_{2i})^2 c_{1i} c_{2i} e^{-((c_{1i}c_{2i})(k-1)\Delta t)}}{(c_{1i}+c_{2i})^4} * \left(1 - 2e^{-((c_{1i}+c_{2i})\Delta t)} + e^{-((c_{1i}+c_{2i})2\Delta t)}\right) =$$

$$= \frac{k_{1i}}{k_{2i}} e^{-(k_{2i}(k-1)\Delta t)} * \left(1 - 2e^{(-k_{2i}\Delta t)} + e^{(-k_{2i}2\Delta t)}\right) \approx$$

$$\approx \frac{(\Delta t)^2 (\lambda_{1i}-\lambda_{2i})^2 c_{1i} c_{2i} e^{-((c_{1i}+c_{2i})(k-1)\Delta t)}}{(c_{1i}+(c_{2i})^2} = (\Delta t)^2 k_{1i} k_{2i} e^{-(k_{2i}(k-1)\Delta t)} \quad.$$

4   Fitting Procedure for the Self-similar Covariance Structure

There are two approaches for fitting the family of MAP to observed data: moment-based approach and likelihood-based approach [31]. The main advantage of moment-based approaches is to reduce computational cost. In such approaches one determines the MAP model parameters in order to fit theoretical moments to empirical ones obtained from observed traffic.

The article [21] proposed a fitting method for a superposition of two-state MAPs (described in Sect. 3) based on Hurst parameter as well as the moments.

For real traffic traces the covariance structure of the counting process is well described by the asymptotic covariance [21]:

$$\mathrm{cov}(k) = \psi_{\mathrm{cov}} k^{-\beta}$$

where ψ_{cov} jest an absolute measure of the variance, $\beta = 2 - 2H$ and k is the lag. The parameters ψ_{cov} and β should be estimated from the real data traces. The objective of the fitting is to achieve:

$$\gamma(k) = \sum_{i=1}^{d} \gamma_i(k) \approx \psi_{\mathrm{cov}} k^{-\beta}$$

where $1 \le k \le 10^n$ and n denotes the number of time scales the model demonstrate self-similar behavior. The fitting procedure requires the following input parameters:

λ^* – the mean rate of the process to be modeled,
n – number of time scales,
d^* – number of active MMPP's,
$H = 1 - \frac{\beta}{2}$ – the Hurst parameter,
ρ – lag 1 correlation.

The modulating parameters of the MMPP's have been chosen logarithmically with a factor a:

$$c_{1i} = c_{2i} = a^{1-i}c_{11}$$

for $i = 1, \ldots, d$. The smallest time scale should relate to the packet level so the fundamental rate has been assumed relative to this time scale: between 1 and 10 [21]. To achieve this, the modulating parameters $c_{11} = c_{21}$ have been initially chosen in the range $[0.25, 0.75]$ (most often to the value 0.4). The parameter a is dependent on the number of active MMPP's and number of time scales:

$$a = 10^{\frac{n}{d-1}} .$$

As was mentioned in Sect. 3, the second-order properties of a superposition of d MMPPs are determined by this three entities: λ_i^*, k_{1i} and k_{2i}. The arrival intensities λ_{1i} and λ_{2i} are only involved in k_{1i} through the quantity:

$$(\lambda_{1i} - \lambda_{2i})^2 .$$

It is only possible to interpret the superposition of d MMPP's as a superposition of d Interrupted Poisson Process (IPP's) and a Poisson process [21]. With this interpretation, the IPP's have arrival intensity:

$$\lambda_i^{\text{IPP}} = \lambda_{1i} - \lambda_{2i} .$$

The Poisson intensity could be determined as:

$$\lambda_p = \sum_{i=1}^{d} \lambda_{2i} .$$

The fitting procedure requires the following steps:

1. give to the variable β the value: $2 - 2H$,
2. give to the number of IPP's d the value: d^*,
3. give to the variable k_{21} the value: $c_{11} + c_{21}$ (see Sect. 3),
4. give to the variable d_0 (the number of ϕ_j's which were set to 0) the value 0,
5. give to the intensity ϕ_d the value 1,
6. give to the variable i the value 1,
7. give to the logarithmic spacing parameter a the value:

$$10^{\frac{n}{d-1}} ,$$

8. give to the variable D the value:

$$a^{i\beta} - \sum_{j=0}^{i-1} (\phi_{d-j})^2 e^{1-a^{-(i-j)}} ,$$

9. **if** $D < 0$ **then:**
>   give to the intensity ϕ_{d-i} the value 0,
>   increment a value d_0 by one,
>   **if** $d^* > d - d_0$ **then:** increment a value d by one and **go to** the step 4,
>   **else:** give to the intensity ϕ_{d-i} the value $\sqrt{D}$,
10. increment a value i by one,
11. **if** $i <> d$ **then: go to** the step 8,
12. **for** $i = 2$ to d **do** $k_{2i} = a^{1-i}k_{21}$,
13. **if not** $(k_{2i} < 1\rho < 0.5)$ **then:** if necessary adjust k_{2i} and/or ρ,
14. give to the variable η the value:

$$\frac{\sqrt{4\rho\lambda^*}}{\sqrt{\sum_{i=1}^{d} \phi_i^2 k_{2i}^{-2} \left((1 - e^{-k_{2i}})^2 - 2\rho(k_{2i} - (1 - e^{-k_{2i}}))\right)}} \; ,$$

15. give to the variable L the value:

$$\eta \sum_{i=1}^{d} \frac{\phi_i}{2},$$

16. **if** $\lambda^* < L$ **then:**
>   give to the Poisson intensity λ_P the value 0,
>   **for** $i = 1$ to d **do:**
>   >   give to the model parameter c_{1i} the value:

$$\frac{L^2}{\lambda^{*2} + L^2} k_{2i} \; ,$$

>   give to the model parameter c_{2i} the value $k_{2i} - c_{1i}$,
>   give to the arrival intensity λ_i^{IPP} the value:

$$\phi_i \frac{\lambda^{*2} + L^2}{\lambda^* \sum_{i=1}^{d} \phi_i},$$

else:
>   give to the Poisson intensity λ_P the value $\lambda^* - L$,
>   **for** $i = 1$ to d **do:**
>   >   give to the model parameters c_{1i} and c_{2i} the value $0.5k_{2i}$,
>   >   give to the arrival intensity λ_i^{IPP} the value $\eta\phi_i$.

5   Results

As was mentioned in Sect. 4 fitting algorithm requires five input parameters. Three of them: λ^* (the mean rate of the process), ρ (lag 1 correlation) and the Hurst parameter H should be estimated from the real data traces. Real traces used in our study comprises of Ethernet traffic data of Bellcore Laboratory [22] and data captured on the gateway of the Institute of Theoretical and Applied Informatics (IITiS) of the Polish Academy of Science in Gliwice (Poland).

Bellcore Laboratory data was already interpreted in multiple studies [6, 32, 25]. A dedicated hardware has been built for measuring each packet arrival and the measurements were performed without losses and with high precision. The large part of collected data is available by Internet [22]. In this study we use the file: OctExt.TL. It contains the first million of external arrivals gathered during 35 hours.

Other data set used in our study has been collected during the whole May 2012 on the Internet gateway of our Institute [23]. The traffic approximately stands for the few dozen office users (researchers), mainly working Monday–Friday 8AM–4PM. During May 1–3, there are national holidays in Poland, so the traffic can be smaller. IP packets were limited to 64 bytes – for most cases the they contain all headers, plus a few bytes of the transport protocol payload. The local DNS traffic is not visible, because of specific setup of the network.

The Hurst parameter was estimated by the *Aggregated Variance* method (see Sect. 2). Figure 1 presents the normalized variance of the aggregated series as a function of time scale in log-log coordinates. The slope of IITiS curve (estimated by the least squares method) is equal to −0.42, which gives the Hurst parameter equal to 0.79. The slope of Bellcore curve is equal to −0.3, which gives the Hurst parameter equal to 0.85. For comparison, the same plot is also drawn for the Poisson process. This line has the slope -1, which gives the Hurst parameter equal to 0.5 (non-self-similar process).

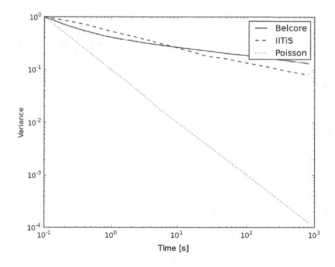

Fig. 1. Variance-time plot; log-log scale (IITiS data, May 2012)

As was mentioned earlier, the superposition of two-state MMPP's can be applied to model traffic exhibiting LRD traffic over a number of time scales. For LRD processes, the autocovariance decays hyperbolically. Any asymptotically second-order self-similar process exhibits LRD properties. Each MMPPs models a specific time scale of data, the volume of traffic modeled by each of the single two-state sources can be associated with the volume of traffic showing variability on a given time scale [21]. Figure 2 shows how the autocorrelations of the three IPP's behave as a function of lag k. The parameters defining this model are:

$$\lambda_1^{IPP} = \lambda_2^{IPP} = \lambda_3^{IPP} = 6.0$$
$$c_{11} = c_{21} = 10^{-2}$$
$$c_{12} = c_{22} = 10^{-4}$$
$$c_{13} = c_{23} = 10^{-6}$$

Figure 3 shows the autocorrelations of the five IPP's. The parameters defining this model are:

$$\lambda_1^{IPP} = \lambda_2^{IPP} = \lambda_3^{IPP} = \lambda_4^{IPP} = \lambda_5^{IPP} = 6.0$$
$$c_{11} = c_{21} = 10^{-2}$$
$$c_{12} = c_{22} = 10^{-\frac{7}{2}}$$
$$c_{13} = c_{23} = 10^{-5}$$
$$c_{14} = c_{24} = 10^{-\frac{13}{2}}$$
$$c_{15} = c_{25} = 10^{-8}$$

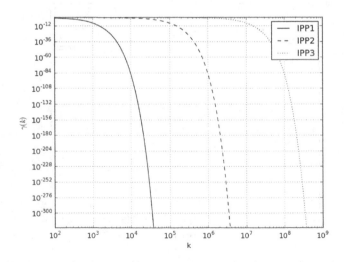

Fig. 2. Autocorrelation of the number of arrivals in a time unit (three IPP's)

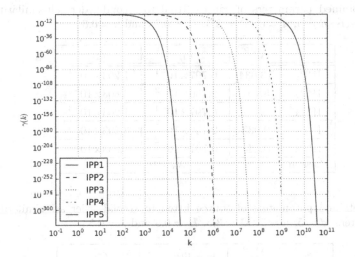

Fig. 3. Autocorrelation of the number of arrivals in a time unit (five IPP's)

Tables 1, 2 and 3 present the parameters obtained from the fitting approach described in Sect. 4. The input parameters defining the asymptotic second-order self-similarity were selected as an example.

Table 1. Obtained parameters of source fitted to second-order self-similarity with input parameters: $d = 5$, $n = 6$, $\lambda^* = 3.5$, $H = 0.6$ and $\rho = 0.6$

	λ_i^{IPP}	c_{1i}	c_{2i}
IPP$_1$	27.646	7.241×10^{-1}	7.590×10^{-2}
IPP$_2$	6.944	2.290×10^{-2}	2.400×10^{-3}
IPP$_3$	1.746	7.241×10^{-4}	7.590×10^{-5}
IPP$_4$	0.434	2.290×10^{-5}	2.400×10^{-6}
IPP$_5$	0.119	7.241×10^{-7}	7.590×10^{-8}
Poisson	$\lambda_p = 0$		

The fitting procedure described in Sect. 4 was also applied to the traces of IP traffic measured at Bellcore and IITiS. The parameters of the superposition of two-state MMPP's was fitted to these obtained from real data traffic. Table 4 gives the parameters defining the model fitted using the set of descriptors obtained from Bellcore trace. The superposition of four MMPP's is sufficient to model asymptotic second-order self-similarity of the counting process over five time-scales. Table 5 gives the parameters obtained for IITiS data traffic traces. For these trace we have used also the superposition of four MMPP's.

Table 2. Obtained parameters of source fitted to second-order self-similarity with input parameters: $d = 5$, $n = 6$, $\lambda^* = 3.5$, $H = 0.75$ and $\rho = 0.6$

	λ_i^{IPP}	c_{1i}	c_{2i}
IPP_1	13.634	6.797×10^{-1}	1.203×10^{-1}
IPP_2	5.701	2.150×10^{-2}	3.803×10^{-3}
IPP_3	2.475	6.797×10^{-4}	1.203×10^{-4}
IPP_4	0.933	2.150×10^{-5}	3.803×10^{-6}
IPP_5	0.540	6.797×10^{-7}	1.203×10^{-7}
Poisson	$\lambda_p = 0$		

Table 3. Obtained parameters of source fitted to second-order self-similarity with input parameters: $d = 5$, $n = 7$, $\lambda^* = 3.5$, $H = 0.9$ and $\rho = 0.5$

	λ_i^{IPP}	c_{1i}	c_{2i}
IPP_1	3.628	5.431×10^{-1}	2.569×10^{-1}
IPP_2	2.571	2.162×10^{-2}	1.022×10^{-2}
IPP_3	1.913	8.60×10^{-4}	4.071×10^{-4}
IPP_4	1.363	3.427×10^{-5}	1.621×10^{-5}
IPP_5	1.424	5.431×10^{-7}	2.569×10^{-7}
Poisson	$\lambda_p = 0$		

Table 4. Obtained parameters of source fitted to the correlation structure of Bellcore data (input parameters: $d = 4$, $n = 5$, $\lambda^* = 6.3$, $H = 0.85$ and $\rho = 0.15$)

	λ_i^{IPP}	c_{1i}	c_{2i}
IPP_1	2.062	4×10^{-1}	4×10^{-1}
IPP_2	1.469	8.618×10^{-3}	8.618×10^{-3}
IPP_3	0.427	1.857×10^{-4}	1.857×10^{-4}
IPP_4	0.602	4×10^{-6}	4×10^{-6}
Poisson	$\lambda_p = 4.020$		

Table 5. Obtained parameters of source fitted to the correlation structure of IITiS data (input parameters: $d = 4$, $n = 5$, $\lambda^* = 9.85$, $H = 0.79$ and $\rho = 0.0213$)

	λ_i^{IPP}	c_{1i}	c_{2i}
IPP_1	1.086	4×10^{-1}	4×10^{-1}
IPP_2	0.508	8.618×10^{-3}	8.618×10^{-3}
IPP_3	0.194	1.857×10^{-4}	1.857×10^{-4}
IPP_4	0.126	4×10^{-6}	4×10^{-6}
Poisson	$\lambda_p = 8.893$		

6   Conclusions

This paper illustrates how a superposition of two-state MMPP's can be fitted to a rate and variance time curve of measured traffic traces. The fitting algorithm for matching asymptotic second-order self-similarity was described in detail. This article also shows how to obtain the set of descriptors from real traffic traces using Bellcore Morristown Laboratory data and IITiS traffic trace. Our future works will focus on the fitting to additional real traffic descriptors besides second-order properties of the counting process. We will also try to use other methods of calculating the Hurst parameter of real data.

Acknowledgements. This research was partially financed by Polish Ministry of Science and Higher Education project no. N N516479640.

References

1. Kleinrock, L.: Queueing Systems, vol. II. Wiley, New York (1976)
2. Becchi, M.: From Poisson Processes to Self-Similarity: a Survey of Network Traffic Models. Technical report, Citeseer (2008)
3. Crovella, M., Bestavros, A.: Self-similarity in World Wide Web Traffic: Evidence and Possible Causes. IEEE/ACM Transactions on Networking 5, 835–846 (1997)
4. Domański, A., Domańska, J., Czachórski, T.: The impact of self-similarity on traffic shaping in wireless LAN. In: Balandin, S., Moltchanov, D., Koucheryavy, Y. (eds.) NEW2AN 2008. LNCS, vol. 5174, pp. 156–168. Springer, Heidelberg (2008)
5. Paxson, V., Floyd, S.: Wide area traffic: the failure of poisson modeling. IEEE/ACM Transactions on Networking 3, 226–244 (1995)
6. Willinger, W., Leland, W.E., Taqqu, M.S.: On the self-similar nature of ethernet traffic. IEEE/ACM Transactions on Networking 2, 1–15 (1994)
7. Garret, M., Willinger, W.: Analysis, modeling and generation of self-similar VBR video traffic. In: ACM SIGCOMM, London, pp. 269–280 (1994)
8. Norros, I.: On the use of fractional Brownian motion in the theory of connectionless networks. IEEE Journal on Selected Areas in Communication 13(6), 953–962 (1995)
9. Taqqu, M.S., Willinger, W., Sherman, R.: Proof of a fundamental result in self-similar traffic modeling. Computer Communication Review 27(2), 5–23 (1997)
10. Mikosch, T., Resnick, S., Rootzen, H., Stegeman, A.: Is Network Traffic approximated by Stable Levy Motion or Fractional Brownian Motion? The Analysis of Applied Probability 12(1), 23–68 (2002)
11. Erramilli, A., Singh, R.P., Pruthi, P.: An Application of Deterministic Chaotic Maps to Model Packet Traffic. Queueing Systems 20(1-2), 171–206 (1995)
12. Gallardo, J.R., Makrakis, D., Orozco-Barbosa, L.: Use a α-stable self-similar stochastic processes for modeling traffic in broadband networks. Performance Evaluation 40(1-3), 71–98 (2000)
13. Harmantzis, F., Hatzinakos, D.: Heavy network traffic modeling and simulation using stable FARIMA processes. In: 19th International Teletraffic Congress, Beijing, China (2005)
14. Laskin, N., Lambadatis, I., Harmantzis, F.C., Devetsikiotis, M.: Fractional Levy Motion and its Application to network traffic modeling. Computer Networks 40(3), 363–375 (2002)

15. Robert, S., Boudec, J.Y.L.: New models for pseudo self-similar traffic. Performance Evaluation 30(1-2), 57–68 (1997)
16. Muscariello, L., Mellia, M., Meo, M., Ajmone Marsan, M., Lo Cigni, R.: Markov models of internet traffic and a new hierarchical MMPP model. Computer Communications 28, 1835–1851 (2005)
17. Clegg, R.G.: Markov-modulated on/off processes for long-range dependent internet traffic. Computing Research Repository, CoRR, arXiv:cs/0610135 (2006)
18. Domańska, J., Domański, A., Czachórski, T.: The Drop-From-Front Strategy in AQM. In: Koucheryavy, Y., Harju, J., Sayenko, A. (eds.) NEW2AN 2007. LNCS, vol. 4712, pp. 61–72. Springer, Heidelberg (2007)
19. Domańska, J., Domański, A., Czachórski, T.: Internet traffic source based on Hidden Markov Model. In: Balandin, S., Koucheryavy, Y., Hu, H. (eds.) NEW2AN/ruSMART 2011. LNCS, vol. 6869, pp. 395–404. Springer, Heidelberg (2011)
20. Domańska, J., Augustyn, D.R., Domański, A.: The choice of optimal 3-rd order polynomial packet dropping function for NLRED in the presence of self-similar traffic. Bulletin of the Polish Academy of Sciences, Technical Sciences 60(4) (2012)
21. Andersen, A.T., Nielsen, B.F.: A Markovian Approach for Modeling Packet Traffic with Long-Range Dependence. IEEE Journal on Selected Areas in Communications 16(5) (1998)
22. Bellcore Morristown Research and Engineering facility traffic traces, http://ita.ee.lbl.gov/html/contrib/BC.html
23. Foremski, P., Callegari, C., Pagano, M.: Waterfall: Rapid identification of IP flows using cascade classification. In: Kwiecień, A., Gaj, P., Stera, P. (eds.) 21th Conference on Computer Networks, CN 2014. CCIS, vol. 431, pp. 14–23. Springer, Heidelberg (2014)
24. Mandelbrot, B., Ness, J.V.: Fractional Brownian Motions, Fractional Noises and Applications. SIAM Review 10 (October 1968)
25. Bhattacharjee, A., Nandi, S.: Statistical analysis of network traffic inter-arrival. In: 12th International Conference on Advanced Communication Technology, USA, pp. 1052–1057 (2010)
26. Cox, D.R.: Long-range dependance: A review. In: Statistics: An Appraisal, pp. 55–74 (1984)
27. Beran, J.: Statistics for Long-Memory Processes. Chapman and Hall (1994)
28. Stallings, W.: High-Speed Networks: TCP/IP and ATM Design Principles. Prentice-Hall (1998)
29. Casale, G.: Building accurate workload models using Markovian Arrival Processes. In: SIGMETRICS 2011, San Jose, USA (June 2011)
30. Fischer, W., Meier-Hellstern, K.: The Markov-modulated Poisson process (MMPP) cookbook. Performance Evaluation 18(2), 149–171 (1993)
31. Okamura, H., Kamahara, Y., Dohi, T.: Estimating Markov-modulated compound poisson processes. In: Valuetools 2007, Nantes, France (October 2007)
32. Domańska, J., Domański, A.: The influence of traffic self-similarity on QoS mechanism. In: International Symposium on Applications and the Internet, SAINT, Trento, Italy (2005)

The Cluster-Based Time-Aware Web System

Krzysztof Zatwarnicki and Anna Zatwarnicka

Department of Electrical, Control and Computer Engineering,
Opole University of Technology, Opole, Poland
{k.zatwarnicki,anna.zatwarnicka}@gmail.com

Abstract. The problem of providing fixed Quality of Web Services (QoWS) is now crucial for further development and application in new areas of internet services. In this paper, we present the MLF (Most Loaded First) adaptive and intelligent cluster-based Web system which provides quality of service on a fixed level. The proposed system keeps the page response time within established boundaries in such a way that with a heavy workload, the page response times, for both small and complex pages, would not exceed the imposed time limit. We show, in experiments conducted with the use of a cluster of real Web servers, that the system is efficient and more effective than other examined systems.

Keywords: quality of web services, guaranteeing web page response time, HTTP request scheduling, request distribution.

1 Introduction

Over the past few years, the Internet has become the most innovative source for information and data. It has evolved from a medium for only privileged users into a medium we could not imagine living without. The rapid development of systems using WWW technology has given rise to the need for research on the effectiveness of the whole system in delivering the necessary content to the user.

Through the many years of the Web development different aspects of QoWS have been noted as the most important. In the early nineties, the most important thing was to properly service HTTP requests, sent by the Web client and deliver the Web page in a response. At the beginning of the 21st century, the Internet began to become more popular and widely used. During that period many popular Web services had problems with simultaneous service of numerous of clients. One of the most prominent ways to evaluate Web systems then was to measure their throughput (number of requests serviced in a time unit). Nowadays, in order to provide appropriate throughput of the system, Web clusters are involved, and the problem of gaining adequate capacity is not a crucial. The challenge now is to service HTTP requests and deliver Web pages to the client within an acceptable amount of time [1].

The problem of servicing HTTP requests, taking the response time in to account was already discussed in our previous work. We have proposed systems minimizing response times in a locally distributed Web cluster system [2], a globally distributed Web cluster system with a broker [3] and a globally distributed

A. Kwiecień, P. Gaj, and P. Stera (Eds.): CN 2014, CCIS 431, pp. 37–46, 2014.

Web cluster system without a broker [4]. In our later work, we dealt with the problem of providing fixed quality of service in Web systems consisting of one Web server [5], a locally distributed cluster of servers [6], and a globally distributed Web cluster with broker [7].

In this paper, we come back to the problem of providing fixed quality of service in a locally distributed cluster of servers and present the MLF system. The system was initially described in [6] together with the results of simulation experiments. This article presents the results of experiments conducted for the MLF system with the use of real Web servers constituting the Web cluster. We analyze these results to determine if the system is equally or more efficient than other systems already known from literature or used in the Internet.

There are many papers on how to provide and guarantee Web service quality. Most of them concern maintaining the quality of the service for individual HTTP requests [8–13]. Very few papers have been dedicated to Web systems guaranteeing to service Web pages within a limited amount of time. In those papers [14, 15], the proposed solution maintains the quality of the Web service only for a limited group of users. Guaranteeing quality of service or providing fixed quality almost always involves rejecting requests of users not belonging to a privileged group. The MLF proposed in this article system not only provides fixed quality of service, keeping the page response time within established boundaries, but also treats all users equally. It should be also noticed that MLF system can keep the quality of service only for the acceptable amount of incoming traffic above which the quality becomes lower than expected.

The paper is divided into four sections. Section 2 presents the MLF system and the methods to schedule and distribute HTTP requests. Section 3 describes the conducted experiments and the results. Finally, Section 4, presents concluding remarks.

2 MLF System

The MLF system consists of: clients sending HTTP requests, MLF Web switch queuing and distributing HTTP requests, Web servers servicing the requests and database servers delivering data to the Web servers (Fig. 1a). The Web switch, the database and the Web server together form the Web cluster. The manner of operation of the Web cluster determines the MLF method. According to the method, clients send HTTP requests to the Web cluster, where the Web switch receives all of the requests, queues them and distributes them among Web servers. The requests are serviced by the Web servers with the assistance of database servers. Responses are delivered back to the Web switch and immediately sent to the clients.

The Web switch controls the operations of the cluster. It especially takes care to deliver the entire Web page to the client within the fixed time t_{max}. To achieve this, the Web switch calculates the term in which each of the incoming requests have to be serviced on the Web server. The requests are queued in the switch according to these calculated terms. When each request leaves the queue, the Web

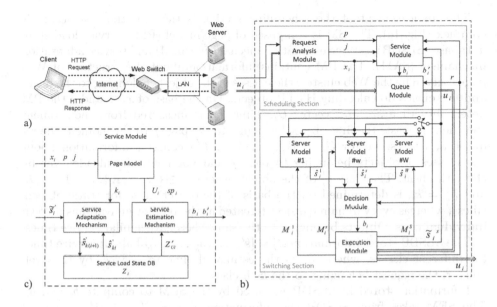

Fig. 1. MLF system: a) Overall view, b) Web switch, c) Service module

switch determines the Web server able to service the request within required time. In the end the request is sent to the chosen Web server.

The Web switch is logically divided into two separate sections (Fig. 1b). The first one is a scheduling section, which is responsible for scheduling requests upon their entrance to the Web cluster. The second section is a switching section, which distributes requests among Web servers.

After receiving the incoming HTTP request u_i, where i is the request index, the Web switch passes it to the Request Analysis Module (RAM). The RAM fetches from the request the following information: address x_i of requested object, identifier j of the client sending the request, and identifier p of the page to which the requested object belongs. The p and j identifiers are included in a *cookie* section of the HTTP request. Upon receiving information from the RAM, a Service Module (SM) computes deadline b_i, indicating the moment the service of the HTTP request has to be started, and term b_i' in which the request service has to be finished. After calculating deadline b_i, the request is passed to the Queue Module (QM) in which the requests are queued according to Earliest Deadline First policy (requests with the shortest deadlines are placed at the beginning of the queue). Not all of the requests are placed in the QM. When the number r of requests serviced concurrently by all of the Web servers in the cluster is smaller than the value $r_{\max}$ then the request u_i is passed directly to the switching section. The value $r_{\max}$ is the lowest number of requests serviced by web servers in the cluster, for which the service reaches the maximal throughput. Requests placed in the QM leave the queue and are passed to the switching section only when $r < r_{\max}$.

The main element of the scheduling section is the SM, those structure is complex (Fig. 1c). The SM is composed of page model (PM), service load state database (SLSDB), service estimation mechanism (SEM) and service adaptation mechanism (SAM). The PM contains information about the structure of Web pages served by the Web cluster, the classes of the objects belonging to the pages and the clients downloading HTTP objects. On the base of x_i, p and j the PM determines time sp_i and vector U_i. Time sp_i is measured from the moment the client requests the first object belonging to the page p to the moment the request u_i arrives. Vector $U_i = [k_i^1, \ldots, k_i^l, \ldots, k_i^L]$ contains information about the classes of objects belonging to page p and not being downloaded yet by the j-th client. The class k_i^l of the object, where $k_i^l \in \{1, \ldots, K\}$, $l = 1, \ldots, L$, and $k_i^1 = k_i$, is determined on the basis of the object's size for static objects (files), whereas every dynamic object (created at the request arrival) has its own individual class. Objects belonging to the same class have similar service times.

The SLSDB stores information $Z_i' = [\hat{s}_{1i}', \ldots, \hat{s}_{ki}', \ldots, \hat{s}_{Ki}']$ about service times for different classes, where $\hat{s}_{ki}'$ is the estimated time to service, by the Web cluster, the request belonging to the k-th class.

Information stored in SLSDB are used by the SEM to compute b_i and b_i'. The SEM takes from SLSDB the information $Z_{Ui}' = [\hat{s}_{k^1 i}', \ldots, \hat{s}_{k^l i}', \ldots, \hat{s}_{k^L i}']$ about service times of objects pointed in the U_i vector. The deadline b_i is calculated in the following way $b_i = \tau_i^{(1)} + \Delta b_i - \hat{s}_{1i}$, where $\tau_i^{(1)}$ is the time of the i-th request's arrival, and $\Delta b_i - \hat{s}_{k^1 i}'(t_{\max} - p_i)/(\lambda \sum_{l=1}^{L} \hat{s}_{k^l i}')$ is the period of time which the request can spend being queued and serviced by the Web server. The λ is a concurrency factor, and it depends on the number of Web objects being downloaded concurrently for given the Web page. According to [16], the value of this factor for average Web pages can be set to $\lambda = 0.267$, or if the WWW pages are not typical, the value can be designated experimentally. The term b' is calculated as follows: $b_i' = \tau_i^{(1)} + \Delta b_i$, and that is the term by which the service of the request has to be finished by the Web server.

The SAM module is responsible for adapting the information Z_i'. After each previously queued request is serviced, the SAM updates the service time for the class the request belonged to. The modification is conducted in the following way $\hat{s}_{k(i+1)}' = \hat{s}_{ki}' + \hat{\eta}(\hat{s}_i - \hat{s}_{ki}')$, where $\hat{s}$ is a measured value of the service time, and $\hat{\eta}$ is an adaptation factor.

After the request u_i leaves the QM it is directed to the switching section of the Web switch. The switching section is composed of: server model modules (SMM), a decision module (DM), and an execution module (EM).

Every SMM is assigned to one Web server working in the cluster. The SMM estimates the service time $\hat{s}_i^w$ of the request for the Web server the module is assigned to, where $w \in \{1, \ldots, W\}$ is the server index, and W is the number of Web servers in the cluster. The service time is computed on the base of information of the requested object class k_i and the load $M_i^w = [e_i^2, f_i^2]$ of the server the SMM is assigned to, where e_i^w is the number of requests being concurrently serviced by the w-th server and f_i^w is the number of dynamic request concurrently serviced.

The DM chooses the Web server to service the request. The decision is made in the following way:

$$z_i = \begin{cases} \min\{w : w \in \{1, \ldots, W\}\} & \text{if } r = r_{\max} \text{ and } \exists_{w \in \{1, \ldots, W\}} \hat{s}_i^w \geq \Delta b_i' \\ w_{\min} : \hat{s}_i^{w_{min}} = \min\{\hat{s}_i^w : w \in \{1, \ldots, W\}\} & \text{in other case} \end{cases} \quad (1)$$

where $\Delta b_i' = \tau_i^{(2)} - b_i'$, and $\tau_i^{(2)}$ is the moment the request u_i leaves the QM. According to the Formula (1) the Web server with the lowest index, which is able to service the request in a time not longer than $\Delta b_i'$, is chosen. If there is no such server, the server offering the shortest service time is chosen. Thanks to this solution, Web servers with the lowest indexes are the most loaded but, still able to service requests, and the servers with higher indexes are unloaded and able to service requests very quickly, when necessary.

When the decision z_i is taken, the EM sends the request u_i to the chosen server. The module also measures the service time $\tilde{s}_i$, and collects the information e_i^s and f_i^s.

The most complex module of the switching section is the SMM. To estimate the service time $\hat{s}_i^w$ of the request, for the given Web server, a neuro-fuzzy model is used (Fig. 2a). Because there are many classes $k_i = 1, \ldots, K$ of requested objects, the same neuro-fuzzy model with different parameters (weights) for each class is used. It can even be said that each of the SMM has K separated neuro-fuzzy networks. All parameters for different classes and networks are stored in the parameter database $Z_i = [Z_{1i}, \ldots, Z_{ki}, \ldots, Z_{Ki}]$, where $Z_{ki} = [C_{ki}, D_{ki}, S_{ki}]$, $C_{ki} = [c_{1ki}, \ldots, c_{lki}, \ldots, c_{(L-1)ki}]$ and $D_{ki} = [d_{1ki}, \ldots, d_{mki}, \ldots, s_{Jki}]$ are parameters of input fuzzy set functions, and $S_{ki} = [s_{1ki}, \ldots, s_{jki}, \ldots, s_{Jki}]$ are parameters of output fuzzy set functions. Input fuzzy set functions are triangular (Fig. 2b) and are denoted as $\mu_{F_{el}}(s)$, $\mu_{F_{fm}}(f_i)$, $l = 1, \ldots, L$, $m = 1, \ldots, M$, whereas ointentfirstutput fuzzy sets functions $\mu_{Sj}(s)$ are singletons (Fig. 2c). The values L and M were chosen experimentally and set to 5, and $J = L \cdot M$.

The service time is calculated in the following way: $\hat{s}_i = \sum_{j=1}^{J} s_{jki} \mu_{R_j}(e_i, f_i)$, while $\mu_{R_j}(e_i, f_i) = \mu_{F_{el}} \cdot \mu_{F_{fm}}(f_i)$. The values of parameters C_{ki}, D_{ki}, S_{ki} are modified in an adaptation process using the Back Propagation Method each time the service of the request on a given server finishes. The parameters of output fuzzy sets are modified as follow: $s_{jk(i+1)} = s_{jki} + \eta_s \cdot (\tilde{s}_i - \hat{s}_i) \cdot \mu_{R_j}(e_i, f_i)$, whereas parameters of input fuzzy sets are computed in following way

$$c_{\phi k(i+1)} = c_{\phi ki} + \eta_c(\tilde{s}_i - \hat{s}_i) \sum_{m=1}^{M} (\mu_{F_{fm}}(f_i) \sum_{l=1}^{L} (s_{((m-1) \cdot L + l)ki} \partial \mu_{F_{el}}(e_i) / \partial c_{\phi ki}))$$

and

$$d_{\gamma k(i+1)} = d_{\gamma ki} + \eta_d(\tilde{s}_i - \hat{s}_i) \sum_{l=1}^{L} (\mu_{F_{el}}(e_i) \sum_{m=1}^{M} (s_{((l-1) \cdot M + m)ki} \partial \mu_{F_{fm}}(f_i) / \partial d_{\gamma ki}))$$

where η_s, η_c, η_d are adaptation ratios, $\phi = 1, \ldots, L-1$, $\gamma = 1, \ldots, M-1$.

3 Testbed and Results of Experiments

In order to evaluate the MLF system simulation experiments were conducted in our previous research [5]. Results of the experiments show that the MLF system can provide higher quality of service than other reference and well known

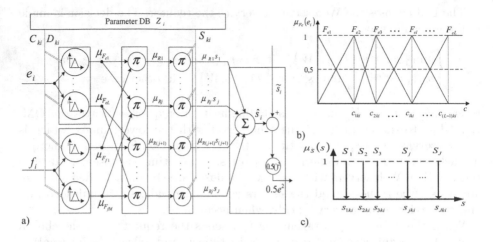

Fig. 2. Server model: a) neuro-fuzzy model, b) input fuzzy sets functions, c) output fuzzy sets functions

distribution methods. The next step to evaluate the system is to conduct experiments with the use of real Web servers and the MLF Web switch. The results of the experiments presented in this article should enable answering the question of whether the proposed system is equally or more effective than other systems already known of from literature or used in real Web switches.

The experiments were conducted using four Web servers, one computer acting as a Web switch and one computer simulating the behavior of Web clients. The first computer with an Intel Core i5-3470 3.3 GHz processor, and Ubuntu 13.04 Desktop operating system, was acting as a generator of the HTTP requests. The second computer, with an Intel Core i7-2670 2.2 GHz processor and Fedora 18 operating system, hosted the Web switch server software. The computers chosen for the Web, and database servers, had the lowest computational power (Intel Celeron 1.7 GHz, Ubuntu 13.04 Server), so it was easy to reach the maximal capacity of the Web servers without overloading other elements of the system. All of the computers were connected through a gigabyte Repotec RP-G3224V network switch.

The Web server hosted five different Web pages. Table 1 presents the structure of the pages. The pages were static and dynamic. All of the pages contained from 10 to 30 embedded objects from 1 to 100 KB in size. Dynamic pages used PHP as the script language generating the content of the page. One of the pages also used SQL requests to the MySQL database, containing 3 related tables of 10 000 rows in size each.

The Web switch server software was written in the C++ language with the use of the *libsoup* [17] and *boost* [18] libraries supporting the supervision of the HTTP requests. The *gcc* compiler was used to create the executable file.

Table 1. Web pages used in the experiments

Name	Type and size of the frame object of the page	Embedded objects number and sizes	MySQL Database
Static 10	Static 1 KB	10, size 1–100 KB, sum 477 KB	—
Static 30	Static 1 KB	30, size 1–100 KB, sum 1.39 MB	—
Dynamic 10	Dynamic, PHP	10, size 1–100 KB, sum 477 MB	—
Dynamic 30	Dynamic, PHP	30, size 1–100 KB, sum 1.39 MB	—
Dynamic MySQL 30	Dynamic, PHP	30, size 1–100 KB, sum 1.39 MB	requests to database containing 3 tables, 10 000 rows each

In order to compare the MLF method with other well known and often applied distribution methods the Web switch had implemented four different scheduling policies:

- MLF,
- LARD (Locality-Aware Request Distribution): one of the best distribution methods taking into account localization of the previously requested object,
- CAP (Content Aware Policy): an algorithm uniformly distributing HTTP requests of different types,
- RR (Round-Robin): an algorithm distributing uniformly all incoming requests.

The software of the HTTP request generator was written in the C++ language with the use of the *LibcURL* library, which enables the creation of HTTP requests and the supervision of the process of sending requests and receiving responses. The request generator generated requests in a similar way to modern Web browsers. It created a given number of virtual clients. Each client, at first, opened a TCP connection to send the requests concerning the frame of the page. After that, it sent in concurrent TCP connections requests concerning objects pointed out in the header of the HTML document. After receiving the frame as a whole, the client opened up to 6 TCP connections to download the embedded objects. Immediately after finishing downloading the Web page as a whole, the client started to download the next page.

The software of the generator was not only generating HTTP request but also collecting information concerning the mean value of request response time and satisfaction. Nowadays the response time is one of the most important measures of the effectiveness of the Web system. Most of the results illustrating the effectiveness of the MLF system in comparison to other systems show the response time in a load function.

The satisfaction allows for determining whether the MLF system is operating as expected and provides the page response time to be no longer then $t_{\max}$ even when the load is high. The satisfaction is often used to evaluate real-time soft systems. It is equal to 1 when the page response time is shorter than t_{max}^s, and decreases to 0 when the time is longer than t_{max}^h (Fig. 3). In all of the experiments, it has been adopted that $t_{max}^s = t_{max} = 2000$ ms and $t_{max}^h = 2t_{max}^s$. Experiments were conducted for various increasing numbers of simulated clients [19].

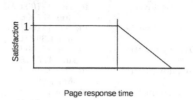

Fig. 3. Satisfaction function

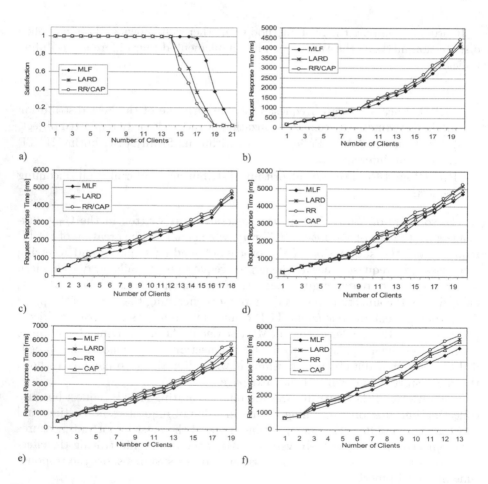

Fig. 4. Results of experiments: a) Satisfaction vs. load for the Web page Static 10; Mean request response time vs. load for Web pages: b) Static 10, c) Static 30, d) Dynamic 10, e) Dynamic 30, f) Dynamic MySQL 30

The Figure 4a presents a diagram of satisfaction vs. number of clients generating HTTP requests. It can be noticed, that when the load is low (the number of clients is between 1 and 14), the satisfaction for all the strategies is very high and equal to 1. This means that almost every Web page was downloaded by the clients with a time no longer than $t_{max}s$. When the load is higher (the number of clients is greater than 15), the satisfaction for LARD and RR/CAP (for static Web pages the RR and CAP algorithms operate in the same way) algorithms significantly decreased. For the MLF method, the satisfaction begins to decrease when the load is very high and equals to 18. That demonstrates that MLF system is able to provide a page response time no longer than t_{max} even when the load is high.

Diagrams b to f in Fig. 4 present the mean request response time vs. the load for different Web pages. In every experiment the mean request response time was lowest for the MLF method. When the load is low, the differences between distribution methods are negligible. However, when the load increases, the request response time for the MLF method also increases but is considerably lower than for other distribution methods. It can be also noticed that the difference between MLF method and the others methods are greater for the dynamic Web pages. The presented results confirm results obtained during simulations [6] therefore it can be concluded that under control of the MLF method the cluster-based Web system is more effective than for other examined methods.

4 Summary

In this paper the HTTP request scheduling and distribution cluster-based Web system providing service at a fixed level, was presented. The proposed MLF system can deliver Web pages to the clients in a time not longer than the Web provider demands. Decision algorithms, used in the MLF method, applies adaptive algorithms using neuro-fuzzy models in its construction. Experiments conducted with the use of real Web servers confirm results obtained during simulations and presented in other articles. A high level of quality can be provided even when the system is heavily loaded. The request response times for individual HTTP requests are considerably lower for the MLF method than for other distribution methods. The results of experiment show that the Web system working under control of the MLF method is more effective than for other examined methods.

References

1. McCabe, D.: Network analysis, architecture, and design. Morgan Kaufmann, Boston (2007)
2. Zatwarnicki, K.: Adaptive control of cluster-based web systems using neuro-fuzzy models. International Journal of Applied Mathematics and Computer Science (AMCS) 22(2), 365–377 (2012)

3. Zatwarnicka, A., Zatwarnicki, K.: Adaptive HTTP request distribution in time-varying environment of globally distributed cluster-based Web system. In: König, A., Dengel, A., Hinkelmann, K., Kise, K., Howlett, R.J., Jain, L.C. (eds.) KES 2011, Part I. LNCS, vol. 6881, pp. 141–150. Springer, Heidelberg (2011)
4. Zatwarnicki, K.: Neuro-Fuzzy Models in Global HTTP Request Distribution. In: Pan, J.-S., Chen, S.-M., Nguyen, N.T. (eds.) ICCCI 2010, Part I. LNCS (LNAI), vol. 6421, pp. 1–10. Springer, Heidelberg (2010)
5. Zatwarnicki, K.: Adaptive Scheduling System Guaranteeing Web Page Response Times. In: Nguyen, N.-T., Hoang, K., Jędrzejowicz, P. (eds.) ICCCI 2012, Part II. LNCS (LNAI), vol. 7654, pp. 273–282. Springer, Heidelberg (2012)
6. Zatwarnicki, K.: Operation of Cluster-Based Web System Guaranteeing Web Page Response Time. In: Bădică, C., Nguyen, N.T., Brezovan, M. (eds.) ICCCI 2013. LNCS (LNAI), vol. 8083, pp. 477–486. Springer, Heidelberg (2013)
7. Zatwarnicki, K.: Guaranteeing quality of service in globally distributed Web system with brokers. In: Jędrzejowicz, P., Nguyen, N.T., Hoang, K. (eds.) ICCCI 2011, Part II. LNCS, vol. 6923, pp. 374–384. Springer, Heidelberg (2011)
8. Abdelzaher, T.F., Shin, K.G., Bhatti, N.: Performance Guarantees for Web Server End-Systems: A Control-Theoretical Approach. IEEE Trans. Parallel and Distributed Systems 13(1), 80–96 (2002)
9. Blanquer, J.M., Batchelli, A., Schauser, K., Wolski, R.: Quorum: Flexible Quality of Service for Internet Services. In: Proc. Symp. Networked Systems Design and Implementation (2005)
10. Harchol-Balter, M., Schroeder, B., Bansal, N., Agrawal, M.: Size-based scheduling to improve web performance. ACM Trans. Comput. Syst. 21(2), 207–233 (2003)
11. Kamra, A., Misra, V., Nahum, E.M.: A Self Tuning Controller for Managing the Performance of 3-Tiered Websites. In: Proc. Workshop Quality of Service, pp. 47–56 (2004)
12. Schroeder, B., Harchol-Balter, M.: Web servers under overload: How scheduling can help. In: 18th International Teletraffic Congress, Berlin, Germany (2003)
13. Olejnik, R.: An Impact of the Nanoscale Network-on-Chip Topology on the Transmission Delay. In: Kwiecień, A., Gaj, P., Stera, P. (eds.) CN 2011. CCIS, vol. 160, pp. 19–26. Springer, Heidelberg (2011)
14. Suchacka, G.z., Borzemski, L.: Simulation-based performance study of e-commerce Web server system – results for FIFO scheduling. In: Zgrzywa, A., Choroś, K., Siemiński, A. (eds.) Multimedia and Internet Systems: Theory and Practice. AISC, vol. 183, pp. 249–259. Springer, Heidelberg (2013)
15. Wie, J., Xue, C.Z.: QoS: Provisioning of client-perceived end-to-end QoS guarantees in Web servers. IEEE Trans. on Computers 55(12) (2006)
16. Zatwarnicki, K., Zatwarnicka, A.: Estimation of Web Page Download Time. In: Kwiecień, A., Gaj, P., Stera, P. (eds.) CN 2012. CCIS, vol. 291, pp. 144–152. Springer, Heidelberg (2012)
17. Libsoup library description, http://developer.gnome.org/libsoup/stable/ (access September 10, 2013)
18. Boost C++ libraries, http://www.boost.org/ (access September 10, 2013)
19. Platek, M.: Guaranteeing quality of service in cluster-base Web system. M.S. thesis, Department of Electroengineering, Automatic Control and Computer Science, Opole University of Technology, Opole, Poland (2013)

Influence of the VM Manager on Private Cluster Data Mining System

Dariusz Czerwinski

Lublin University of Technology, 38A Nadbystrzycka Str, 20-618 Lublin, Poland
d.czerwinski@pollub.pl

Abstract. In this paper the comparative analysis of the impact of the virtual machine manager on the Private Cluster Data Mining System performance was shown. The idea of the research is comparison of the performance of Data Mining System under different VM's and different operating systems. In this article the results obtained in a test environment, based on the Ubuntu Desktop 13.10 x64 used as host OS and Cloudera Hadoop distribution used as personal cluster guest OS, were shown. The main focus of the research is the hypervisor impact on the typical operations in data mining system, such as parallelized calculation, file system operations and the use of CPU resources.

Keywords: computer networks, data mining system, virtualization, Cloudera Hadoop.

1 Introduction

Data mining systems play a very important role in the IT industry. They are used in many fields related to science and also are applied as a commercial solutions. An example of such a system is Hadoop – a data mining project developed by the Apache Foundation. Hadoop project is part of the Cloudera distribution. It consists of software and a set of libraries for creating tasks using the Map Reduce paradigm [1–5]. MapReduce is useful tool to perform big data analysis in the cloud or cluster environment.

One of the basic assumptions that have been adopted by the developers of the Hadoop project, is the simplicity of programming data mining methods. This allows for a convenient implementation of a wide range algorithms useful in many applications. Applications written in the Java language have direct access to the Hadoop API, but Hadoop also has features that allow for implementations written in any programming language [6].

At present, many organizations that need to process large amounts of data are using the MapReduce. Google first started using MapReduce before 2004 [1]. The company Yahoo has the largest Hadoop cluster that uses more than 42 000 nodes [7]. Amazon uses Hadoop as part of their cloud services. Hadoop is used by such entities as: Facebook, last.fm, New York Times, AOL Advertising, NAVTEQ, Samsung, TrendMicro, and many others, the list of which can be

A. Kwiecień, P. Gaj, and P. Stera (Eds.): CN 2014, CCIS 431, pp. 47–56, 2014.

found on the Hadoop project website [2]. Decentralised calculations are widely used in the IT industry as also as science and their projects are extremely diversified [8].

Against this background, the distinguishing features of the Hadoop project are: [9]

– availability – Hadoop can run as large clusters of generally available machines in cloud computing systems, such as Amazon Elastic Compute Cloud (EC2) as well as small clusters or private clouds,
– reliability – Hadoop is designed to work on PC and was therefore designed with the assumption of frequent hardware failures,
– scalability – Hadoop exhibits linear scalability in situations where there is a need to increase the number of nodes in the cluster,
– simplicity – Hadoop allows the user in a simple and efficient way to implement the method of parallel calculations.

The idea of Hadoop cluster action, which performs its tasks in the cloud computing environment, is shown in Fig. 1. In the simplest case, the Hadoop cluster is a set of dedicated computers that are connected in computer network, usually realized in a single location. It is also possible to configure a Hadoop cluster that the machines are in different locations. Storage and processing of all data is performed in the "cloud" of machines. Different users, both in remote locations as well as local, can submit jobs to the Hadoop cluster manager.

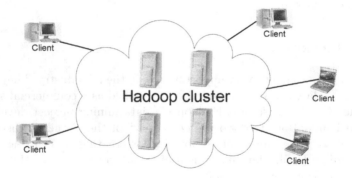

Fig. 1. Idea of the Hadoop cluster, services available analogous to the cloud availability

2   Tests of the Private Cluster

Cloudera system allows building structures of both computing clusters as well as private clouds. Thanks to this, users of the system use the resources in the same way as public cloud resources offered by Amazon. This allows for easy integration of this solution in a hybrid cloud projects [10–12].

Hadoop, which is part of the Cloudera distribution consists of the following elements [2, 9, 13]:

- Slave Node (SN) – it is a physical resource (usually a single host), where the tasks are running. This node can operate in either data collection mode (called DataNode) and the accepting tasks mode (called TaskTracker). It is possible to configure the nodes so that they operate only as nodes collecting data or as nodes that support commissioned tasks.
- Master Node (MN) – in small clusters, this device has four functions: a node that manages tasks (called JobTracker) in the cluster, the node that accepts the task (called TaskTracker), node handling directory tree in the HDFS file system (called NameNode) but it does not store the data and the node storing and replicating data in the HDFS (called DataNode).
- HDFS File System (Hadoop Distributed File System) – is crucial to store the data in cluster. HDFS is highly resistant to damage, so it can be implemented even on PC hardware. HDFS file system provides high throughput access to application data.

2.1 Test Platform

Developed Testbed provided the base for performance evaluation of data mining system in the private cluster structure under different virtual hypervisors. This cluster consisted of three PCs that are running Ubuntu Desktop 13.10 64-bit operating system with latest updates. Studies based on the so constructed private cluster are realized in two steps. The first step was to assess the impact of VM hypervisor on performance of data mining system when the host operating system was LINUX family. The results of this phase are presented in this article. In the next stage, the tests for other host operating systems from the Microsoft Windows family are planned.

Taking this into account, as a basis for the implementation of virtualization, three different packages were selected: VMware (implemented using VMware Player v.6.0.1-1379776), VirtualBox package (implemented on the basis of VirtualBox 4.2.16 for LINUX hosts amd64) and packet KVM with Qemu 1.0.1 (KVM with virtio implementation). Cluster members are installed and configured on PC computers equipped with a dual-core processors from Intel and AMD with support for hardware virtualization, 4 GB RAM, disk storage capacity of 250 GB and a Gigabit Ethernet network interfaces. In such environment, the virtual guest machine was started with Centos 5.3 system on each PC. The guest OS include an implementation of Cloudera's Distribution Including Apache Hadoop (CDH) version CDH3U3 [14].

Testbed architecture is shown in Fig. 2. The individual components of hardware test structure of the system have been assigned to Hadoop as follows (Fig. 2a):

- PC1 – cluster controller, the parent node MN (CPU – AMD Athlon X2 240, 2 cores), 4 GB RAM, 250 GB HDD SATA, host system Ubuntu Desktop 13.10 x64, guest OS Cloudera (Centos 5.3 x64),
- PC2 – node controller SN (CPU – AMD Phenom X2 555, 2 cores), 4 GB RAM, 250 GB HDD SATA, host system Ubuntu Desktop 13.10 x64, guest OS Cloudera (Centos 5.3 x64),

– PC2 – node controller SN (CPU – Intel C2Duo P8700, 2 cores), 4 GB RAM, 250 GB HDD SATA, host system Ubuntu Desktop 13.10 x64, guest OS Cloudera (Centos 5.3 x64).

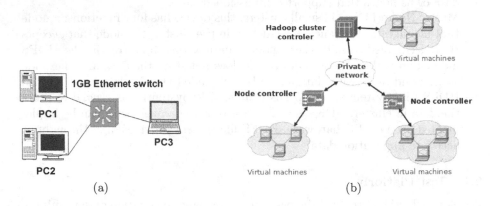

Fig. 2. The architecture of the testbed: (a) physical, (b) logical

All nodes were defined as a virtual machines run under different VM hypervisors and their configuration was as follows:

– CPU: host CPU,
– RAM: 1512 MB of host RAM,
– HDD: 10 GB of host disk space,
– NIC: host network interface (bridged mode).

In the case of Virtual Box and VMware, the Guest Additions and VMware Tools had been installed on guest operating system respectively.

2.2 Benchmarks

Testing of the Hadoop cluster can be focused on: unit testing, integration testing and performance testing. Unit testing verifies the function of each node in isolation. Integration tests verify whether the system is working as a whole. Both, unit and integration testing should be done on pre-production stage. Performance testing, can be focused on two areas: application code optimization and cluster performance optimization [15–18]. Tests and their results carried out in the public cloud were described in [20]. The author used different virtual machines configuration and Hadoop wordcount test to obtain the information on vertical and horizontal scaling. Hadoop benchmarks carried out with different VM hypervisors and on one physical machine were described in [18, 19]. Hypervisors, which have been tested, were: KVM, Xen, OpenVZ and commercial one (name not specified). Described in [15–20] Hadoop tests were conducted for HDFS file system and MapReduce algorithm performance. The *TestDFSIO* benchmark and Hadoop *wordcount, grep, K-Means Clustering, Hivebench* tests were executed for HDFS file system and MapReduce algorithm respectively.

The contribution of this paper is to provide some results to easier choice of the hypervisor for a particular Hadoop cluster workload. In this work the results of the tests directed on the private cluster performance were presented. An unmodified Hadoop implementation has been run on the cluster described in the above section. Three different benchmarks of Hadoop found in Cloudera distribution where chosen: *mapredtest, mrbench* and *DFSCIOTest. TeraSort* benchmark was not run due to the lack of space on HDFS file system. Additionally the Dhrystones and Linpack benchmarks were conducted.

Measurements were carried out in three different test configurations marked as "VMware", "Virtual Box" and "KVM", where the Cloudera OS run properly in the following virtual machines: VMware, Virtual Box and KVM. During testing, operating systems, host and guest were not loaded with additional processes. Hosts worked in a separate local area network where no additional network traffic, except that which occurred between cluster nodes, exists. In order to investigate the influence of the hypervisor a few selected types of tests were performed to obtain meaningful results of individual solutions. Selected tests are:

- Generic map/reduce load generator test run with command:
  `hadoop jar hadoop-test*.jar mapredtest 100 1000000`,
- A map/reduce benchmark that can create many small jobs run with command: `hadoop jar hadoop-test*.jar mrbench`,
- HDFS read test run with command: `hadoop jar hadoop-test*.jar DFSCIOTest -read -nrFiles 3 -fileSize 100`,
- HDFS write test run with command: `hadoop jar hadoop-test*.jar DFSCIOTest -write -nrFiles 3 -fileSize 100`,
- Dhrystones benchmark (No Opt – no optimization at compile time, OPT3 – enabled optimization at compile time),
- Linpack benchmark (No Opt – no optimization at compile time, OPT3 – enabled optimization at compile time).

Each test consisted of 50 measurements, for which the average values and standard deviations were calculated. In the case of MapReduce tests, they were run with addition *time* command before *hadoop* to collect the time taken by program in real, user and kernel mode.

3   Experimental Resuls

Based on the presented above private cluster, the tests to examine the impact of virtual machine hypervisor on data mining system performance were performed. The results of mapredtest were presented in Fig. 3.

The given time is a time taken by program in real mode. The shortest time was obtained under KVM hypervisor, in the case of VMware hypervisor the time grows only about 1.76 %. The slowest solution was obtained under Virtual Box hypervisor and comparing to KVM the difference is about 5.1 %.

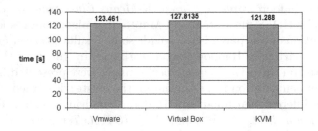

Fig. 3. MapReduce test results obtained with *mapredtest*

Second performed test was mrbench, with its default parameters (*numRuns*=1, *maps*=2, *reduces*=1). The results of the tests are shown in Fig. 4. It can be seen that for KVM and VMware hypervisors perform almost the same. The Virtual Box hypervisor was approximately 9.2 % slower compared to previous ones.

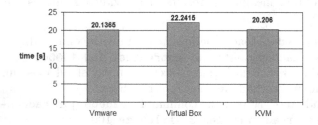

Fig. 4. MapReduce test results obtained with *mrbench*

In order to complement and deepen the above-mentioned conclusions, the assessment of an additional parameter, so-called system overhead, has been done. The system overhead was defined as a percentage value given by proportion *kernel mode time/user mode time·*100 %. The results of the comparison of system overhead for each virtual machine hypervisor was shown in Fig. 5.

It may be noted, that the smallest overhead during both MapReduce tests has the KVM hypervisor, while for hypervisors VMware and Virtual Box is larger even up to 45 % in the case of the last one. VMware hypervisor performs almost the same as KVM, however its system overhead parameter value is over twice in *mapredtest* and almost twice in *mrbench* test.

Next performed tests where focused on cluster HDFS file system performance. Test where achieved with *DFSCIOTest* benchmark run under different virtual machine hypervisors. The scheduled task for that benchmark consist the 3 files, each in 100 MB size. The size of the files was limit due to availability of space on the hard drive. HDFS file system benchmark results were shown in Fig. 6 for read and Fig. 7 for write tests. The performance of the throughput HDFS file system under three different hypervisors is almost the same during read benchmarks.

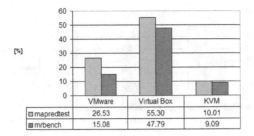

	VMware	Virtual Box	KVM
mapredtest	26.53	55.30	10.01
mrbench	15.08	47.79	9.09

Fig. 5. System overhead comparison for *mapredtest* and *mrbench* tests (lower is better)

The average IO rate during read benchmark has the lowest value for KVM hypervisor and is about 30 % lower than in the case of VMware and Virtual Box. During the write benchmarks the situation is reversed, the KVM hypervisor performance is the highest one and VMware is comparable with it (Fig. 7). The Virtual Box hypervisor write test value was the lowest one (throughput and average IO rate), however the difference, comparing to best ones, is about 6 %.

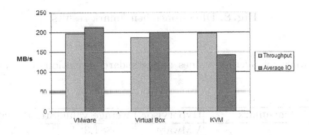

Fig. 6. HDFS read test results (higher is better)

Another benchmark used during the testing of the private cluster was Dhrystones. Figure 8 presents a summary of the results obtained for this test. The standard deviations for this test are summarized in Table 1. On the basis of research results and their analysis (average values with standard deviations do not overlap) can be stated that the highest performance for the Dhrystones benchmark has been obtained for the KVM hypervisor. It can be also noticed that the performance of other hypervisors is close to best one.

The last test was the Linpack benchmark. Figure 9 and Table 2 show a summary of the results obtained for this test. The standard deviations are small and it is clear that the highest performance in this test was obtained for the KVM hypervisor.

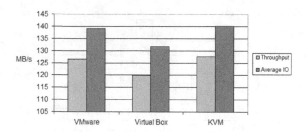

Fig. 7. HDFS write test results (higher is better)

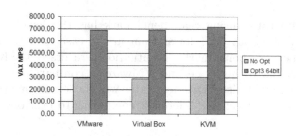

Fig. 8. Dhrystones benchmark results

Table 1. Summary of average values and standard deviations of measurements for Dhrystones

Benchmark	Hypervisor	Average value	Std. deviation
Dhrystones Opt3 64bit	VMware	6881.384	50.63
	Virtual Box	6865.16	83.9
	KVM	7182.66	63.72
Dhrystones No Opt	VMware	2975.97	36.11
	Virtual Box	2903.91	32.16
	KVM	3035.36	29.36

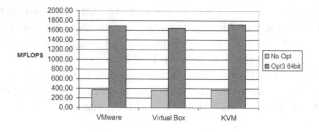

Fig. 9. Linpack benchmark results

Table 2. Summary of average values and standard deviations of measurements for Linpack

Benchmark	Hypervisor	Average value	Std. deviation
	VMware	1695.04	9.48
Linpack Opt3 64bit	Virtual Box	1647.80	15.70
	KVM	1723.57	12.05
	VMware	367.29	1.91
Linpack No Opt	Virtual Box	358.90	3.67
	KVM	373.77	1.94

4   Conclusions

The article compares the impact of virtual machine hypervisor on performance of data mining system on the example of Cloudera Hadoop distribution. The effect of VM hipervisora on the common operations performed in the private Hadoop cluster was examined. This cluster used standard PC computers, the host operating system was Ubuntu Desktop 13.10 x64 and running a Cloudera Hadoop x64 distribution. Based on these results it can be concluded that, hypervisors VMware and KVM offer time of the test tasks in Hadoop cluster on the same level while the implementation Virtual Box is a little bit slower.

To capture the differences in the impact of hypervisors the further performance tests have been carried out. The interesting aspect is the comparative evaluation of guest operating system overhead that runs virtually. For KVM hypervisor system overhead has the lowest value equal about 10 %. For the other two solutions, this value was above 15 % and 50 % respectively for VMware and Virtual Box. Therefore, one would expect better performance of the Hadoop cluster that run under control of KVM. This hypothesis was confirmed in subsequent tests. Hadoop cluster achieved best performance results in almost all benchmarks when it was running under the KVM hypervisor. However the results of VMware manager were very close to it.

The study showed that there is a virtual machine hypervisor impact on the performance of data mining systems. The size and nature of the impact will depend on the type of tasks that run in the private Hadoop cluster. Best results, in most cases, where obtained with KVM hypervisor.

More generally, this work focused on influence of VM manager on running MapReduce workloads in private cluster environment. Experimental results show that the differences are noticeable which can be also seen in other works [18–20]. However there is a need of further research which hypervisor to use for particular type user applications.

References

1. Dean, J., Ghemawat, S.: MapReduce: Simplified Data Processing on Large Clusters (December 2013), http://research.google.com/archive/mapreduce.html
2. Welcome to Hadoop Apache (December 2013), http://hadoop.apache.org
3. Borzemski, L.: Data Mining in Evaluation of Internet Path Performance. In: Orchard, B., Yang, C., Ali, M. (eds.) IEA/AIE 2004. LNCS (LNAI), vol. 3029, pp. 643–652. Springer, Heidelberg (2004)
4. Gruca, A., Czachórski, T., Kozielski, S. (eds.): Man-Machine Interactions 3. AISC, vol. 242. Springer, Heidelberg (2014)
5. Bembenik, R., Skonieczny, L., Rybiński, H., Niezgodka, M. (eds.): Intelligent Tools for Building a Scient. Info. Plat. SCI, vol. 390. Springer, Heidelberg (2012)
6. Yao, K.-T., Lucas, R.F., Ward, C.E., Wagenbreth, G., Gottschalk, T.D.: Data Analysis for Massively Distributed Simulations. In: Interservice/Industry Training, Simulation, and Education Conference IITSEC 2009 Paper No. 9350 (2009)
7. Hadoop Blog (December 2013), http://developer.yahoo.com/blogs/hadoop/
8. Krauzowicz, Ł., Szostek, K., Dwornik, M., Oleksik, P., Piórkowski, A.: Numerical Calculations for Geophysics Inversion Problem Using Apache Hadoop Technology. In: Kwiecień, A., Gaj, P., Stera, P. (eds.) CN 2012. CCIS, vol. 291, pp. 440–447. Springer, Heidelberg (2012)
9. Lam, C.: Hadoop in Action. Manning Publications Co. (2011)
10. Nurmi, D., Wolski, R., Grzegorczyk, C., Obertelli, G., Soman, S., Youseff, L., Zagorodnov, D.: The Eucalyptus Open-source Cloud-computing System. In: 9th IEEE/ACM International Symposium on Cluster Computing and the Grid (CCGRID), pp. 124–131 (2009)
11. Donnelly, P., Bui, P., Thain, D.: Attaching Cloud Storage to a Campus Grid Using Parrot, Chirp, and Hadoop. In: IEEE Second International Conference on Cloud Computing Technology and Science (CloudCom), pp. 488–495 (2010)
12. Czerwiński, D., Przyłucki, S., Sawicki, D.: Comparison of cloud and virtualization systems. Drives and Control 13(11), 93–103 (2011)
13. Cooper, B.: The Prickly Side of Building Clouds. IEEE Internet Computing 14(6), 64–67 (2010)
14. CDH Version and Packaging Information – Cloudera Support (December 2013), https://ccp.cloudera.com/display/DOC/CDH+Version+and+Packaging+Information
15. Wheeler, T.: Testing Hadoop Applications. In: O'Reilly Strata Conference, New York (October 2012)
16. White, T.: Hadoop: The Definitive Guide, MapReduce for the Cloud. O'Reilly Media (2009)
17. Holmes, A.: Hadoop in Practice. Manning Publications Co. (October 2012)
18. Li, J., Wang, Q., Jayasinghe, D., Park, J., Zhu, T., Pu, C.: Performance Overhead Among Three Hypervisors: An Experimental Study using Hadoop Benchmarks. In: IEEE International Congress on Big Data, pp. 9–16 (June/July 2013)
19. Yang, Y., Xiang, L., Xiaoqiang, D., Chengjian, W.: Impacts of Virtualization Technologies on Hadoop. In: Third International Conference on Intelligent System Design and Engineering Applications, Hong Kong, pp. 846–849 (January 2013)
20. Qingye, J.: Virtual Machine Performance Comparison of Public IaaS Providers in China. In: 2012 IEEE Asia Pacific Cloud Computing Congress, pp. 16–19 (2012)

JavaScript Frameworks and Ajax Applications

Adam Domański[1], Joanna Domańska[2], and Sebastian Chmiel[1]

[1] Institute of Informatics,
Silesian Technical University,
Akademicka 16, 44-100 Gliwice, Poland
adamd@polsl.pl
[2] Institute of Theoretical and Applied Informatics
Polish Academy of Sciences
Baltycka 5, 44-100 Gliwice, Poland
joanna@iitis.gliwice.pl

Abstract. JavaScript frameworks are useful in the development of interactive web pages. Ajax technology supports a dialog with the WWW server. This article compares performance and functionality of the Ajax libraries for major Web browsers. The size of the libraries, their load time and execution time is compared. The article also evaluates the Document Object Model (DOM) support provided by selected libraries.

Keywords: Ajax, DOM, JavaScript, web applications.

1 Introduction

Ajax technology is an asynchronous communication between the browser and the server [1, 2]. This technology is currently standard in dynamic websites. It is used by such companies as Google, Facebook, YouTube and many others. This solutions improves the usability of web applications. Ajax gives the programmer more capabilities and allows to create applications with the same level of complexity as a normal window-based programs [3–6]. This techology becomes more and more popular and lot of new frameworks have been created.

The purpose of this paper is to compare the most popular Javascript software libraries supporting the Ajax technology. The article presents a functionality and a performance of the most popular frameworks. During the test such indicators as: size of library, loading time, execution time of the Ajax requests and event handling were considered.

This paper shows how different libraries behave in the specific cases and how to select the library to meet the programmer needs.

This paper is organized as follows. Section 2 gives a brief description of the Ajax technology and presents some popular Javascript libraries. Section 3 describes the experiments and gives some numerical results. Section 4 concludes this article.

A. Kwiecień, P. Gaj, and P. Stera (Eds.): CN 2014, CCIS 431, pp. 57–68, 2014.
© Springer International Publishing Switzerland 2014

2   Ajax Technology

In this section the Ajax (Asynchronous JavaScript And XML) technology and some popular Javascript libraries were briefly presented.

Ajax is not a new technology. It is a combination of several existing solutions [7–9]. The main idea of this technology is a completely new approach of communication between browser and server [10]. This technology provides a mechanism which allows to download new web page content using JavaScript without reloading the WWW page. In the standard solution, the browser sends the request to the server and the server sends another page in response [11]. Figure 1 presents such situation. Downloading the whole pages is sometimes unnecessary and redundant. In some situations we need to load only a small part of the page (i.e. the panel side management). In the case of Ajax technology it is possible to send asynchronous request to download only interesting content from the server [11–13].

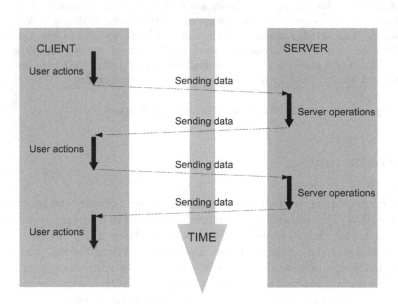

Fig. 1. Scheme of synchronous communication between the client and the web server

Figure 2 presents the asynchronus comunication between browser and server. The user event started the new content downloading procedure using JavaScript. The Ajax engine sends a request to the server (the web page has not been reloaded) and waits for new content (in background). After successful new content downloading, Ajax engine processes collected data and publish them on the site.

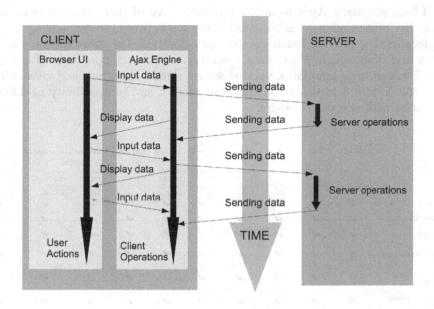

Fig. 2. Asynchronous requests to the server

An asynchronous communication with the server uses a special object called *XMLHttpRequest* [8]. Originally this object had a different name. It was created with Internet Explorer 5.0 and defined as an *ActiveX* controler [11]. Because other browsers began to use the similar object and called it the *XMLHttpRequest* from IE 7.0 the controler *ActiveX* was deprecated.

Ajax technology uses four data formats to exchange data with the server [14]:

- XML (eXtensible Markup Language),
- plain text,
  URL   data are transmitted in the URL
  (eg. *file.php?var1=value1&var2=value2*),
- JSON (JavaScript Object Notation).

Ajax can be applied in many solutions. Some of them are considered as essential of the Web 2.0 Internet [15], some are an unnecessary additions. The Ajax technology is suitable for presented below mechanisms:

- forms validations,
- autocomplete and word completion,
- edit in place – clicking on an element causes the appearance of the edit box with the ability to save the new value,
- backlight changes – suitable for services, which update the content from time to time without refreshing the page,
- slow loading – content from the top of the page are downloaded and shown at once.

There are many Ajax technology libraries. All of them provide basic functionality of managing asynchronous connection between browser and server, but generally they provide much greater opportunities. The most spectacular are: drag and drop, edit in place, data validation supporting, possibility of using defined widgets and templates, various widget skins, animations and visual effects [16–25]. Figure 3 presents the the functionality and compatibility of different libraries with selected browsers.

	DHTMLX	Dojo	Enyo	Ext JS	jQuery	MochiKit	MooTools	Prototype	Script.aculo.us	Yahoo UI
Browser support	IE 6+ Firefox 3+ Safari 4+ Opera 10.5+ Chrome 3+	IE 6+ Firefox 3.6+ Safari 5+ Opera 10.5+ Chrome 13+	IE 8+ Firefox 4+ Safari 5+ Opera 10+ Chrome 10+	IE 6+ Firefox 1.5+ Safari 3+ Opera 9+ Chrome 3+	IE 9+ Firefox 2+ Safari 3+ Opera 9+ Chrome 1+	IE 6+ Firefox 2+ Safari 2+ Opera 8.5+ Chrome 10+	IE 6+ Firefox 2+ Safari 3+ Opera 9+ Chrome 1+	IE 6+ Firefox 1.5+ Safari 2.0.4+ Opera 9.25+ Chrome 1+	IE 6+ Firefox 1.5+ Safari 3+ Opera 10+ Chrome 1+	IE 6+ Firefox 3+ Safari 4+ Opera 10+ Chrome 10+
DOM	✔	✔	✔	✔	✔	✔	✔	✔	✔	✔
JSON Format	✔	✔	✔	✔	✔	✔	✔	✔	✔	✔
AJAX	✔	✔	✔	✔	✔	✔	✔	✔	✔	✔
Button backward	✘	✔	✔	✔	✔	✔	✔	✔	✔	✔
Simple visual effects	✔	✔	✔	✔	✔	✔	✔	✔	✔	✔
Adanced visual effects	✘	✔	✔	✔	✔	✔	✔	✘	✔	✔
Additional widgets	✔	✔	✔	✔	✔	✔	✔	✘	✔	✔
Drag and drop	✔	✔	✔	✔	✔	✔	✔	✔	✔	✔
Edition in place	✔	✔	✔	✔	✔	✔	✔	✔	✔	✔
Forms validation	✔	✔	✔	✔	✔	✔	✔	✔	✔	✔
Pages templates	✔	✔	✔	✔	✔	✘	✔	✘	✘	✔

Fig. 3. Comparison of the functionality of selected libraries based on information from manufacturers websites

3   Performance of the Ajax Libraries

In this section the comparing of the performance of selected libraries were carried out. Durring the tests we checked:

- sizes of libraries in the basic version (Ajax-enabled DOM events and parsing the data) and in the expanded versions (graphics effect, forms and data validation),
- libraries load times in primary and extended versions,
- an execution time of the XMLHttpRequest method ("GET" or "POST"). This test was performed in the case of three typical feedback formats: plain text, XML, and JSON,
- searching and editing elements in the DOM model.

Each of described bellow tests (except library size checking) was conducted for five types of browser:

- Internet Explorer 10.0.9200,
- Firefox 23.0.1,
- Safari 5.1.7,
- Opera 12.16,
- Chrome 29.0.1547.

3.1 Library Size

We analysed the empty web page size with the loaded library. We considered: library with basic modules (functions responsible for the asynchronous communication with the server, DOM handling functions, events handling and parsing data), library with extended version (graphics effects handling and forms validation). The page size was measured using the Firefox plugin called "Firebug".

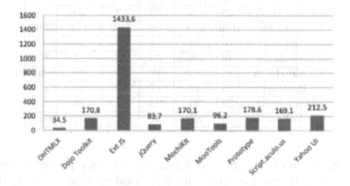

Fig. 4. Libraries sizes (basic versions) [KB]

We concluded that the worst case was Ext JS library. This is due to number of additional mechanisms and widgets builded into the base library file. The library authors did not divide modules, so it is necessary to load the entire library to handle basic functionality. Better results were obtained for the DHTMLX, jQuery and MooTools libraries. This is due to code optimization (unnecessary whitespace were removed, variable names have been changed and comments have been removed). Such optimization can save 80 % of the initial size of the code.

3.2 Library Load Time

We have measured the library loading time. The code from the previous test was used. The time was measured using a JS functions.

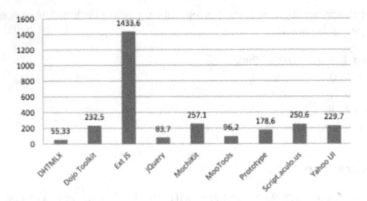

Fig. 5. Libraries size (extended versions) [KB]

Table 1. Library loading times (basic versions) [ms]

	IE	Firefox	Safari	Opera	Chrome	Avg.
DHTMLX	16	18	4	3	3	8.8
Dojo Toolkit	26	45	28	27	19	29
Ext JS	198	292	183	184	174	206.2
jQuery	20	29	14	16	11	18
MochiKit	49	35	23	21	11	27.8
MooTools	46	45	17	22	20	30
Prototype	26	45	17	22	20	24.8
Script.aculo.us	39	39	23	21	26	29.6
YahooUI	59	87	47	97	32	64.4

The results (Tables 1, 2) are similar to the results obtained in the previous test. This is due to the fact that the page loading time depends on its size. The difference in the time of loading of libraries in different browsers is not always the fault of the browser. Often this is due to the fact that the JavaScript language has worked differently in different browsers (usually older ones), and the authors of libraries provided a proper work in all browsers.

3.3 Sending and Receiving Ajax Request

In this subsection the most important mechanism of the Ajax technology were rated. There were no delays on the line during the client-server data transfer because the tests were carried out on the local server. Figure 6 presents the code of the experiment.

For each Ajax task the method for sending data has to be specified (GET or POST). The GET method sends the variables and their values in the page URL. The POST method sends variables in the header of HTTP. The experiment allowed to specify the transmitted data format (plain text, XML, JSON). The file content of all free formats is the same. This aproach allows to compare different

Table 2. Library loading times (extended versions) [ms]

	IE	Firefox	Safari	Opera	Chrome	Avg.
DHTMLX	4	22	9	4	3	8.4
Dojo Toolkit	33	47	31	53	23	37.4
Ext JS	198	292	183	184	174	206.2
jQuery	20	29	14	16	11	18
MochiKit	66	38	28	29	15	35.2
MooTools	46	45	17	22	20	30
Prototype	26	37	20	23	18	24.8
Script.aculo.us	76	58	33	47	32	49.2
YahooUI	56	99	49	87	32	64.6

formats for the same content. Text was endorsed with HTML tags (to be ready to display on the page). Its size was 4 125 bytes. In the case of the XML the file size was 4 157 bytes. Although the XML procesing is very simple but its additional tags occupies more space than other formats. The last variant was the JSON data format. The advantage of this format is a small data size (only 4 002 bytes) and the fact that this format is directly imported as an object in JavaScript. The common results are presented in Table 3 and Table 4.

The results of this test show that the data format and the transfer method do not play a significant role in the length of transfer time. The data processing is also fast. The differences between formats are insignificantly small. In most cases, one can notice better results for JSON (the size of the transmitted content is smallest). The fastest libraries in this case were DHTMLX, Dojo Toolkit and MooTools. The obtained times were less than 10 milliseconds.

3.4   Document Object Model Support

The data after downloading from the server are usually processed and displayed on the website. The object model of the web page called the DOM is used for webpage modifications. This subsection shows the performance of searching and editing elements in this model. Tested library mostly have their own mechanisms for handling DOM model. We evaluated the three most frequently performed operations:

- searching of elements (by different classifiers),
- modification of elements content,
- creating of new elements.

The single test was repeated 100 000 times due to short time of single test. Some libraries did not have its own mechanism for handling DOM. The tests for them are omitted and the values are indicated by 0.

For first case (searching the elements according to their attributes "id"), most libraries had short seek time. The worst of the tested libraries were MooTools

Table 3. Ajax request (method GET) [ms]

		IE	Firefox	Safari	Opera	Chrome	Avg.
DHTMLX	Plain text	2	11	4	4	10	6.2
	XML	1	17	4	9	8	7.8
	JSON	10	15	6	7	10	9.6
Dojo Toolkit	Plain text	3	9	7	4	5	5.6
	XML	4	6	6	5	5	5.2
	JSON	17	6	7	4	6	8
Ext JS	Plain text	6	35	14	6	42	20.6
	XML	47	39	14	6	45	22.2
	JSON	7	35	15	6	43	21.2
jQuery	Plain text	14	12	8	4	12	10
	XML	15	18	10	8	12	12.6
	JSON	14	17	10	5	11	11.4
MochiKit	Plain text	9	19	11	14	19	14.4
	XML	16	16	10	8	12	12.6
	JSON	16	16	10	12	18	14.4
MooTools	Plain text	13	9	11	4	10	9.4
	XML	13	9	9	5	12	9.6
	JSON	12	8	9	4	10	8.6
Prototype	Plain text	14	21	9	4	15	11.2
	XML	15	16	9	7	19	13.2
	JSON	14	12	9	4	18	11.4
Script.aculo.us	Plain text	15	20	9	5	13	12.4
	XML	19	13	10	7	13	12.4
	JSON	17	9	8	6	12	10.4
YahooUI	Plain text	3	21	8	14	22	16.6
	XML	13	21	8	15	23	16
	JSON	12	21	8	15	23	15.8

Table 4. AJAX request (method POST) [ms]

		IE	Firefox	Safari	Opera	Chrome	Avg.
DHTMLX	Plain text	5	19	5	7	8	8.8
Dojo Toolkit	Plain text	5	5	7	6	6	5.8
Ext JS	Plain text	8	28	14	9	46	2.1
jQuery	Plain text	9	16	11	8	13	11.4
MochiKit	Plain text	10	17	7	10	17	12.2
MooTools	Plain text	8	8	7	8	12	8.6
Prototype	Plain text	9	13	9	9	16	8.4
Script.aculo.us	Plain text	8	15	11	9	14	11.4
YahooUI	Plain text	7	21	9	18	22	15.4

```
<! DOCTYPE html>
<html>
<head>
   <meta charset="utf-8">
   <title> [ LIBRARY NAME] - [TEST NAME] < / title>
< / head>
<body>
   <h1 id="test_name"> [ LIBRARY NAME] - [TEST NAME] < / h1 >
   <div id="result_div"> [ RESULTS ] < / div >
   <script> [TIME FUNCTION - Loading ] < / script >
   <script>
      var resultDiv = document.getElementById (" result_div ");
      [ START TIME ]
      var test = fTestTimeCounter ();
      test.start ();
      new Ajax.Request ( [URL ] , {

      method: [ POST / GET ]
         parameters: {
            [ SENDING PARAMETERS ]
         }
         [ REQUEST SUCCESS ]
         onSuccess : function ( data) {
            var response = [DATA PROCESSING TEXT / XML / JSON ]
            [TIME STOP]
            alert ( test.stop () + ' ms ');
            [ DISPLAYING RESULTS ]
            resultDiv.innerHTML = " <pre> " + response + " < / pre> " ;
         }
         [ REQUEST FAILURE ]
         onFailure : function () {
            alert (' Something went wrong ...');
         }
      } ) ;
   < / script >
< / body>
< / html >
```

Fig. 6. AJAX requests – sending and receiving

i Yahoo UI. This situation is presented in Fig. 7. The searching element using identifier is most popular in JavaScript hence for all libraries the seek time was short.

Figure 8 presents the results achieved during the modification of the DOM elements. The libraries: Prototype, Script.aculo.us, and Yahoo UI changed slowly the values of the DOM elements. Very probably they do not use the prepared JavaScript function. The worst result 458 ms corresponds to 0.0458 ms for a single modification.

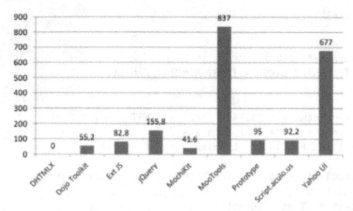

Fig. 7. Searching the element in the DOM according to the value attribute "id" [ms]

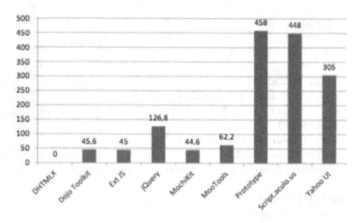

Fig. 8. Modification the element in the DOM [ms]

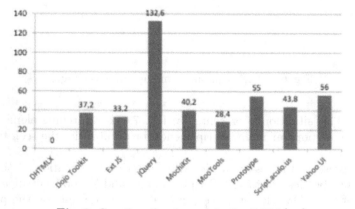

Fig. 9. Creating the element in the DOM [ms]

Creation of the DOM elements takes much more time than modifying their contents (Fig. 9). At the time of the creation two operations are performed: the creation of a new element and assign it to the specified parent. The obtained results are largely comparable. Only library JQuery works a few times slower (this time is highest in the case of Internet Explorer).

4 Conclusions

There is a large choice of available libraries and most of them have additional functionality. These additions can simplify the work programmer, but may be also an unnecessary ballast. Selected libraries can be divided into two categories: simple and complex. MochiKit and Prototype belong to the first type. They focus on the core functionality. The others hace additional modules with advanced operations: drag and drop functionality, advanced service forms with the data validation, edit in place and autocomplete. Most of them have the ability of creating visual effects.

The Open Source libraries were selected to tests. Their functionality were nearly identical hence the choice of appropriate library can be difficult. We assumed that the important factors are size and performance of library. Large library size negatively influence the page load time. The Ext JS library had the biggest size. However, it was characterized by good performance. The main element of Ajax speed communication with the server was alse examined. We obtained the best results for DHTMLX, dojo Toolkit and MooTools. In the case of transfer methods ("GET" and "POST") and data formats (plain text, XML, and JSON) performance is relatively similar. Small acceleration was noted in the case of the JSON. fIn this case the size of transfering data was the smallest.

The DOM operating tests confirmed the biggest differences between the studied libraries. Although single call lasted only milliseconds, this functionality may have great importance for large applications . The libraries: Dojo, MochiKit, Ext JS, Prototype, and Script.aculo.us were the fastest in the case of the test of seraching the structure of the DOM by identifier. Based on the test results the weaknesses of libraries Prototype, Script.aculo.us and Yahoo UI can be determined. The DOM elements modifying time was worst in the case of this libraries. In the case of inserting new elements the worst was jQuery library.

During our researches we noticed the differences in the quality of the available documentation. Some libraries had a well written tutorials with examples (Dojo Toolkit, jQuery, MooTools, Prototype). The other forced to longer search. Considering all the experiments, the best choice is the library Dojo Toolkit. Slightly worse are the prototype, Script.aculo.us, Ext JS and jQuery. and at the very end were MooTools and Yahoo UI. The worst results were obtained for MooTools and Yahoo UI.

References

1. Duda, C., Frey, G., Kossmann, D., Matter, R., Zhou, C.: AJAX Crawl: Making AJAX Applications Searchable. In: IEEE 25th International Conference on Data Engineering, ICDE (2009)

2. Tian, X.: Extracting Structured Data from Ajax Site. In: 2009 First International Workshop on Database Technology and Applications (2009)
3. Han, W., Di, L., Zhao, P., Li, X.: Using Ajax for desktop-like geospatial web application development. In: 2009 17th International Conference on Geoinformatics (2009)
4. Zepeda, J., Chapa, S.: From Desktop Applications Towards Ajax Web Applications. In: 4th International Conference on Electrical and Electronics Engineering, ICEEE (2007)
5. Weiguo, H., Liping, D., Peisheng, Z., Xiaoyan, L.: Using Ajax for desktop-like geospatial web application development. In: 2009 17th International Conference Geoinformatics (2009)
6. Dong, S., Cheng, C., Zhou, Y.: Research on AJAX technology application in web development. In: 2011 International Conference E-Business and E -Government, ICEE (2011)
7. Bruce, P.W.: Ajax Hacks Tips and Tools for Creating Responsive Web Sites. O'Reilly Media (2006)
8. Lauriat, S.M.: Advanced Ajax: Architecture and Best Practices. Prentice Hall (2008)
9. Lin, Z., Wu, J., Zhang, Q., Zhou, H.: Research on Web Applications Using Ajax New Technologies. In: International Conference on MultiMedia and Information Technology, MMIT 2008 (2008)
10. Jingjing, L., Chunlin, P.: jQuery-based Ajax general interactive architecture. In: 2012 IEEE 3rd International Conference on Software Engineering and Service Science, ICSESS (2012)
11. Vossen, G., Hagemann, S.: Unleashing Web 2.0: From Concepts to Creativity. Morgan Kaupmman (2007)
12. Matthijssen, N., Zaidman, A., Storey, M., Bull, I., van Deursen, A.: Connecting Traces: Understanding Client-Server Interactions in Ajax Applications. In: 2010 IEEE 18th International Conference on Program Comprehension, ICPC (2010)
13. Matthijssen, N., Zaidman, A.: FireDetective: understanding ajax client/server interactions. In: 2011 33rd International Confer. Software Engineering, ICSE (2011)
14. Zhang, X., Wang, H.: AJAX Crawling Scheme Based on Document Object Model. In: 2012 Fourth International Conference on Computational and Information Sciences, ICCIS (2012)
15. Al-Tameem, A.B., Chittikala, P., Pichappan, P.: A study of AJAX vulnerability in Web 2.0 applications. In: First International Conference on Applications of Digital Information and Web Technologies, ICADIWT 2008 (2008)
16. Powers, S.: Adding Ajax. O'Reilly Media (2007)
17. The DHTMLX library website, http://dhtmlx.com/
18. The Dojo Toolkit library website, http://dojotoolkit.org/
19. The Ext JS library website, http://www.sencha.com/products/extjs
20. The jQuery library website, http://jquery.com/
21. The library MochiKit website, http://mochi.github.io/mochikit/
22. The MooTools library website, http://mootools.net/
23. The Prototype library website, http://prototypejs.org/
24. The Script.aculo.us library website, http://script.aculo.us/
25. The Yahoo UI Library website, http://yuilibrary.com/

Communication-Aware Algorithms for Target Tracking in Wireless Sensor Networks

Bartłomiej Płaczek

Institute of Computer Science, University of Silesia,
Będzińska 39, 41-200 Sosnowiec, Poland
placzek.bartlomiej@gmail.com

Abstract. This paper introduces algorithms for target tracking in wireless sensor networks (WSNs) that enable reduction of data communication cost. The objective of the considered problem is to control movement of a mobile sink which has to reach a moving target in the shortest possible time. Consumption of the WSN energy resources is reduced by transferring only necessary data readings (target positions) to the mobile sink. Simulations were performed to evaluate the proposed algorithms against existing methods. The experimental results confirm that the introduced tracking algorithms allow the data communication cost to be considerably reduced without significant increase in the amount of time that the sink needs to catch the target.

Keywords: wireless sensor networks, data collection, object tracking.

1 Introduction

Wireless sensor networks (WSNs) can be utilized as target tracking systems that detect a moving target, localize it and report its location to the sink. So far, the WSN-based tracking systems have found various applications, such as battlefield monitoring, wildlife monitoring, intruder detection, and traffic control [1,2].

This paper deals with the problem of target tracking by a mobile sink which uses information collected from sensor nodes to catch the target. Main objective of the considered system is to minimize time to catch, i.e., the number of time steps in which the sink reaches the moving target. Moreover, due to the limited energy resources of WSN, also the minimization of data communication cost (hop count) is taken into consideration. It is assumed in this study that the communication between sensor nodes and the sink involves multi-hop data transfers.

Most of the state-of-the-art data collection methods assume that the current location of the target has to be reported to sink continuously with a predetermined precision. These continuous data collection approaches are not suitable for developing the WSN-based target tracking applications because the periodical transmissions of target location to the sink would consume energy of the sensor nodes in a short time. Therefore, the target tracking task requires dedicated algorithms to ensure the amount of data transmitted in WSN is as low as possible.

A. Kwiecień, P. Gaj, and P. Stera (Eds.): CN 2014, CCIS 431, pp. 69–78, 2014.
© Springer International Publishing Switzerland 2014

Intuitively, there is a trade-off between the time to catch minimization and the minimization of data communication cost. In this study two algorithms are proposed that enable substantial reduction of the data collection cost without significant increase in time to catch. The introduced communication-aware algorithms optimize utilization of the sensor node energy by selecting necessary data readings (target locations) that have to be transmitted to the mobile sink. Simulation experiments were conducted to evaluate the proposed algorithms against state-of-the-art methods. The experimental results show that the presented algorithms outperform the existing solutions.

The paper is organized as follows. Related works are discussed in Sect. 2. Section 3 contains a detailed description of the proposed target tracking methods. The experimental setting, compared algorithms and simulation results are presented in Sect. 4. Finally, conclusion is given in Sect. 5.

2   Related Works

In the literature, there is a variety of approaches available that address the problem of target tracking in WSNs. However, only few publications report the use of WSN for chasing the target by a mobile sink. Most of the previous works have focused on delivering the real-time information about trajectory of a tracked target to a stationary sink. This section gives references to the WSN-based tracking methods reported in the literature that deal explicitly with the problem of target chasing by a mobile sink. A thorough survey of the literature on WSN-based object tracking methods can be found in references [1,3].

Kosut et al. [4] have formulated the target chasing problem, which assumes that the target performs a simple random walk in a two-dimensional lattice, moving to one of the four neighbouring lattice points with equal probability at each time step. The target chasing method presented in [4] was intended for a system composed of static sensors that can detect the target, with no data transmission between them. Each static sensor is able to deliver the information about the time of the last target detection to the mobile sink only when the sink arrives at the lattice point where the sensor is located.

A more complex model of the WSN-based target tracking system was introduced by Tsai et al. [5]. This model was used to develop the dynamical object tracking protocol (DOT) which allows the WSN to detect the target and collect the information on target track. The target position data are transferred from sensor nodes to a beacon node, which guides the mobile sink towards the target. A similar method was proposed in [6], where the target tracking WSN with monitor and backup sensors additionally takes into account variable velocity and direction of the target.

In this paper two target tracking methods are proposed that contribute to performance improvement of the above-mentioned target tracking approaches by reducing both the time to catch (i.e., the time in which mobile sink can reach the target) and the data communication costs in WSN. In this study, the total hop count is analysed to evaluate the overall cost of communications, however it should be noted that different metrics can also be also used, e.g., number of data transfers to sink, number of queries, number of transmitted packets, and energy consumption in sensor nodes.

The introduced algorithms provide decision rules to optimize the amount of data transfers from sensor nodes to sink during target chasing. The research reported in this paper is a continuation of previous works on target tracking in WSN, where the data collection was optimized by using heuristic rules [7] and the uncertainty-based approach [8]. The algorithms proposed in that works have to be executed by the mobile sink. In the present study the data collection operations are managed by distributed sensor nodes.

To reduce the number of active sensor nodes the proposed algorithms adopt the prediction-based tracking method [9]. According to this method a prediction model is applied, which forecasts the possible future positions of the target. On this basis only the sensor nodes expected to detect the target are activated at each time step.

3   Proposed Methods

In this section two methods are proposed that enable reduction of data transfers in WSN during target tracking. The WSN-based target tracking procedure is executed in discrete time steps. At each time step both the target and the sink move in one of the four directions: north, west, south or east. Their maximum velocities (in segments per time step) are assumed to be known. Movement direction of the target is random. For sink the direction is decided on the basis of information delivered from WSN. During one time step the sink can reach the nearest segments (x_S, y_S) that satisfy the maximum velocity constraint: $|x_S - x'_S| + |y_S - y'_S| \leq v_{\max}$, where coordinates (x'_S, y'_S) describe previous position of the sink. Sink moves into segment (x_S, y_S) for which the Euclidean distance $d[(x_S, y_S), (x_D, y_D)]$ takes minimal value. Note that (x_D, y_D) are the coordinates of target that were lately reported to the sink.

Let (x_C, y_C) denote coordinates of the segment where the target is currently detected. The sensor node that detects the target will be referred to as the target node. According to the proposed methods the information about target position is transmitted from the target node to the sink only at selected time steps. If this information is transmitted then the destination coordinates at sink (x_D, y_D) are updated, i.e, $(x_D, y_D) = (x_C, y_C)$. It means that the current position of the target is available for sink only at selected time steps. In remaining time periods the sink moves toward the last reported target position, which is determined by coordinates (x_D, y_D).

Hereinafter, symbol $dir(x, y)$ will be used to denote the direction chosen by sink when moving toward segment (x, y). At each time step, the coordinates (x_D, y_D) and (x_C, y_C) are known for the target node. Therefore, the target node can determine the direction which will be chosen by the sink in both cases: if the current target position is transmitted to the sink and if the data transfer is skipped.

According to the first proposed method, the coordinates (x_C, y_C) are transmitted to the sink only if $dir(x_D, y_D) \neq dir(x_C, y_C)$, i.e., if the direction chosen on the basis of coordinates (x_D, y_D) is different than the one selected by taking into account the current position (x_C, y_C).

In the second proposed method, the target node evaluates probability $P[dir]$ that the move of sink in direction dir will minimize its distance to the segment in which the target will be caught. The target coordinates (x_C, y_C) are transferred to the sink only if the difference $P[dir(x_C, y_C)] - P[dir(x_D, y_D)]$ is above a predetermined threshold. To evaluate probabilities $P[dir]$, the target node determines an area where the target can be caught. This area is defined as a set of segments:

$$A = \{(x, y) : t_T(x, y) \leq t_S(x, y)\} , \tag{1}$$

where $t_T(x, y)$ and $t_S(x, y)$ are the minimum times required for target and sink to reach segment (x, y).

Let $(x_S, y_S)_C$ and $(x_S, y_S)_D$ denote the segments into which the sink will enter at the next time step if it will move in directions $dir(x_C, y_C)$ and $dir(x_D, y_D)$ respectively. In area A two subsets of segments are distinguished: subset A_C that consists of segments that are closer to $(x_S, y_S)_C$ than to $(x_S, y_S)_D$ and subset A_D of segments that are closer to $(x_S, y_S)_D$ than to $(x_S, y_S)_C$:

$$A_C = \{(x, y) : (x, y) \in A \wedge d[(x, y), (x_S, y_S)_C] < d[(x, y), (x_S, y_S)_D]\} , \tag{2}$$

$$A_D = \{(x, y) : (x, y) \in A \wedge d[(x, y), (x_S, y_S)_D] < d[(x, y), (x_S, y_S)_C]\} . \tag{3}$$

On this basis the probabilities $P[dir]$ are calculated as follows:

$$P[dir(x_C, y_C)] = \frac{|A_C|}{|A|} , \quad P[dir(x_D, y_D)] = \frac{|A_D|}{|A|} , \tag{4}$$

where $|\cdot|$ denotes cardinality of the set.

The operations discussed above are illustrated by the example in Fig. 1, where the positions of target and sink are indicated by symbols "T" and "S" respectively. Velocity of the target is 1 segment per time step. For sink the velocity equals 2 segments per time step. Gray color indicates the area A in which the sink will be able to catch the target. The direction $dir(x_C, y_C)$ is shown by the arrow with number 1 and $dir(x_D, y_D)$ is indicated by the arrow with number 2, thus $(x_S, y_S)_C = (1, 3)$ and $(x_S, y_S)_D = (3, 1)$. Subset A_C includes gray segments that are denoted by 1. The segments with label 2 belong to A_D. In the analyzed example $|A| = 82$, $|A_C| = 44$, and $|A_D| = 31$. According to Equation (4) $P[dir(x_C, y_C)] = 0.54$, $P[dir(x_D, y_D)] = 0.38$ and the difference of these probabilities equals to 0.16.

If the first proposed method is applied for the analysed example then the data transfer to sink will be executed, since $dir(x_D, y_D) \neq dir(x_C, y_C)$, as shown by the arrows in Fig. 1. In case of the second method, the target node will send the coordinates (x_C, y_C) to the sink provided that the difference of probability (0.16) is higher than a predetermined threshold. The threshold value should be

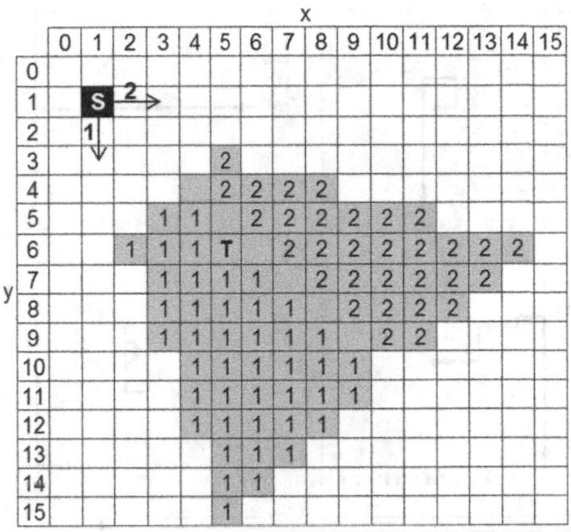

Fig. 1. Example of $P[dir]$ calculations

interpreted as a minimum required increase in the probability of selecting the optimal movement direction, which is expected to be obtained after transferring the target position data.

4 Experiments

Experiments were performed in a simulation environment to compare performance of the proposed methods against state-of-the-art approaches. The comparison was made by taking into account two criteria: time to catch and hop count. The time to catch is defined as the number of time steps in which the sink reaches the moving target. Hop count is used to evaluate the cost of data communication in WSN.

4.1 Experimental Setting

In the experiments, it was assumed that the monitored area is a square of 200 x 200 segments. Each segment is equipped with a sensor node that detects presence of the target. Thus, the number of sensor nodes in the analysed WSN equals 40 000. Communication range of each node covers the eight nearest segments. Maximum velocity equals 1 segment per time step for the target, and 2 segments per time step for the sink.

Experiments were performed using simulation software that was developed for this study. The results presented in Sect. 4.3 were registered for 10 random tracks of the target (Fig. 2). Each simulation run starts with the same location of both the sink (5, 5) and the target (100, 100). During simulation the hop counts are calculated assuming that the shortest path is used for each data transfer to sink, the time to catch is measured in time steps of the control procedure. The simulation stops when target is caught by the sink.

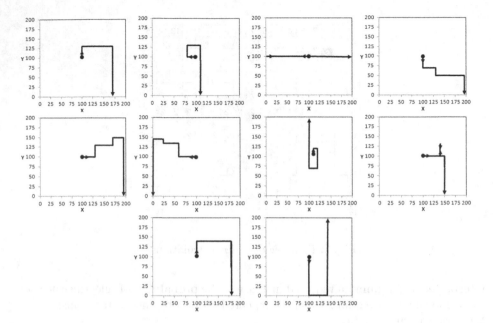

Fig. 2. Simulated tracks of target

4.2   Compared Algorithms

In the present study, the performance is analysed of four WSN-based target tracking algorithms. Algorithms 1 and 2 are based on the approaches that are available in literature, i.e. the prediction-based tracking and the dynamical object tracking. These algorithms were selected as representative for the state-of-the-art solutions in the WSN-based systems that control the movement of a mobile sink which has to reach a moving target. The new proposed methods are implemented in Algorithms 3 and 4. The pseudocode in Table 1 shows the operations that are common for all the examined algorithms. Each algorithm uses different condition to decide if current position of the target will be transmitted to the sink (line 6 in the pseudocode). These conditions are specified in Table 2.

For all considered algorithms, the prediction-based approach is used to select the sensor nodes that have to be activated at a given time step (t). Prediction of the possible target locations is based on a simple movement model, which takes into account the assumptions on target movement directions and its maximum velocity. If for previous time step ($t-1$) the target was detected in segment (x'_C, y'_C), then at time step t the set of possible target locations M can be determined as follows:

$$M = \{(x,y) : |x - x'_C| + |y - y'_C| \le v_{\max}\} , \tag{5}$$

where $v_{\max}$ is the maximum velocity of target in segments per time step.

Algorithm 1 uses the prediction-based tracking method without any additional data transfer condition. According to this algorithm, the target location

Table 1. Pseudocode for WSN-based target tracking algorithms

1	repeat
2	at target node do
3	determine M
4	collect data from each node$(x, y) : (x, y) \in M$
5	determine (x_C, y_C)
6	if *condition* then communicate (x_C, y_C) to the sink
7	set node(x_C, y_C) to be target node
8	at sink do
9	if data received from target node then $(x_D, y_D) := (x_C, y_C)$
10	move toward (x_D, y_D)
11	until $(x_S, y_S) = (x_C, y_C)$

Table 2. Compared algorithms

Alg.	Method	Data transfer condition
1	Prediction-based tracking	None
2	Dynamical object tracking	$(x_S, y_S) = (x_D, y_D)$
3	Proposed method #1	$dir(x_D, y_D) \neq dir(x_C, y_C)$
4	Proposed method #2	$P[dir(x_C, y_C)] - P[dir(x_D, y_D)] > threshold$

is reported to the sink at each time step. Sensor nodes for all possible target locations $(x, y) \in M$ are activated, and the discovered target location is transmitted to the sink. An important feature of Algorithm 1 is that the information about current target position is delivered to the sink with the highest available frequency (at each time step of the tracking procedure).

Algorithm 2 is based on the tracking method which was proposed for the dynamical object tracking protocol. According to this approach sink moves toward location of so-called beacon node (x_D, y_D). A new beacon node is set if the sink enters segment (x_D, y_D). In such case, the sensor node which currently detects the target in segment (x_C, y_C), becomes new beacon node and its location is communicated to the sink. When using this approach, the cost of data communication in WSN can be reduced because the data transfers to sink are executed less frequently than for the prediction-based tracking method. The proposed communication-aware tracking methods are applied in Algorithm 3 and Algorithm 4 (see Table 2). Details of these methods were discussed in Sect. 3.

4.3   Simulation Results

Simulation experiments were carried out in order to determine time to catch values and hop counts for the compared algorithms. As it was mentioned in Sect. 3, the simulations were performed by taking into account ten different tracks of the target. Average results of these simulations are shown in Fig. 3. It is evident that the best results were obtained for Algorithm 4, since the objective is to minimise both the time to catch and the hop count. It should be noted that Fig. 3. presents the results of Algorithm 4 for different threshold values. The relevant threshold values between 0.0 and 0.9 are indicated in the chart by the decimal numbers. According to these results, the average time to catch increases when the threshold is above 0.2. For the threshold equal to or lower than 0.2 the time to catch takes a constant minimal value. The same minimal time to catch is obtained when using Algorithm 3, however in that case the hop count is higher than for Algorithm 4.

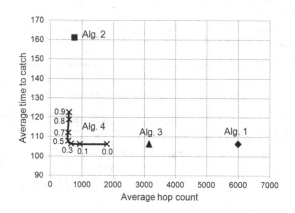

Fig. 3. Average time to catch and hop count for compared algorithms

In comparison with Algorithm 1 both proposed methods enables a considerable reduction of the data communication cost. The average hop count is reduced by 47 % for Algorithm 3 and by 87 % for Algorithm 4 with threshold 0.2. Algorithm 2 also reduces the hop count by about 87 % but it requires much longer time to catch the target. The average time to catch for Algorithm 2 is increased by 52 %.

Detailed simulation results are presented in Fig. 4. These results demonstrate the performance of the four examined algorithms when applied to ten different tracks of the target. The threshold value in Algorithm 4 was set to 0.2. The shortest time to catch was obtained by Algorithms 1, 3 and 4 for all tracks except the 5th one. In case of track 5, when using Algorithm 4 slightly longer time was needed to catch the target. For the remaining tracks the three above-mentioned algorithms have resulted in equal values of the time to catch. In comparison with Algorithm 1, the proposed algorithms (Algorithm 3 and Algorithm 4) significantly reduce the data communication cost (hop count) for all analysed cases. For each considered track Algorithm 2 needs significantly longer time to reach the moving target than the other algorithms. The hop counts for Algorithm 2 are close to those observed in case of Algorithm 4.

According to the presented results, it could be concluded that Algorithm 4, which is based on the proposed method, outperforms the compared algorithms. It enables a significant reduction of the data communication cost. This reduction

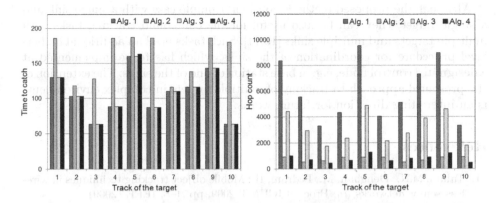

Fig. 4. Time to catch and hop counts for different tracks of target

is similar to that obtained for Algorithm 2. Moreover, the time to catch for Algorithm 4 is as short as in case of Algorithm 1, wherein the target position is communicated to the sink at each time step.

5 Conclusion

The cost of data communication in WSNs has to be taken into account when designing algorithms for WSN-based systems due to the finite energy resources and the bandwidth-limited communication medium. In order to reduce the utilization of WSN resources, only necessary data shall be transmitted to the sink. This paper is devoted to the problem of transferring target coordinates from sensor nodes to a mobile sink which has to track and catch a moving target. The presented algorithms allow the sensor nodes to decide when data transfers to the sink are necessary for achieving the tracking objective. According to the proposed algorithms, only selected data are transmitted that can be potentially useful for reducing the time in which the target will be reached by the sink.

Performance of the proposed algorithms was compared against state-of-the-art approaches, i.e., the prediction-based tracking and the dynamical object tracking. The simulation results show that the introduced algorithms outperforms the existing solutions and enable substantial reduction in the data collection cost (hop count) without significant decrease in the tracking performance, which was measured as the time to catch.

The present study considers an idealistic WSN model, where the information about current position of target (x_C, y_C) is always successfully delivered through multi-hop paths to the sink and the transmission time is negligible. In order to take into account uncertainty of the delivered information, the precise target coordinates (x_C, y_C) should be replaced by a (fuzzy) set. Relevant modifications of the presented algorithms will be considered in future experiments.

Although the proposed methods consider a simple case with a single sink and a single target, they can be also useful for the compound tracking tasks with multiple targets and multiple sinks [10,11]. Such tasks need an additional higher-level procedure for coordination of the sinks, which has to be implemented at a designated control node, e.g., a base station or one of the sinks. The extension of the presented approach to tracking of multiple targets in complex environments is an interesting direction for future works.

References

1. Tahan, M.N., Dehghan, M., Pedram, H.: Mobile object tracking techniques in wireless sensor networks. In: Proc. of ICUMT 2009, pp. 1–8. IEEE (2009)
2. Płaczek, B.: Uncertainty-dependent data collection in vehicular sensor networks. In: Kwiecień, A., Gaj, P., Stera, P. (eds.) CN 2012. CCIS, vol. 291, pp. 430–439. Springer, Heidelberg (2012)
3. Bhatti, S., Jie, X.: Survey of target tracking protocols using wireless sensor network. In: Proc. of ICWMC 2009, pp. 110–115. IEEE (2009)
4. Kosut, O., Lang, T.: The nose of a bloodhound: target chasing aided by a static sensor network. In: Proc. of ACSSC 2007, pp. 376–380. IEEE (2007)
5. Tsai, H.-W., Chu, C.-P., Chen, T.-S.: Mobile object tracking in wireless sensor networks. Computer Communications 30(8), 1811–1825 (2007)
6. Bhuiyan, M.Z.A., Wang, G., Wu, J.: Target tracking with monitor and backup sensors in wireless sensor networks. In: Proc. of ICCCN 2009, pp. 1–6. IEEE (2009)
7. Płaczek, B., Bernaś, M.: Optimizing Data Collection for Object Tracking in Wireless Sensor Networks. In: Kwiecień, A., Gaj, P., Stera, P. (eds.) CN 2013. CCIS, vol. 370, pp. 485–494. Springer, Heidelberg (2013)
8. Płaczek, B., Bernaś, M.: Uncertainty-based information extraction in wireless sensor networks for control applications. Ad Hoc Networks 14C, 106–117 (2014)
9. Samarah, S., Al-Hajri, M., Boukerche, A.: An energy efficient prediction-based technique for tracking moving objects in WSNs. In: Int. Conf. on Communications ICC 2011, pp. 1–5. IEEE (2011)
10. Schenato, L., Oh, S., Sastry, S., Bose, P.: Swarm coordination for pursuit evasion games using sensor networks. In: Proc. of ICRA 2005, pp. 2493–2498. IEEE (2005)
11. Zheng, J., Yu, H., Zheng, M., Liang, W., Zeng, P.: Coordination of multiple mobile robots with limited communication range in pursuit of single mobile target in cluttered environment. Journal of Control Theory and Applications 8(4), 441–446 (2010)

MAGANET – On the Need of Realistic Topologies for AMI Network Simulations*

Sławomir Nowak[1], Mateusz Nowak[1], and Krzysztof Grochla[2]

[1] Institute of Theoretical and Applied Informatics, Polish Academy of Science
Bałtycka 5, 44-100 Gliwice, Poland
{s.nowak,mateusz}@iitis.pl
[2] Proximetry Poland Sp. z o.o.,
Al. Rozdzienskiego 91, 40-203 Katowice, Poland
kgrochla@proximetry.com

Abstract. The article proposes MAGANET (Map-based Generator of AMI Network Topology), new map-based topology generator, designated for Advanced Metering Infrastructure (AMI) networks as well as for the urban wireless sensor networks (WSNs). The correct, reliable representation of network topology is crucial for the correct performance evaluation on large multihop wireless networks. Most of the current work us different random topologies. In the article we describe proposed new tool and its features. The comparison of the properties of topology provided by MAGANET for three different generation modes: uniform random, grid and map based is also presented.

Keywords: topology generator, AMI, advanced metering infrastructure, smart metering, sensor networks.

1 Introduction

The Advanced Metering Infrastructure (AMI) refers to the measurement and collection system that consist of meters, communication network and data acquisition system. The AMI system exchange information regarding operation of meters at customer site, recording consumption of electric energy, gas or water. The AMI networks are gathering data from fixed nodes, which transmit the measurements to one or multiple gateways. Nodes (meters) may also work as the intermediate communication nodes. The smart metering infrastructure, typically installed in urban areas, may consist even of millions of devices, covering large area and big number of buildings. In the majority of AMI solutions the communication takes place in wireless way, however network of electricity-only meters may use Power Line Communication (PLC) technology.

AMI networks require new dedicated solutions, both in hardware level and in software level, particularly new transmission protocols. Evaluation of new

---

* The work is partially supported by NCBIR INNOTECH Project K2/HI2/21/1 84126/NCB R/13: The effective management of Telecommunication Networks consisting of millions of devices (beneficiary Proximetry Poland Sp. z o.o.).

A. Kwiecień, P. Gaj, and P. Stera (Eds.): CN 2014, CCIS 431, pp. 79–88, 2014.

network protocols in multi-thousand nodes network is a significant problem. Analytical models (e.g. diffusion approximation [1]) or discrete event simulation models of multihop wireless networks [2] are used, as it is hardly possible to make experiments in real network. In order to evaluate new algorithms and protocols, researchers need realistic models of network topologies to be used in the simulation environment. The clustering of nodes in specific areas of the network is of particular importance. Thus simplified assumptions on deployment of network nodes, taken for simulation studies, studies can lead to erroneous conclusions as to the effectiveness of developed solutions and protocols.

The rest of the paper is organized as follows: in the next section we present the overview of available network generators (Internet, wireless, PLC AMI). Next we introduce MAGANET (Map-based Generator of AMI Network Topology), a new tool for AMI topology generation, which makes use of real-world building arrangement information and we summarize its properties. In Section 4 we present sample topolo-gies generated and their properties. We conclude the paper with a short summary in Sect. 5.

2   Related Work

Network topology generation involves producing synthetic topologies which imitate the characteristics of the real networks. The generation of realistic network topologies is an important research problem.

Particularly many studies were carried out to reflect the actual Internet topology and large effort has been put on the Internet topology generators. The output of Internet topologies generators is a connection graph. Contemporary tools take into account different topological coefficients and complex properties e.g. heavy tailed distributions. Popular generators are BRITE [3] and GT-ITM [4]. GT-ITM generates a connected random graph in which node is considered as transit domain. Each transit domain is grown to contain other connected random graphs, expanding on n-levels. BRITE utilized the better attachment model to generate power-law networks, allowing to create Internet topologies with heavy tail distribution, resulting in realistic hierarchical model of the interconnected network nodes. Similarly INET [5] produces AS level Internet topologies according to power-law with the accurate degree distribution.

Wireless AMI networks are special case of wireless sensor networks (WSNs). As the WSN is a relative new concept only few articles on their topologies are available. The general remarks about sensor placement are presented in [6]. The known wireless and WSNs networks topology generators are able to generate topologies with miscellaneous statistical parameters and neighborhood graphs on the basis on randomly generated positions of the network nodes and their transmission range. Attaraya [7] is a topology generator and event-driven simulator of topology control algorithms, dedicated to sensor networks. The generator enables the user to deploy the nodes randomly with uniform or normal distribution, or in a regular grid. It supports different models of energy consumption. The operator can also place the nodes manually. Attaraya allows for a graphical preview of topology and results of algorithms.

The Topo_gen [8] is a simple tool for sensor network topology generation with limited capabilities.It is adaptable and can be easily ported to support different protocols. Topo_gen allows the creation of topology files for EmSIM and ns-2. It does not take into account realistic sensor network placement, except for random node distribution and basic clusterization.

Another generator of WSN topologies is GenSeN [9], designed to use with ns-2 simulator. It allows deployment based on a set of strategies, including random, grid and different types of manual deployment. Nodes can be place also by a method representing the dropping from a higher point, e.g. from helicopter, or with use of propellant.

The application of the tools listed above in the study of AMI networks is, however, limited. The AMI networks do not use full TCP/IP stack of network protocols, so their logical topology may differ from IP network topology, wired or wireless. Network nodes are installed in buildings or in their direct neighborhood. The number of nodes in each building is proportional to number of flats or offices in it. Thus the AMI nodes deployment is closely related to the arrangement of buildings, their area, function, relative positions and other urban planning features. The resulting topology is not similar to topology of other types of networks. Therefore it is not easy to use any existing network topology generators, designed for non-AMI networks, to obtain reliable AMI topology. If we consider specific features found in real wireless AMI network, e.g. long links (wired or radio relays), the problems arise.

Some existing study of AMI networks, along with topology generator [10] assumes use of Power Line Communication (PLC) technology, instead of wireless communi-cation. It presents the conclusion, that Smart Grids build over PLC technology have "small world" properties, but PLC based systems are out of the scope of the article.

3   The Need of Realistic Topology Generator

Authors of papers related to AMI networks usually tend to use network topologies far from real deployment (having random uniform distribution, e.g. [11] or a grid deployment, e.g. [12]). However, the need of using realistic topologies increases among researchers. The study [13] indicate the necessity to incorporate urban planning features into AMI topologies. Some authors, like [14], make use of real networks deployment data, which are hardly available for the community. On the other hand, the wireless multihop networks (WMN) study [15] shows that use of realistic, buildings dependent topologies give much better results than uniform random topology. The authors use NPART, topology generation software, allowing, however, only genera-tion of topologies similar to real networks placed in two particular cities, leading to not universal tool. NPART generator is based on Freifunk WMN Project [15]. Similarity of WMN project and AMI networks is weak, as WMNs are created in different way, which result in different topologies. The goal of Freifunk project is enabling users' access to the Internet, therefore the topology has different properties. In particular AMI network nodes are placed in every building, flat or office. For WMN the density of network nodes across buildings is random, and parts of network need not to be fully connected. Therefore, comparison between topologies generated by NPART and AMI would not be reliable. The AMI topology generator should have the following properties:

- reflects the specificity of the AMI network (closely related to the arrangement of buildings, their area, function, relative positions and other urban planning features) and allows analysis of selected areas;
- allows high scalability (various sizes of networks);
- allows visualization of topologies;
- export/import topology;
- allows for preliminary analysis of the topological metrics.

Since none of the known tools has the properties listed above, the MAGANET (Map-based Generator of AMI Network Topology) generation tool was developed. It imports maps of buildings from a particular area of the world, indicated by geographical coordinates, and deploys the nodes inside buildings according to set of configuration parameters (Fig. 1). The nodes are uniformly distributed within the buildings. The map data are fetched from OpenStreetMap, free to use and available under Open Database Licence (ODbL) [16]. By choosing representative areas, it allows to obtain a topology representing various housing types, e.g. urban areas or rural areas.

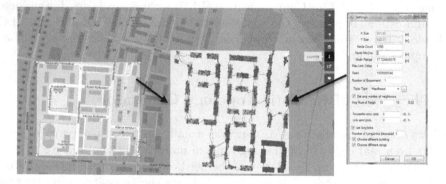

Fig. 1. The network topology is based on specific area and set of configuration parameters. Map image © OpenStreetMap contributors.

The MAGANET generates a graph of Layer II connectivity between the nodes. Currently the tool assumes a planar (2D) topology, based on simple model of radio propagation – "disk model" where only radio parameter is the maximum range allowing for communication [17]. The model can be extended to support more complex signal propagation models, e. g. taking into account already available or additional information (building walls, height of buildings, terrain etc.). Sample map and the resulting topology of connections are presented on Fig. 1.

The resulting connectivity graph may be inconsistent, so optionally "long links" may be introduces (Fig. 2). The "long links" are links between specific nodes (one to one) with longer range. These links correspond to radio relays or cable connections used by generator for assuring connectivity in the whole network.

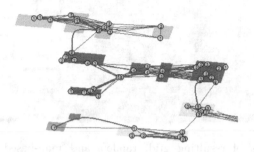

Fig. 2. The enlarged part of a sample topology generated. Different colors denote buildings in separate areas, connected via a long-links (indicated by the arcs).

The generation parameters include the number of nodes placed on the map and the transmission range of a node, or alternatively the average number of neighbors per node (the range is then calculated automatically). For comparative evaluations the generator allows also to create uniform random or grid topology. We plan to add additional deployment strategies in the future.

The MAGANET can perform some topological analysis on resulting networks, calculating the subset of well-known topology metrics [18]: Clustering coefficient, Shared neighbors distribution, Average path length and path length distribution and neighborhood connectivity. The listed metrics can be saved in a CSV format. The resulting topology (neighborhood graph) can be exported as neighbor lists in text file or XML and used as input for simulation studies with popular simulation software (e.g. ns-3, OMNeT++).

4   Example of AMI Network Topology

Depending on the specific area, city or urban type, the meter network topology and metrics representing its properties can be different. The scope of AMI topologies consists e.g. of a high, medium and low density urban, rural, industrial areas (according to [16]). For each case a set of topologies can be generated and analyzed in order to find and describe average and extreme values.

We present here an example of map based AMI network topology and compare the topology metrics to topology based on uniform random and grid based deployment, which are often being used as referenced topologies in research on AMI networks (see Fig. 3).

We considered the medium density urban area of size aprox. 300 x 500 meters, consisting of about 50, 1–4 storey multi-family houses (a part of the Trynek district, Gliwice, Poland).We assumed deploying of 1050 meters (aprox. 10 meters

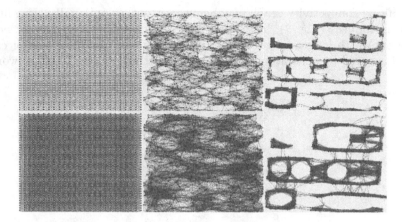

Fig. 3. The figures of resulting grid, random and map-based topologies (upper: 20 neighbors average, lower: 40 neighbors average)

per floor in each building). The results are compared to corresponding (the same size and number of meters) random and grid topologies. The radio range was selected to obtain the given average number of neighbors. The relation between radio range and average number of neighbors is presented on Fig. 4. Than the results were compared to the specific values of average number of neighbors (10–50, step 5).

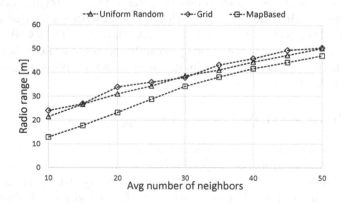

Fig. 4. The relation between radio range and average number of neighbors

The popular metrics to compare topologies is clustering coefficient (Fig. 5(a)). The map-based topology clustering coefficient is a bit higher than in the random topology and significantly higher than in case of grid based topology. However on that basis to draw we cannot have far-reaching conclusions, as has been studied only a sample topology. It is clear that for other areas that metrics can have very different values. However, preliminary evaluations show, that in most cases map-based topologies reveals greater tendency for clustering and larger values of the coefficient, what is consistent with the remark in [3] that smart metering systems often exhibit the "small world" properties.

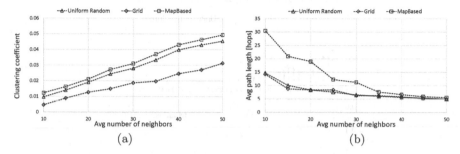

(a)   (b)

Fig. 5. Clustering coefficient (a) and average path length (b) as the function of average number of neighbors

Map-based topologies have longer paths than random topologies. This effect de-creases with increasing communication range. It is related to the increasing number of connections between buildings. The images of topologies on Fig. 3 are the intuitive confirmation of this fact. This property needs to be taken into consideration while designing the management algorithms for AMI networks and the network structure. Longer transmission paths make the dynamic routing protocols ineffective and influence on the transmission parameters. The average path length between random two nodes is presented on Fig. 5(b).

More differences between the topologies show distributions of the metrics' values. The results below show examples for average neighbors equal to 20. The distribution of shared neighbors (the probability of two nodes to have given number of common neighbors) may have a significant impact on the performance of protocols dedicated to sensor networks, e.g. defining a logical topology for multicast communications. Grid topology, due to its regularity, has only certain values. Random distribution is steady while the map-based topology fluctuates and exhibits the "long tail" property, which is not present in the network topology generators available in the literature. The shared neighbors' distribution is presented on Fig. 6. Note that most the probability value for 0 shared neighbors (which is most probable) is not shown on the chart for a clear presentation of the remaining values. So the sum of the distribution is lower than 1 is in this case.

The distribution of number of neighbors is important as larger number of direct neighbors increase the probability of interferences and collisions. Some routing mechanisms are also based on the local information exchange. The differences in topologies are clearly visible. The direct neighbor's distribution has the "long tail" representing some non-zero probability of having very high number of common neighbors. The result is presented on Fig. 7. Note that the scale

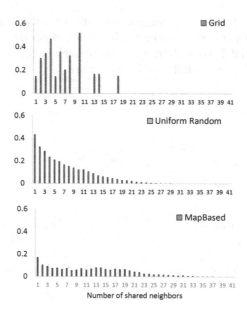

Fig. 6. The shared neighbors distributions

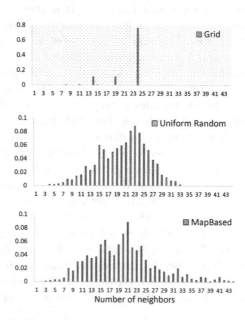

Fig. 7. The average number of neighbors distributions

for the grid based topology is different as the distribution is concentrated in few characteristics values. In comparison to random or grid based topologies, the map-based topology also the path distribution is significantly different and has the "long tail" property. The average path distribution is presented on Fig. 8.

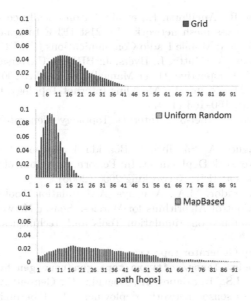

Fig. 8. The average path length distributions

5   Summary

The proposed MAGANET network topology generator allows generating the topologies based on the maps of buildings representing the wireless AMI network structure. On the basis of MAGANET generator the statistical properties of real (or close to real) AMI networks topologies can be collected, suitable for direct use in modeling, or as input data to adequate random generators. It is possible to create a Layer II topology having parameters corresponding to very specific areas and make a detailed analysis of a planned, real applications.

The comparison to the random topologies shows that the AMI networks have important statistical properties that need to be considered in the per-performance evaluation and design of networking protocols. The long tail of shared neighbor distribution and average path length distribution shows that the map based topologies differ from the results provided by network topology generators used so far. MAGANET can also represent different types of areas: industrial buildings, high-density housing typical for downtown, low-density suburb housing, dispersed development in rural areas. The MAGANET is available at http://modeling.iitis.pl.

References

1. Czachórski, T., Grochla, K., Pekergin, F.: Diffusion approximation model for the distribution of packet travel time at sensor networks. In: Cerdà-Alabern, L. (ed.) Wireless and Mobility 2008. LNCS, vol. 5122, pp. 10–25. Springer, Heidelberg (2008)

2. Tragos, E., Bruno, R., Ancillotti, E., et al.: Automatically configured, optimised and QoS aware wireless mesh networks. In: 21st IEEE International Symposium on Personal, Indoor and Mobile Radio Communications (PIMRC), Istanbul (2010)
3. Medina, A., Lakhina, A., Matta, I., Byers, J.: BRITE: Universal Topology Generation from a User's Perspective. (User Manual) BU-CS-TR-2001-003 (2005)
4. Calvert, K., Doar, M., Zegura, E.: Modeling Internet Topology. IEEE Transactions on Communications, 160–163 (1997)
5. Winick, J., Jamin, S.: Inet-3.0: Internet Topology Generator. CSE-TR-456-02 (2002)
6. Camilo, T., Rodrigues, A., Sá Silva, J., Boavida, F.: Lessons Learned from a Real Wireless Sensor Network Deployment. In: Performance Control in Wireless Sensor Networks Workshop at IFIP Networking 2006 (2006)
7. Wightman, P., Labrador, M.A.: Atarraya: A Simulation Tool to Teach and Research Topology Control Algorithms for Wireless Sensor Networks. In: ICST 2nd International Conference on Simulation Tools and Techniques, SIMUTools 2009 (2009)
8. I-LENSE Topology Generator topo_gen,
   http://www.isi.edu/ilense/software/topo_gen/topo_gen.html
9. Camilo, T., Silva, J.S., Rodrigues, A., Boavida, F.: Gensen: A topology generator for real wireless sensor networks deployment. In: Obermaisser, R., Nah, Y., Puschner, P., Rammig, F.J. (eds.) SEUS 2007. LNCS, vol. 4761, pp. 436–445. Springer, Heidelberg (2007)
10. Zhifang, W.: Generating Statistically Correct Random Topologies for Testing Smart Grid Communication and Control Networks. IEEE Transactions on Smart Grid 1(1) (2010), doi:10.1109/TSG.2010.2044814
11. Wang, D., Zhifeng, T., Jinyun, Z., Alhussein, A.: RPL Based Routing for Advanced Metering Infrastructure in Smart Grid. Mitsubishi Electric Research Laboratories, Inc. (2010)
12. Park, J., Lim, Y., Moon, S.-J., Kim, H.-K.: A scalable load-balancing scheme for advanced metering infrastructure network. In: Cho, Y., Gantenbein, R.E., Kuo, T.-W., Tarokh, V. (eds.) RACS, pp. 383–388. ACM (2012)
13. A proposal for Smart Metering Networking Solution. White Paper, Albentia Systems, ALB-W012-000en (2012)
14. Cespedes, S., Cardenas, A.A., Iwao, T.: Comparison of Data Forwarding Mechanisms for AMI Networks. In: 2012 IEEE PES Innovative Smart Grid Technologies, ISGT (2012), doi:10.1109/ISGT.2012.6175683
15. Milic, B., Malek, M.: NPART-node placement algorithm for realistic topologies in wireless multihop network simulation. In: Proceedings of the 2nd International Conference on Simulation Tools and Techniques (2009)
16. Open Street Map homepage, http://www.openstreetmap.org
17. Ben Hamida, E., Chelius, G., Gorce, J.M.: Impact of the Physical Layer Modeling on the Accuracy and Scalability of Wireless Network. Simulation 85, 574–588
18. Newman, M.E.J.: The structure and function of complex networks. SIAM Review 45, 167–256 (2003)

Topology Properties of Ad-Hoc Networks with Topology Control

Maciej Piechowiak[1] and Piotr Zwierzykowski[2]

[1] Kazimierz Wielki University, Bydgoszcz, Poland
[2] Poznan University of Technology, Poznan, Poland
mpiech@ukw.edu.pl, pzwierz@et.put.poznan.pl

Abstract. This article provides a detailed analysis and a description of the implementation of methods for the topology generation in ad-hoc networks with the application of topology control protocols. Network topology planning and performance analysis are crucial challenges for wire and wireless network designers. They are also involved in the research on routing algorithms and protocols for ad-hoc networks. The article focuses on a determination of ad-hoc network parameters and discusses their influence on the properties of networks topologies. The article also proposes a new ad-hoc topology generator. In addition, it is worth to emphasize that the generation of realistic network topologies makes it possible to construct and study routing algorithms, protocols and traffic characteristics for ad-hoc networks.

Keywords: ad-hoc networks, topology control, network topology.

1 Introduction

Ad-hoc networks are sets of nodes that form temporary networks without any additional infrastructure and no centralized control. The nodes in an ad-hoc network can represent end-user devices such as smart phones, or laptops in traditional networks. In some measurement systems nodes can represent autonomous sensors or indicators. These nodes generate traffic to be forwarded to some other nodes (unicast) or a group of nodes (multicast). Due to a dynamic nature of ad-hoc networks, traditional network routing protocols are not viable. Thus, nodes act both as the end system (transmitting and receiving data) and the router (allowing traffic to pass through), which results in multihop routing. Networks are *in motion*, i.e. nodes are mobile and may go out of range of other nodes in the network [1].

In some measurement systems nodes can represent autonomous sensors or indicators. Wireless networks can be used to collect sensor data for data processing for a wide range of applications, such as tensor systems, air pollution monitoring, etc. Nodes in these networks generate traffic to be forwarded to some other nodes (unicast) or a group of nodes (multicast) [2,3].

Modeling the energy consumption of nodes in ad-hoc networks is one of the most important problems faced by their designers. Ad-hoc networks are usually constructed from devices of different types. Due to their diversity, in the

A. Kwiecień, P. Gaj, and P. Stera (Eds.): CN 2014, CCIS 431, pp. 89–98, 2014.

literature the emphasis is only on optimizing energy absorbed by transceivers, although it represents only 15 % to 35 % consumed energy in nodes communication using wireless cards in the 802.11 standard [4].

An analysis of topology generation and implementation methods are necessary steps in the testing process of ad-hoc network. Creation of a universal tool requires an adoption of realistic network models and design methods for topology generation that reflect characteristics of existing networks and are based on the solutions proposed in the literature.

The article discusses the basic topology control protocols for ad-hoc networks and analyzes their impact on the basic topology parameters of ad-hoc networks, such as the average node degree, diameter and edge density.

The article is divided into the following parts: Section 2 defines topology control mechanisms and basic parameters describing network topology. In Section 3, the network topology generator proposed by the authors is presented. The results of the simulation of the implemented topology control protocols along with their interpretation are described in Sect. 4. Section 5 concludes the article.

2 Topology Control

The efficient use of energy resources available to ad-hoc and sensor network nodes is one of the fundamental tasks for network designers [5]. Reduction of the energy consumed by radio communication is an important issue.

Topology control is the art of controlling decision-making mechanisms of network nodes, taking into account their transmission range, that aims at a generation of networks with specific properties, while maintaining the lowest energy requirements of nodes and the maximum throughput of the network.

Topology control mechanisms are used to ensure that certain parameters in the whole network are secure. Decisions in nodes are made locally to achieve a global goal. Both centralized and distributed techniques of topology control can be classified as topology control mechanisms.

2.1 Protocols of Distributed Topology Control

A practical approach to topology control requires a creation of distributed protocols that operate locally, without the knowledge of the global state of the network, and generate topologies close to the optimal.

Topology graphs should provide desirable properties of a network using symmetric edges and should be consistent (if these properties are satisfied in the graph of the maximum power that contains the edges resulting from the maximum transmit power of the nodes) [6]. It is desirable then to build a graph of the least degrees of nodes, which reduces the probability of interference in the network. It is also desirable to create optimal topology based on inaccurate information. Providing accurate information on the nodes is often too expensive, because it requires GPS receiver in each node of the network.

Topology control protocols based on the knowledge of the position of the nodes (called *location-based topology control*) are based on the assumption of available information to the nodes with a very precise location of the neighboring nodes. The easiest way to satisfy this condition is to equip the nodes with GPS receivers, which are expensive, but provide reliable and accurate information. An alternative solution is to use techniques that make an approximation of the position based on messages received from its neighbors possible. A few nodes equipped with a GPS receiver communicating with neighboring nodes may enable them to calculate position. This solution is less expensive to implement, but is associated with the generation of additional traffic on the network [7].

LMST protocol (*Local Minimum Spanning Tree*) calculates the local approximation of the minimum spanning tree [8]. It is performed in three, or optionally four, stages.

The first stage is the exchange of information. All nodes send messages to their visible neighbors containing their identities and locations (visible neighbor nodes that are within range when transmitting at the maximum power).

In the second stage of topology creation, each node performs locally Prim's algorithm [9] taking their Euclidean length of edge as cost – the minimum spanning tree $T_u = (VN_u, E_u)$ contains all visible neighbors of node u (VN_u) in the max-power graph $G_\varepsilon = (N, V_\varepsilon)$. Then, each node defines a set of neighbors.

The node v is treated as a neighbor of node u ($u \rightarrow v$) if a node v is within range of node u and is available in one step in a minimum spanning tree computed in this node $T_u = (VN_u, E_u)$:

$$u \longrightarrow v \Longleftrightarrow (u, v) \in E_u \ . \tag{1}$$

A set of neighbors of node u is defined as:

$$N(u) = \{v \in VN_u | u \longrightarrow v\} \ . \tag{2}$$

Network topology defined in the LMST protocol is represented by a directed graph $G_{\text{LMST}} = (N, E_{\text{LMST}})$, where directed edge $(u, v) \in E_{\text{LMST}}$ exists only if $u \longrightarrow v$.

In the last (required) step of the protocol, power levels of signals required for the communication with neighboring nodes are calculated. This can be obtained by measuring the power of incoming messages sent to the nodes in the first stage of protocol with the maximum power received from the visible neighbors.

The fourth (optional) step creates a topology with symmetric links. This is achieved either by replacing the asymmetric edges of symmetric ones or by removing asymmetric edges. Figure 1 presents the steps of generating network topology with an application of the LMST model for exemplary node placements, while Fig. 2(a) shows an example topology with 75 nodes generated by the LMST protocol.

DistRNG protocol (*Distributed Relative Neighborhood Graphs*) [10] constructs a RNG graph built on a set of nodes N that has an edge between a pair of nodes $u, v \in N$ if and only if there is a node $w \in N$ such that:
    $\max\{\delta(u, w), \delta(v, w)\} \leq \delta(u, v)$.

The DistRNG protocol uses the concept of *coverage area*. If node v is a neighbor of node u, the coverage area of node v: $Cov_u(v)$ is defined as the clipping

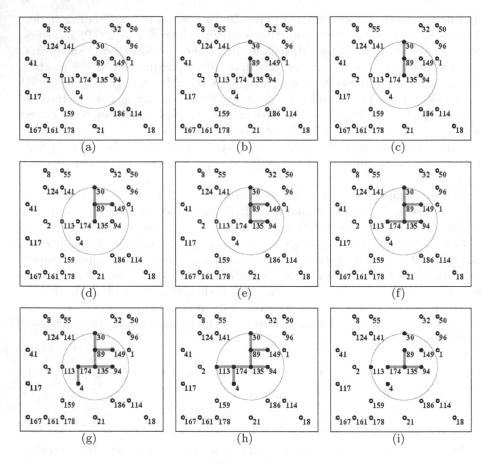

Fig. 1. The steps for generating network topology with an application of the LMST model for exemplary node deployments

plane with the center at node u and width $a\hat{u}b$, where a and b are the points of intersection of the circles with the radius $\delta(u, v)$ and midpoints in the nodes of u and v. The coverage area of a neighboring node v for node u is shown in Fig. 3. The total coverage area of node u is the sum of the areas of all of its neighbors. Figure 4 presents the steps for generating network topology with an application of the DistRNG model for exemplary node placements, while Fig. 2(b) shows an example topology with 75 nodes generated by the DistRNG protocol.

2.2 Graph Model and Network Parameters

The ad-hoc network is represented by an undirected, connected graph $G = (V, E)$, where V is a set of nodes and E is a set of links. The existence of the link $e = (u, v)$ between node u and v entails the existence of the link $e' = (v, u)$ for any $u, v \in V$ (corresponding to two-way links in communications networks). In

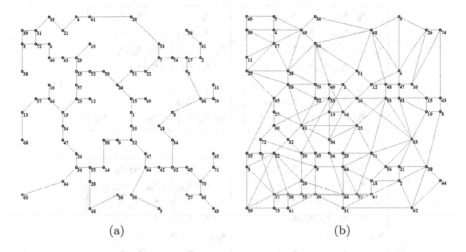

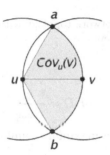

Fig. 2. Exemplary ad-hoc networks generated with LMST (a) and DistRNG (b) topology control methods ($n = 75$)

Fig. 3. The coverage area of neighboring node v for a node u

the most common power-attenuation model, the power needed to support a link $e = (u,v)$ is $p(e) = ||u,v||^{\beta}$, where $||u,v||$ is the Euclidean distance between u and v, and β is a real constant between 2 and 5 dependent on the wireless transmission environment (*path loss model*) [5].

To evaluate different structures of ad-hoc networks it is important to define the basic parameters describing network topology:

- *average node degree* [11]:

$$D_{\mathrm{av}} = \frac{2k}{n} \qquad (3)$$

where n – number of nodes, k – number of links,
- *hop-diameter* [11] – the length of the longest shortest path between any two nodes; the shortest paths are computed using *hop count* metric,

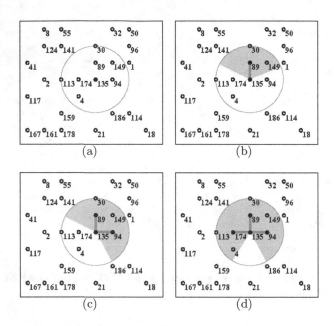

Fig. 4. The steps for generating network topology with an application of the DistRNG model for exemplary node placements

— *edge density*:

$$ED = \frac{k}{E_{\max}} \tag{4}$$

where $E_{\max} = \frac{n(n-1)}{2}$ is the maximum number of edges in fully connected graph with n nodes.

3   Network Topology Generator

The topology generator for ad-hoc networks was created based on the structure and the methods that support the process of topology generation of the BRITE application [12]. Its flexibility and functionality to generate the topology of wired networks was preserved. Its capabilities were additionally extended by creating new classes supporting the process of generation of ad-hoc network topologies [13].

The BRITE generator was equipped with tools needed to generate the topologies according to the two basic topology control protocols described in Sect. 2. Protocols based on the knowledge of the position and direction were selected. These protocols are widely used in existing ad-hoc networks and their usefulness in the simulation of theoretical network models is beyond dispute. Implementation of distributed protocols is associated with a relatively high computational complexity and, consequently, with significant power requirements from the processor and memory demands from the generator. Each node in the network has limited knowledge about the entire network topology. For this reason, a creation of optimal topology is generally not possible in realistic scenarios, hence reflecting this problem in generative models is desirable.

During application development, additional classes extending the functionality of the generator were created. The purpose of these structures was to represent ad-hoc network basis in a format determined by the BRITE application. In this way, the application was extended by additional tools that mainly supported the visualization of network topologies and the presentation of data obtained in the simulation.

Java classes used to generate the topology of ad-hoc networks (*DistRNG*, *LMST* and *KNeigh*) are derived from *AdhocModel* class that, just as *ASModel* and *RouterModel* classes, is derived from the abstract *Model* class [12]. The *AdhocModel* class stores the values of the variables used by all models of ad-hoc networks generation – *RangeMax*, *bwMin*, *bwMax*, *bwSpread* and provides a method of *PlaceNodes* responsible for the deployment of nodes within the plane as well as *AssignBW* that assigns metrics to edges in the final stage of topology generation. The *PlaceNodes* method was taken from the classes created for the generation of router-level and autonomous systems topologies. Nodes were deployed with a normal distribution, though the choice of heavy-tail distribution was omitted.

The *AssignBW* method, responsible for the imposition of attributes on the edges, was changed. In wireless networks there are other factors affecting the bandwidth and latency as compared to wired networks. The bandwidth of the link does not depend on the type of transmission medium, that is the same throughout the network, but on its length instead.

The bandwidth in generated ad-hoc networks is first allocated on the basis of the input parameter *bwMin* and *bwMax* using a linear function that assigns a specific minimum bandwidth to edges of length equal to the radio range of the source node and the maximum bandwidth to edges with a length of 0. Randomness, that can partly compensate for inaccuracies of generative methods resulting from the adoption of a very simple model of propagation, is ensured by the implementation of the *bwSpread* parameter introducing random scattering of bandwidth *bw*. It is obtained by a linear function within the range $[bw \cdot (1 - bwSpread), bw \cdot (1 + bwSpread)]$.

4   Simulation Results

A comparative analysis of the most important parameters of the topology generated by the implemented method were conducted. The topologies generated by models based on the DistRNG and LMST protocols and situated in the square plane with a side length of $Size = 50$ were compared. Nodes in all models assumed the value of the maximum transmission range of $RangeMax = 10$.

Distributed topology control protocols do not guarantee the consistency of the generated graph. Calculations of topologies diameters were performed only for nodes forming coherent graphs.

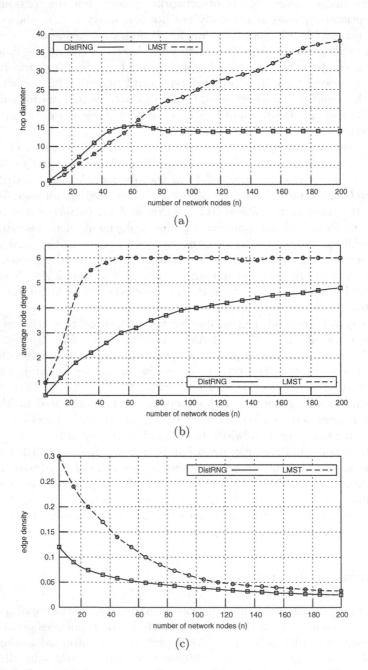

Fig. 5. Hop-diameter (a), average node degree (b) and edge density (c) versus to the number of network nodes

Hop-diameter determines the maximum number of hops in the transmission along the shortest path between any pair of nodes in the network. Its value is important, the broadcast traffic, among others. A low value of the hop-diameter has an impact on a reduction of the level of traffic on the network. The theoretical value of hop-diameter tends to infinity. In realistic scenarios, however, the density of nodes within the plane is limited and depends on the maximum number of nodes that can be deployed in a given area.

The lowest values of the hop-diameter parameter are observable in the networks generated with an application of the DistRNG protocol, whereas the highest values were obtained in the LMST protocol. The DistRNG protocol stood out offering a faster stability of this parameter (constant values were obtained for a smaller number of network nodes).

The average node degree (D_{av}) in the graph allows the level of interference in the network to be estimated [14]. The higher the average degree of logical nodes in the graph, the greater the probability of interfering transmissions. The networks generated with the LMST protocol are characterized by higher values of the average node degree. In the networks above 50 nodes, the value is fixed at 6.

The limitation of the range of nodes makes the edge density values low regardless of the density of nodes. The lowest values of this parameter are observable for the topologies generated with the application of the DistRNG protocol (Fig. 5(c)).

The analysis of the results shows that the best performance is achieved by using the generation model based on the DistRNG protocol. The choice of the generative method must also take into account the technological capacity to implement them.

5 Conclusions

The article discusses the basic topology control protocols for ad-hoc networks and analyzes their impact on the basic parameters of ad-hoc networks. The results show a significant influence of adopted topology control protocol on the parameters of ad-hoc network topology. The authors believe that the variation of the results will have an impact on the efficiency of unicast and multicast routing algorithms[1].

The article also presents an original topology generator of ad-hoc networks. The proposed topology generator is an indispensable tool for researchers and designers of ad-hoc networks that allows ad-hoc topology features to be simulated and analyzed. Further, it makes it possible to improve the process of designing and testing routing protocols dedicated particularly for ad-hoc networks.

The conducted research presented in the article will be used by the authors in the routing algorithms evaluation. Further investigations will include a greater number of network parameters (i.e., the clustering coefficient) and parameters directly related to the routing algorithms (both unicast and multicast).

---

[1] In earlier works [15], the authors showed the influence of the average node degree on the cost of trees constructed by multicast routing algorithms.

References

1. Santi, P.: Mobility Models for Next Generation Wireles Networks: Ad Hoc, Vehicular, and Mesh Networks. John Wiley and Sons, Chichester (2012)
2. Wireless sensors and integrated wireless sensor networks. Frost & Sullivan Technical Insights (2004)
3. Głąbowski, M., Musznicki, B., Nowak, P., Zwierzykowski, P.: An algorithm for founding Shortest Path Tree using Ant Colony Optimization metaheuristic. In: Choras, R.S. (ed.) Image Processing and Communications Challenges 5. AISC, vol. 233, pp. 317–326. Springer, Heidelberg (2014)
4. Gomez, K., Riggio, R., Rasheed, T., Miorandi, D., Chlamtac, I., Granelli, F.: Analysing the Energy Consumption Behaviour of WiFi Networks. In: Proceedings of IEEE Greencom 2011, pp. 98–104 (2011)
5. Santi, P.: Topology Control in Wireless Ad Hoc and Sensor Networks. John Wiley and Sons, Chichester (2005)
6. Rajaraman, R.: Topology Control and Routing in Ad Hoc Networks: A Survey. ACM SIGACT News 33, 60–73 (2002)
7. Santi, P.: Topology Control in Wireless Ad Hoc and Sensor Networks. ACM Computing Surveys 37, 164–194 (2005)
8. Li, N., Hou, J., Sha, L.: Design and Analysis of an MST-based Topology Control Algorithm. In: INFOCOM 2003, Twenty-Second Annual Joint Conference of the IEEE Computer and Communications, pp. 1702–1712 (2003)
9. Prim, R.: Shortest Connection Networks and Some Generalizations. Bell Systems Tech. J. 36, 1389–1401 (1957)
10. Borbash, S., Jennings, E.: Distributed Topology Control Algorithm for Multihop Wireless Networks. In: Proceedings of 2002 World Congress on Computational Intelligence (WCCI 2002), pp. 355–360 (2002)
11. Zegura, E.W., Calvert, K.L., Bhattacharjee, S.: How to Model an Internetwork. In: IEEE INFOCOM 1996, pp. 592–602 (1996)
12. Medina, A., Lakhina, A., Matta, I., Byers, J.: BRITE: An Approach to Universal Topology Generation. In: IEEE/ACM MASCOTS, pp. 346–356 (2001)
13. Zamożniewicz, A.: Methods for generating topologies of ad-hoc networks. MSc thesis, Poznan University of Technology (2009) (in Polish)
14. Riggio, R., Rasheed, T., Testi, S., Granelli, F., Chlamtac, I.: Interference and Traffic Aware Channel Assignment in WiFi-based Wireless Mesh Networks. Ad Hoc Networks, 864–875 (2010)
15. Piechowiak, M., Stasiak, M., Zwierzykowski, P.: Analysis of the Influence of Group Members Arrangement on the Multicast Tree Cost. In: 5th Advanced International Conference on Telecommunications AICT 2009, Venice, Italy, pp. 429–434 (2009)

Modelling and Simulation Analysis of Some Routing Schemes for Mobile Opportunistic Networks

Jerzy Martyna

Institute of Computer Science, Faculty of Mathematics and Computer Science,
Jagiellonian University, ul. Prof. S. Lojasiewicza 6, 30-348 Cracow, Poland

Abstract. Opportunistic networking explores the data transmission of small mobile devices such as mobile phones, Personal Digital Assistants (PDAs), etc. Performance modelling and simulation analysis of these networks is an essential ingredient of opportunistic network research. In this paper, the modelling of two classes of routing protocols – context-oblivious and context-aware – is considered. The main performance measured are derived for both classes of opportunistic routing protocols. The simulation results show that the model is a promising framework for mobile opportunistic networks.

Keywords: mobile opportunistic networks, wireless communications, performance evaluation.

1 Introduction

Opportunistic networks [1,2] are an extension of Mobile Ad Hoc Networks (MANETs). They are characterised by a very sparse density of nodes and link instability. For this reason, opportunistic networks spend most of their time in isolation, without any wireless communication with other nodes. During a contact or communication, opportunity nodes exchange the messages and decide to forward some of them to another nodes. Therefore, opportunistic networks must be delay- and disruption-tolerant networks [3].

In opportunistic networks mobility is used as a technique to provide communication between disconnected "groups" of nodes. In a number of opportunistic networks, the mobile nodes coexist with a few fixed nodes. All vehicular networks such as DieselNet [4] belong to these networks. From an architectural point of view, these networks are hybrid networks, sharing features with MANET. Part of the vehicular networks is formed by vehicle-to-vehicle links, and the occasionally occurring vehicular to roadside unit or road traffic hotspot link. Recently, a new type of these networks has expressed a great interest in being used, namely MANET of Unmanned Aerial Vehicles (UAVs). They can improve many homeland defence systems, as well as the detection of dangerous materials, infrastructure and perimeter surveillance, etc. [5].

It is important to note that efficient use of a mobile opportunistic network is dependent on the routing protocols. The design of efficient routing protocols for

A. Kwiecień, P. Gaj, and P. Stera (Eds.): CN 2014, CCIS 431, pp. 99–107, 2014.

mobile opportunistic networks is a difficult task due to the absence of knowledge about network topology. The routing efficiency of these networks is dependent on knowledge of network topology [6]. Therefore, a key piece of knowledge to design efficient routing protocols is the context information about the destination. It represents the current working address and the probability of meeting with other places or nodes. Two main classes of routing protocols are identified: context-oblivious and context-aware routing protocols.

The first class of routing protocols for opportunistic networks, referred to as context-oblivious, includes the Epidemic Routing Protocol [7,8]. This protocol is used when context information about other nodes in the network is not available. Unfortunately, it generates high overhead and provides an effective congestion control mechanism.

The second group of routing protocols for opportunistic networks, called context-aware routing protocols, is characterised by using all types of context information for making routing decisions. The context information allows us to improve routing performance. Among others, the 2-Hops Protocol described by Groenevelt et al. [9] belongs to this group of protocols for opportunistic networks.

2   Model of Opportunistic Networks

In this section, we present a model of mobile opportunistic networks.

We assume that the mobile network is modelled as a time-varying graph $G_T \triangleq \{V, E_T\}$, where V is identified by the first positive integers $\mid n \mid$. The set E_T represents the edges at time t. We suppose that all nodes are deployed uniformly at random in region R. Each node is able to monitor and cover circular region of radius r. It is equal to the sensing range that is centred at its position. We can formulate the following definitions:

Definition 1. *A static point $(x, y) \in R$ to be covered if there exists at least one node at a distance less than or equal to r from (x, y), and let $C(R)$ be the sub-region of R formed by the covered points.*

Definition 2. *The sensing coverage of node is equal to the ratio of the area $C(R)$ to the area of R.*

It is obvious that the network designer's goal is to maximise sensing coverage, which ideally should be 1. Considering the effect of mobility on sensing coverage, we can give the following definition:

Definition 3. *Let mobile sensing coverage $MC(R) = MC(R, t)$ be the set of points in R that are covered by at least one node in at least one time instant in the interval $[0, t)$.*

Then, we have $MC(R) \geq C(R)$. On the other hand, MC(R) can possibly be much larger than $C(R)$ if t is relatively large and/or movement speed is relatively high. We assume that after initial deployment, the nodes of mobile network start moving independently of each other according to the following rules. The direction of movement is chosen in the $[0, 2\pi]$ interval according to a distribution $f_\Theta(\theta)$. Hence, the movement speed v_s is chosen in the interval $[0, v_{\max}]$ according to the distribution $f_V(v)$. The assumption is valid for all nodes in the mobile network. This implies that the nodes are assumed to be sufficiently far away from the boundary of R. It means that the border effect can be ignored.

Based on the observation above, Liu [10] has shown the following theorem.

Theorem 1. *Let λ be the average node density in R, assuming that the nodes move according to the mobility model described above, and let v denote the expected node speed as dictated by the speed distribution $f_V(v)$. Then, the sensing coverage and the mobile sensing coverage are as follows:*

$$C(R) = 1 - e^{-\lambda \pi r^2} , \tag{1}$$

$$MCR(R, t) = 1 - e^{-\lambda(\pi r^2 + 2r\overline{v}r)} . \tag{2}$$

Movement of nodes is modelled using the whale mobility model [11]. This model was derived from the observation of radio-equipped whales. It has been observed that whale movement is influenced by three main factors: migration, grouping with other whales, and the direction of the nearest feeding grounds. Thus, the mobility of whales, treated here as nodes, is modelled as the result of the sum of three weighted vectors, representing the contribution to migration movement, grouping, and feeding respectively. The movement status of each node is updated every time step t, upon which new speed and direction values are generated for the next time step.

For all mobile networks, we can define the encounter graph (ER).

Definition 4. *Each mobile network is represented by a node in the ER graph, and an edge is added between two nodes if the two corresponding mobile networks have encountered each other at least once during the studied trace period.*

3 New Metrics of Performance Effectiveness for Mobile Opportunistic Networks

In this section, we introduce several performance metrics for mobile opportunistic networks.

The characteristic of the ER graphs is given by the following metrics:

Definition 5. *[12]. The clustering coefficient is defined as*

$$CC = \frac{\sum_{n=1}^{M} CC(n)}{M} \tag{3}$$

where

$$CC(n) = \frac{\sum_{a,b \in N(n)} I(a \in N(b))}{|N(n)| \cdot (|N(n)| - 1)} . \tag{4}$$

It is assumed that $N(n)$ is the set of neighbours of node n in the ER graph and $N(n)$ is its cardinality. $I(.)$ is the indicator function and N is the total number of nodes in the graph. The coefficient CC shows a higher tendency that the neighbours of a given node are also neighbours to each other, or there is a heavy cliquishness in the relationship between mobile opportunistic network formed through encounters. On the other hand, the clustering coefficient is the average ratio of neighbours of a given node that are also neighbours of one another.

We assumed that in the whale mobility model each ER graph is assigned to a logical "center", which determines the movement of the graph. In particular, the movement is modelled as a sequence of vectors $\mathbf{V}_i(t), \mathbf{V}_i(t+1), \ldots$ corresponding to the speed $v_i(t)$ and direction $d_i(t)$ of the logical "center" during time interval $[t, t+1), [t+1, t+2), \ldots$, respectively. Assuming that the whale mobility pattern belongs to the stationary Gauss-Markov processes [13,14], we have

$$v_i(t+1) = \alpha_i \cdot v_i(t) + (1 - \alpha_i)\bar{v}_i + V_{\text{rand}}\sqrt{1 - \alpha_i^2} \tag{5}$$

and

$$d_i(t+1) = \alpha_i \cdot d_i(t) + (1 - \alpha_i)\bar{d}_i + D_{\text{rand}}\sqrt{1 - d_i^2} \tag{6}$$

where $\bar{v}_i$ (resp., $\bar{d}_i$) are constants representing the average speed (resp., direction of movement), V_{rand} (resp., D_{rand}) is a normal variable with zero mean and standard deviation σ_i ($\sigma_i > 0$) of i-th logical "center" of nodes.

We introduce the metric of node moving in accordance with the direction of cluster movement. We use the well known cosine similarity metric [15], which measures similarity between A_k node of cluster A and logical "center" C_i as the cosine of the angle $\angle A_k CC_i$ between the vectors corresponding to A_k and C_i belonging to the cluster $\mathcal{C}$. Formally

Definition 6. *Given two m-dimensional vectors A_k ($A_k \in \mathcal{C}$) and C_i the cosine similarity metric, denoted $\Theta(A_k, C_i)$, is defined as follows*

$$\Theta(A_k, C_i) = \cos(\angle A_k C_i) = \frac{A_k \cdot C_i}{\| A_k \| \| C_i \|} \tag{7}$$

where $\| x \|$ represents the length of vector x.

If $\Theta(A_k, C_i) = 1$, we havve the accordance between the cluster and the node movement directions. If $\Theta(A_k, C_i) = 0$, we have non-accordance between both movement directions.

The next definition presents strongly connected graph.

Definition 7. *If graph $G_T = \{V, E_T\}$ is directed, we say that G_T is strongly connected if for any two nodes $u, w \in E_T$ there exists a path from u to w and a path from w to u in G_T.*

We note that the links in the ER graphs are directed. Only some of the ER graphs are strongly connected graphs. When we consider strongly connected graphs, we can use the strongly connected index based on the introduced definition of the strongly connected clustering coefficient of a node, namely:

Definition 8. *Strongly connected clustering coefficient of a node is given by*

$$SCC(n) = \frac{\sum_{A_k \in S(n)} \sum_{A_j \in S(n)} I(A_k \in S(A_j))}{|S(n)| \cdot (|S(n)| - 1)} \tag{8}$$

where $S(n)$ is the set of chosen nodes with node n will maintain links.

The strongly connected clustering coefficient is the average ratio of the connected nodes of a node that also include each other as a connected node.

To describe the degree of disconnectivity of the ER graph we use the metric defined as follows.

Definition 9. *The degree of disconnectivity of the ER graph is given by*

$$DISC = \frac{\sum_{n=1}^{M}(M - |SCC(n)|)}{M(M-1)} \tag{9}$$

where $SCC(n)$ is the set of strongly connected nodes that are in the same ER graph with node n.

This metric shows the percentage of unreachable nodes starting from a given node in the ER graph.

4   Simulation Results

We have implemented two routing protocols: the Epidemic routing protocol belonging to the context-aware routing protocols and the 2-Hops routing protocol as the context-oblivious routing protocol in the Opportunistic Network Environment (ONE) simulator [16] released under GPLv3 license [17]. The topology awareness mechanism was implemented in the Epidemic routing protocol. The use of topology awareness mechanism allows us to get the link state updates. We are taking two simulation scenarios with mobile opportunistic networks and the static opportunistic network. The parameters for the simulation are given in Table 1.

The goal of the simulation is to compare the clustering coefficient and disconnectivity degree for both protocols. Our objective is also to compare the quotient of the numbers of delivered and formed messages for both protocols. The quotient shows the quality of the routing protocol: the greater quotient indicates the worst property of the protocol.

Figure 1a shows the clustering coefficient versus the percentages of strongly connected nodes. We found that the Epidemic routing protocol is better than the 2-Hops protocol. Figure 1b demonstrates the change of disconnectivity degree (DISC) by taking various percentage of strongly connected nodes for both routing protocols.

Figures 2a and 2b show the quotient of the number of delivered and formed messages in the static opportunistic network with the Epidemic and 2-Hops

Table 1. Parameters used in simulation

Parameter	Value
Number of nodes	10
Mobility	Random Waypoint Model
Speed	0.5–1.5 m/s (uniform distr.)
Simulation time	8 000 s
Transmission range	20 m
Area	1 km × 1 km
Buffer size	10 MB
Payload size	200 KB – 2 MB
Load	40–50 Bytes/s/node

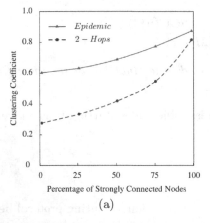
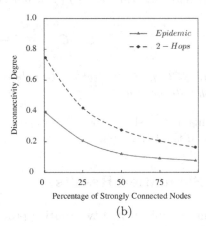

Fig. 1. Clustering coefficient (a) and disconnectivity degree (b) versus the percentages of strongly connected nodes

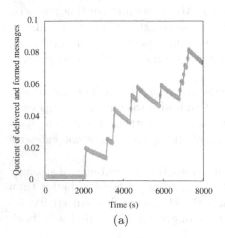
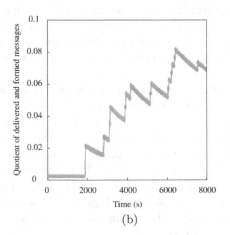

Fig. 2. Quotient of the number of delivered and formed messages in the static opportunistic network with the Epidemic (a) and 2-Hops protocols (b), respectively

protocols respectively. From the graphs, it can be seen that both protocols have similar properties.

Figures 3a and 3b show the same networks with a speed equal to 10 m/s. We can see that the Epidemic routing protocol is minimally better then the 2-Hops protocol. It is caused by a smaller value of the obtained quotient of the delivered and formed messages.

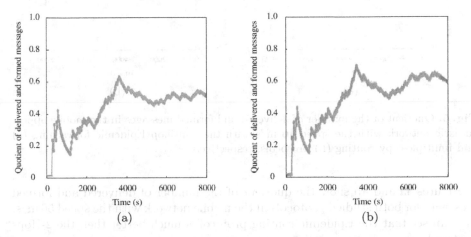

Fig. 3. Quotient of the number of delivered and formed messages in the mobile opportunistic network with the speed 10 m/s with the Epidemic (a) and 2-Hops (b) protocols, respectively

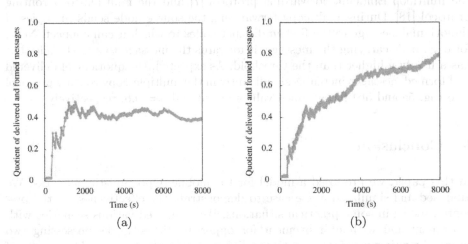

Fig. 4. Quotient of the number of delivered and formed messages in the mobile opportunistic network with the speed 10 m/s with the Epidemic (a) and 2-Hops (b) protocols, respectively

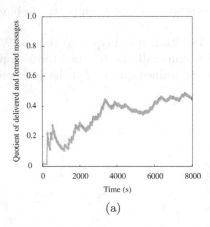
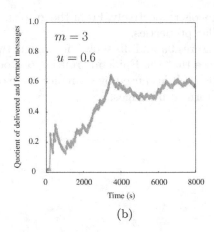

Fig. 5. Quotient of the number of delivered and formed messages in the mobile opportunistic network with the speed 10 m/s with the multihop Epidemic forwarding (a) and multiple-copy routing (b) protocols, respectively

Figures 4a and 4b show the quotient of the number of delivered and formed messages for both studied protocols in the mobile network with the speed 50 m/s. We can see that the Epidemic routing protocol is much better then the 2-Hops protocol.

Figures 5a and 5b show an interesting result, when we increase the number of copies forwarded to different nodes. The problem is how many copies should be sent to other nodes? In this scenario, the simulation results were obtained using the multihop Epidemic forwarding protocol [7] and the multiple-copy routing protocol [18]. During multicopy forwarding, the source node sends m copies of the original message to the first m distinct nodes to which it can connect. Next, for each node carrying the message, it forwards the message to a next node that has a utility u higher than the threshold. As expected, the quotient of delivered and formed messages increases significantly in the multiple-copy routing protocol (see Fig. 5a and 5b) for assumed values $m = 3$ and $u = 0.6$, respectively.

5   Conclusion

In this paper, we proposed a model for the mobile opportunistic networks. We analysed and simulated a use case to demonstrate the effectiveness of the presented model in some practical situations. We compared various scenarios with movement and without movement for opportunistic networks possessing two types of routing protocols, context-oblivious and context-aware. The results of the experiments in the simulation scenario show that both the analytical and simulation models achieve high performance in all conditions.

Future work will introduce some extensions to the simulation program, as well as a new set of experiments. The simulation programme extensions will include additional variables such as denial probability, integration delay, etc. We will also research varying movement patterns for the mobile networks.

References

1. Pelusi, L., Passarella, A., Conti, M.: Opportunistic Networking: Data Forwarding in Disconnected Mobile Ad Hoc Networks. IEEE Communication Magazine 44(11), 134–141 (2006)
2. Martyna, J.: Performance Modeling of Opportunistic Networks. In: Kwiecień, A., Gaj, P., Stera, P. (eds.) CN 2013. CCIS, vol. 370, pp. 240–251. Springer, Heidelberg (2013)
3. Farell, S., Cahill, V.: Delay- and Disruption-Tolerant Networking. Artech House, London (2006)
4. GroupPR (2007), http://prisms.cs.umass.edu/dome/umassdieselnet
5. Han, Z., Swindleburst, A., Liu, K.: Optimization of MANET Connectivity via Smart Deployment/Movement of Unmanned Air Vehicles. IEEE Trans. on Vehicular Technology 58(7), 3533–3546 (2009)
6. Jain, S., Fall, K., Patra, R.: Routing in a Delay Tolerant Network. In: Proc. of the 2004 ACM Conf. on Applications, Technologies, Architectures, and Protocols for Computer Communications (SIGCOMM), pp. 145–158. ACM, New York (2004)
7. Vahdat, A., Becker, D.: Epidemic Routing for Partially Connected Ad Hoc Networks. Technical Report CS-2000-06, Computer Science Department, Duke University (2000)
8. Zhang, X., Neglia, G., Kurose, J., Towsley, D.: Performance Modeling of Epidemic Routing. Computer Networks 51(10), 2867–2891 (2007)
9. Groenevelt, R., Nain, P., Koole, G.: The Message Delay in Mobile Ad Hoc Networks. Performance Evaluation 62, 210–228 (2005)
10. Liu, B., Bras, P., Dousse, O., Nain, P., Towsley, D.: Mobility Improves Coverage of Sensor Networks. In: Proc. of the ACM Int. Conf. on Mobile Ad Hoc Networking and Computing (MobiHoc), pp. 300–308 (2005)
11. Small, T., Haas, Z.: The Shared Wireless Infostation Model – A New Ad Hoc Networking Paradigm (Or Where There is a Whale, There is a Way). In: Proc. of ACM MobiHoc, pp. 233–244 (2003)
12. Albert, R., Barbasi, A.: Statistical Mechanics of Complex Networks. Review of Modern Physics 74(1), 47–97 (2002)
13. Papoulis, A.: Probability, Random Variables, and Stochastic Processes, 3rd edn. McGraw-Hill, New York (1991)
14. Liang, B., Haas, Z.: Predictive Distance-based Mobility Management for PCS Networks. In: Proc. of the INFOCOM 1999, vol. 3, pp. 1377–1384 (1999)
15. Deza, M.M., Deza, E.: Encyclopedia of Distances. Springer, Berlin (2009)
16. Keränen, A., Ott, J., Kärkkäinen, T.: The ONE Simulator for DTN Protocol Evaluation. In: Proc. of the 2nd International Conference on Simulation Tools and Techniques (SIMUTools), pp. 1–10 (2009)
17. The ONE (Opportunistic Network Environment Simulator) (2014), http://www.netlab.tkk.fi/tutkimus/dtn/theone/
18. Spyropoulus, T., Psounis, K., Baghavendra, C.S.: Efficient Routing in Intermittently Connected Mobile Networks: The Multiple-copy Case. IEEE/ACM Trans. on Networking (TON) 16(1), 77–90 (2008)

Modelling of Half-Duplex Radio Access for HopeMesh Experimental WMN Using Petri Nets

Remigiusz Olejnik

West Pomeranian University of Technology, Szczecin
Faculty of Computer Science and Information Technology
ul. Żołnierska 49, 71-210 Szczecin, Poland
r.olejnik@ieee.org

Abstract. The article presents a problem of shared access to a radio channel in an experimental Wireless Mesh Network. Half-duplex radio access algorithm used by the radio modules used in the network has been shown, then the access problem has been thoroughly defined. Finally a shared radio channel access problem has been modelled using Petri nets.

Keywords: Wireless Mesh Network, Petri net, wireless networks.

1 Introduction

According to Akyildiz [1] "a Wireless Mesh Network (WMN) consists of mesh routers and clients where mesh routers have minimal mobility and form a mesh of self-configuring, self-healing links among themselves". The article presents a half-duplex radio access algorithm used by the radio modules in the experimental HopeMesh wireless mesh network along with proposal of its modelling using Petri nets.

HopeMesh experimental Wireless Mesh Network is composed of simple nodes based on AVR ATmega162 microcontroller with external 62256 SRAM memory chip that offers additional 32 KiB and HopeRF RFM12B radio module. Available memory can keep routing data for a maximum number of 2838 nodes – one entry in the network routing table needs of a total 11 bytes. The network stack for the network has been described in another paper in this volume.

2 Half-Duplex Radio Access

2.1 Problem Definition

The implementation outlined in [2] used an identical (physically) blocking implementation in order to send or receive data via the RFM12B hardware module. Listing 1.1 shows the algorithm used for sending data.

A. Kwiecień, P. Gaj, and P. Stera (Eds.): CN 2014, CCIS 431, pp. 108–117, 2014.

Listing 1.1. Sender routine for the RFM12B hardware module [2]

```
void rfTx(uint8_t data)
{
        while(WAIT_NIRQ_LOW());
        rfCmd(0xB800 + data);
}
```

This implementation physically blocks the main loop the same way as the UART algorithm shown in [2]. In this case the algorithm does not wait for the status of an internal register to send data but rather waits for the external nIRQ pin from the RFM12B hardware module to go low. The official "RFM12B programming guide" [3] also proposes a physically blocking algorithm.

The goal of this paper is to show improvement of the algorithm in a similar fashion as the UART algorithm. The nIRQ pin of the RFM12B was connected to the INT0 pin of the ATmega162 microprocessor allowing to execute the SIG_INTERRUPT0 interrupt service routine asynchronously. But it turned out that the implementation could not be reused at all. The RFM12B radio hardware imposes the following algorithmic challenges for the driver implementation:

- **Single interrupt request for multiple events.** The RFM12 radio module uses only one nIRQ pin in order to generate an interrupt for the following events [4]:
  - The TX register is ready to receive the next byte (RGIT).
  - The RX FIFO has received the preprogrammed amount of bits (FFIT). The state management has to be implemented in software otherwise the current state of operation (sending or receiving) is undefined.
- **Half-Duplex operation.** The RFM12 radio module only allows either to receive or to send data at a time but not simultaneously.

The operation of the RFM12B driver algorithm was abstracted as a (proto)thread. Interestingly enough the thread has a state modelled as a state machine depending whether it receives or sends data. The following states are valid:

- **RX:** the receiving state – the thread (logically) blocks until a complete packet has been received. Whether a packet is complete or not depends on the upper network stack layers.
- **TX:** the sending state – the thread (logically) blocks until a complete packet has been sent. Again the upper network stack layers decide whether the transmission is complete or not.

The abstract algorithm is shown in Algorithm 1. Receiving data is *non-deterministic*. A packet can arrive at any time and thus the invocation of the SIG_INTERRUPT0 interrupt service routine. Therefore the algorithm sets the RX state as the *default* state for the radio thread. After receiving the whole packet the driver has to signal the receiver thread that it can process the packet.

Sending data on the other hand is *deterministic*. When a user hits the Enter key via the UART module a packet can be constructed. A sender thread

Algorithm 1. RFM12B driver thread algorithm

---

while *true* do
    if state is RX then
        receive data
        signal completion to receiver thread
    else if state is TX then
        send data
        signal completion to sender thread
    end if
    set state to RX
end while

---

has to inform the radio driver thread to change its state to TX and wait until the packet has been fully transmitted. It was implemented as a concurrency problem between three threads and a single resource:

- **Sender Thread:** inside the main loop – wants to acquire the control over the radio module until the transmission of a packet is complete.
- **Receiver Thread:** inside the main loop – wants to acquire the control over the radio module until the reception of a packet is complete.
- **Radio Thread (ISR):** also wants to acquire the control over the radio module until the packet reception is complete if it is in the RX state or until the packet transmission is complete if it is in the TX state.
- **Single resource:** in this case is the radio module; only one thread at a time can own the radio hardware resource.

The question is who controls the state of the radio thread and who and when acquires and releases the lock on the single resource (the radio module). This is rather a complicated algorithm which needs further research and investigation. The solution to this problem is presented in the paper.

2.2   Petri Net Model

The purpose of the model is to validate the correct behaviour of the complete algorithm. The most common models for parallel processing are: **Process calculus**, **Actor model** and **Petri nets**.

The Petri net [5, 6] model has been chosen, well known from modelling of RTOS systems [7]. This solution allows also for a visual modelling of the concurrent algorithm. Many different formal definitions for Petri nets exists. According to Peterson [5] a Petri net is composed of:

- A set of Places $P = \{p_1, p_2, \ldots p_n\}$.
- A set of Transitions $T = \{t_1, t_2, \ldots t_m\}$.
- An input function I who maps from a transition t_j to a collection of input places $I(t_j) = \{p_0, p_1, \ldots, p_i\}$.
- An output function O who maps from a transition t_j to a collection of output places $O(t_j) = \{p_0, p_1, \ldots, p_o\}$.

The dynamic behaviour of a Petri net can be modelled using marking. A marked Petri net contains one or more tokens which can only reside inside places (not transitions). Peterson [5] expresses the marking of a Petri net as a vector $\mu = \mu_1, \mu_2, \ldots, \mu_n$ which stores the number of tokens for each place p_i in a Petri net. The marking of a Petri net is not constant through the life-cycle but rather changes over time. The change of a marking will be expressed as a token movement from a place "A" p_a to a place "B" p_b. This abstraction allows to animate the change of Petri net marking by moving tokens from one place to another.

The symbols used in the paper are presented in Figure 1.

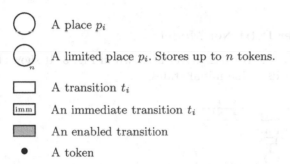

A place p_i

A limited place p_i. Stores up to n tokens.

A transition t_i

An immediate transition t_i

An enabled transition

A token

Fig. 1. Petri net symbols

But how can this abstraction be used to model concurrency? The key aspect is that a marking of a Petri net at a discrete point of time simply expresses the current state of the complete system. The location of a token (the place it currently resides in) defines the system state. A concurrently running system includes multiple states, one for each concurrently running module as was shown above. Thus a concurrent system includes multiple locations in which tokens can reside.

Regarding the initial problem we can as an example define two concurrently running modules which have to share a common resource. The following states can be defined:

- **Module 1 state.** The location of this token abstracts the current state of the first module.
- **Module 2 state.** The location of this token abstracts the current state of the second module.
- **Lock state.** The two modules both have critical section of code which must run mutually exclusive because they share a common resource. A lock (mutex) has to be introduced. The mutex token location represents the state of the mutex lock.

The common Petri net model for mutual exclusion as proposed by Peterson [5] can be seen in Fig. 2. It was used for the development of the algorithm.

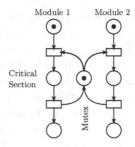

Fig. 2. Mutual exclusion Peterson Petri net model

2.3   RFM12 Driver Petri Net Model

The final algorithm for the RFM12 driver is shown in Fig. 3. It shows the three above mentioned threads in the initial states:

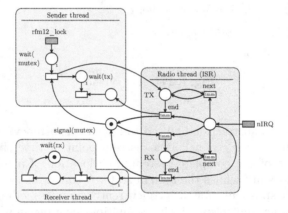

Fig. 3. Half duplex algorithm modeled as a Petri net

- **Sender thread.** The sender thread includes an always enabled transition which emits a token whenever the user prompts to send a message. When this happens it waits until it can acquire the control over the radio module in order to send a complete packet.
- **Receiver thread.** The receiver thread by default always logically blocks in a waiting state until a packet reception is complete.
- **Radio ISR thread.** The radio thread is in the RX (or idle) state by default. All transitions in the radio thread are immediate transitions. This is because the interrupt service routine itself cannot be interrupted which is a constraint defined by the hardware of the used CPU. A context switch to other threads therefore can only then happen when there are no enabled transitions in the ISR.

The Petri net behaviour is used to validate the correctness of the presented algorithm.

Data Reception. Whenever the radio module detects a valid sync pattern it will fill data into its FIFO buffer. When the reception of one byte is complete the radio hardware pulls the nIRQ pin low triggering the ISR. The model shown in Fig. 4 shows this behaviour as an infinitely enabled transition labelled as "nIRQ" which can fire at any non-deterministic time. The following Fig. 4 and the corresponding events describe the behaviour when a new byte is received by the radio.

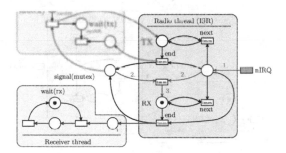

Fig. 4. nIRQ triggers reception

1. The nIRQ transition fires the ISR. An interrupt occurs caused by the radio module.
2. The radio thread being in the RX state by default tries to acquire the lock (mutex) on the radio and succeeds. Since the mutex is free the interrupt source *must* be caused by the reception of a byte.
3. The radio thread can begin its critical section by taking the received byte and delegating it to the upper network layers. The radio thread stays in the RX state and blocks the radio by not releasing the mutex.

Since the radio module has acquired the lock on the radio a sender thread will have to wait until the mutex will be released. All following nIRQ interrupt sources therefore also *must* be caused by a reception of a next byte which is shown in Fig. 5.

1. The nIRQ transition fires the ISR. An interrupt occurs caused by the radio module.
2. The radio thread is in the RX state and already acquired the lock (mutex) on the radio. It is ready to receive the next byte and delegates it to the upper network layers.
3. The upper network layers did not indicate that the reception is complete so the radio thread stays in the RX state.

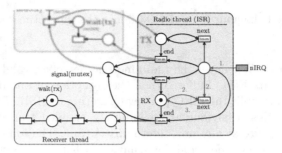

Fig. 5. Next byte reception

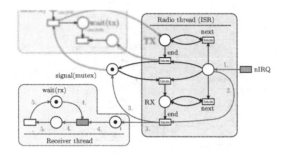

Fig. 6. Final byte reception

Still no sender thread will be able to acquire the mutex lock. The above described reception of bytes happen until the upper network layers detect the end of a frame or packet which is described in Fig. 6.

1. The nIRQ transition fires the ISR. An interrupt occurs caused by the radio module.
2. The upper network layers decide that the reception is complete. The mutex on the radio can be released.
3. After releasing the mutex the receiver thread is signalled by the radio thread that the reception of a packet is complete. The signal itself is emitted by upper network layers who need to wait for the receiver thread to process the incoming packet.

   Note that this is a limited place (currently with the limit of one token). The maximum number of token corresponds to the maximum number of packets which have to be buffered by the network stack. If the receiver thread is too slow to process incoming packets all further incoming packets will be dropped.
4. The receiver thread now changes its state. From a waiting state it switches to a processing state where it processes the received packet and i.e. displays it on the console.
5. The receiver thread finally switches back to a waiting state in order for the next packet to arrive.

Data Transmission. The data transmission in contrast to data reception is triggered by the sender thread. Thus Figure 7 shows an infinitely enabled transition in the sender thread which triggers a token (a packet to be sent) whenever the prompts a new send command. Figure 7 describes the behaviour when a new packet is sent by the sender thread.

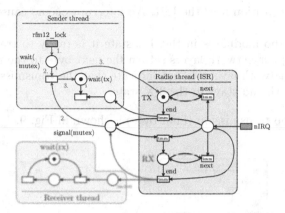

Fig. 7. Sender thread triggers transmission

1. The sender thread tries to send a new packet by changing its state to a mutex waiting state. It waits for the mutex to become free in order to acquire the lock on the radio module.
2. If the radio module becomes free the mutex can be acquired. The radio module transmitter hardware and the radio module transmitter FIFO is enabled.
3. Because the sender thread took over control it resets the radio driver thread to the new state TX. Afterwards the sender thread puts itself into a waiting state until the packet transmission is complete.

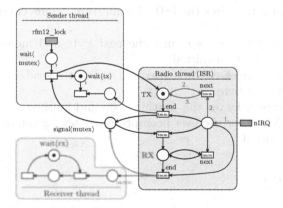

Fig. 8. Next byte transmission

The sender thread is in a different state now. It waits for the transmission of the complete packet. The transmission of the first (or next) byte is triggered now by the radio module itself. Whenever the transmission FIFO of the radio is ready the radio module hardware triggers nIRQ and the ISR fires as shown in Fig. 8.

1. The nIRQ transition fires the ISR. An interrupt occurs caused by the radio module.
2. Since the radio module is in the TX state it is ready to transmit the next byte. The upper network layers return the next byte to the radio driver.
3. The upper network layers do not indicate that the transmission is complete so the radio thread stays in the TX state.

The end of the transmission (final byte) is shown in Fig. 9.

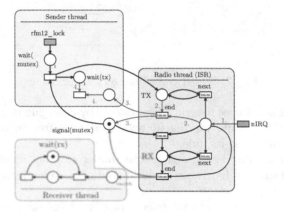

Fig. 9. Final byte transmission

1. The nIRQ transition fires the ISR. An interrupt occurs caused by the radio module.
2. The upper network layers return the next byte and indicate that this is the last byte to be transmitted.
3. The radio thread releases the mutex and signals the sender thread the transmission of the last byte.
4. The sender thread leaves the waiting state and finishes the transmission of the packet. Whenever the user prompts a new send command a new packet can be transmitted.

3   Conclusion

The paper presents the problem of shared access to radio module in HopeMesh experimental WMN. The original algorithm used for RFM12B access has been presented, then Petri net model along with mutual exclusion problem has been introduced. Finally the RFM12B driver has been successfully modelled with Petri nets – all possible states for data reception and transmission: triggering, sending/receiving of bytes and sending/receiving of final bytes have been shown. The algorithms has been used for the implementation of data link layer as a part of the network stack of HopeMesh experimental WMN.

Acknowledgments. I would like to thank my graduate student, Sergiusz Urbaniak, who implemented my preliminary ideas of RFM12B and AVR ATmega16 based wireless mesh network in his master dissertation [8].

References

1. Akyildiz, I.F., Wang, X., Wang, W.: Wireless Mesh Networks: a Survey. Computer Networks 47(4), 445–487 (2004)
2. Olejnik, R.: An Experimental Wireless Mesh Network Node Based on AVR ATmega16 Microcontroller and RFM12B Radio Module. In: Kwiecień, A., Gaj, P., Stera, P. (eds.) CN 2010. CCIS, vol. 79, pp. 96–105. Springer, Heidelberg (2010)
3. Datasheet: RFM12B programming guide. HOPE Microelectronics (2011)
4. Datasheet: Si4421 Universal ISM Band FSK Transceiver. Silicon Labs (2008)
5. Peterson, J.L.: Petri net theory and the modeling of systems. Prentice-Hall (1981)
6. Murata, T.: Petri Nets: Properties, Analysis and Applications. Proceedings of the IEEE 77(4), 541–580 (1989)
7. Rzońca, D., Trybus, B.: Hierarchical Petri Net for the CPDev Virtual Machine with Communications. In: Kwiecień, A., Gaj, P., Stera, P. (eds.) CN 2009. CCIS, vol. 39, pp. 264–271. Springer, Heidelberg (2009)
8. Urbaniak, S.: Communication algorithms and principles for a prototype of a wireless mesh network. Master thesis, West Pomeranian University of Technology, Szczecin (2011)

Study of Internet Threats and Attack Methods Using Honeypots and Honeynets

Tomas Sochor and Matej Zuzcak

University of Ostrava, Ostrava, Czech Republic
tomas.sochor@osu.cz
http://www1.osu.cz/home/sochor/en/

Abstract. The number of threats from the Internet has been growing in the recent period and every user or administrator should protect against them. For choosing the most suitable protection the detailed information about threats are required. Honeypots and honeynets are effective tools for obtaining details about current and recent threats. The article gives an introduction into honeypots and honeynets and shows some interesting results from initial 3-months period of the implementation of a small honeynet made of 3 Dionaea and one Kippo low-interaction honeypots. Basic conclusions regarding the amount of currently actively spread malware and their type are formulated.

Keywords: computer attack, Dionaea, honeynet, honeypot, Internet threat, Kippo, low-interaction honeypot, malware.

1 Honeypot Application Introduction

Despite growing popularity of honeypot and honeynet application only few detailed statistical reports from their application are publicly available. The main reason is probably the risk of misuse of such reports by potential attackers as well as financial aspect. Among few exceptions there are DenyHosts [1], Dshield.org [2], Honey Pot Project [3], Shadowserver [4] and HoneyMap [5]. All the mentioned project have a common drawback that is the lack of data available for third-party analyses. Some of them are not up-to-date as well. In the recent past there was a project called honeynet.cz publishing a lot of reports thus being widely appreciated in Europe and worldwide but it is no longer available.

The aim of the study described here was to identify suitable honeypot implementations (with focus to low-interaction server honeypots) and test them in practical implementation of our own honeynet. Subsequent evaluation of gathered data is inevitable part of the study as well. The main focus of the study is the protection of local IP-based networks using honeypots.

2 Honeypot and Honeynet Classification

Honeypots are classified into various categories and not all publications apply the same approach. The first and the most obvious classification is based on

A. Kwiecień, P. Gaj, and P. Stera (Eds.): CN 2014, CCIS 431, pp. 118–127, 2014.
© Springer International Publishing Switzerland 2014

the activity of the honeypot. Therefore honeypots can be either *passive* that simulates a server by offering some (vulnerable) services and is just waiting for an interaction from an attacker. The other approach (*active* honeypot) means that the honeypot simulates a client actively looking for available services. When an active honeypot finds a server then it interacts with the server simulating a client software like (vulnerable) www browser, e-mail client etc.).

Honeynet is a logical network of several honeypots (either connected to a single physical network or to multiple networks interconnected using the Internet). The purpose on honeynets is to improve monitoring of detected threats and to explore them in a more efficient way. The typical implementation of a honeynet for research purposes is shown in Fig. 1.

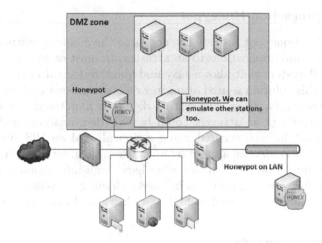

Fig. 1. Example of possible research honeypot implementation in a production network

Honeypots are commonly classified into two main categories according to the level of their interaction (in accordance to [6], Chap. 1.4.2):

– low-interaction honeypots,
– high-interaction honeypots.

2.1 Low-Interaction Honeypots

Low-interaction honeypots are based on network services emulation. Their aim is to achieve the active connection by an attacker. After the attacker's successful connection the honeypot will perform a predefined action (e.g. it reacts by displaying a predefined banner message, it downloads a malware specimen intended to store on the attacked server by the attacker etc. Low-interaction honeypots offer a limited scope of activities that can be performed and limited number of active connection as well. On the other hand it present very safe solution because an attacker is not offered with the whole operation system but they have an access to an emulated service only. This service is modified so that the attacker has a feeling that they attack a real running service with productive value. The main aim of such honeypot is to gain as much information about an attacker as possible.

This type of honeypots is the most frequently used as production honeypots, their administration is much easier than high-interaction honeypots, their operating costs are much lower as well, and the their application does not pose any risk. On the other hand they can catch only automated attacks that have been known in advance. Their typical use is for gathering statistical data, mapping threats, to distract attackers from production systems, or for fine-tuning of rules for safety tools.

2.2 High-Interaction Honeypot

High-interaction honeypot provide the access to the whole operating system including services and applications to an attacker. In most cases it is implemented as a virtualized system that allows easy and quick restore after possible modification. When this solution is used also some risks are present and their operation requires more administration. The system should be monitored in a detailed way so that any activity of an attacker could be recorded, analyzed and evaluated. The system must be well secured using a firewall and an IPS system that is adapted so that it cannot be abused by any attacker e.g. for a targeted attack or its integration into a botnet. On the other hand high-interaction honeypots are very beneficial for research, they can be used to identify zero-day vulnerabilities, to reveal a new malware as well as highly sophisticated targeted attacks.

2.3 Shadow Honeypots

Shadow honeypots present a specific subgroup of production (low-interaction) honeypots. The operate in combination with an ADS (Anomaly Detection System). If the ADS detects certain anomalies in the incoming network traffic, it is subsequently redirected to the honeypot instead of its blocking. The traffic is further analyzed on the honeypot. The legitimate traffic is segmented from a traffic containing certain anomalies this way. The benefit of this solution is in the fact that it can detect new or highly sophisticated attacks focused to real production systems.

3 Research Methods

There are various possible approaches applicable to the study of honeypots but the direct measurement of number of attacks and their nature prevails. One of recent studies [7] is a nice example of this approach that is applied by our study, too. The results presented in the paper [7] confirm the majority of our conclusions. Also other recent studies (e.g. [8] are in accordance to our results where applicable.

Two types of low-interactive honeypots, i.e. Kippo [9] and Dionaea [10], were used for gathering data about threats. Kippo is a low-interaction honeypot written in Python that emulates a SSH shell. It was inspired by the Kojoney honeypot. It contains false file system that is able to emulate adding and deleting files. It was inspired by debian Linux OS. The user can download new files (e.g. using wget command) and use cat command, too. The downloaded files are recorded. The most of Linux commands and tools are not implemented (emulated) however and if an attacker tries to use them an error message is produced. All attacks are logged with proper timestamps.

Dionaea was configured to support all emulated services available (including SMB, http, ftp, tftp, SIP, and database services of MS SQL and MySQL). Certain additional modules (e.g. VirusTotal, sandboxes, XMPP) were also used but with no influence to the measurements. After data gathering they were processed and statistics were prepared in the form of tables and diagrams (see below in the Sect. 5).

The reason for using two honeypots is the following. The Dionaea honeypot emulates the vulnerabilities of Windows operating system, especially focusing the SMBD protocol. The results from Dionaea are affected by the fact that it cannot emulate other services at the required level of quality, however.

On the other hand attackers focusing to Linux presented also an object of interest. Their common modus operandi is that they penetrate the system via a weak SSH password. This is what Kippo honeypot emulates: namely such system that looks like self-contained application. In Kippo only the SSH service (port 22) is emulated and after system penetration only activities in the system are analyzed. But thanks to the fact that Kippo is a low-interaction honeypot, the attacker is allowed to do hardly anything. No command is executed it the real system, some of them are available as dummy command while many others produce an error message. The measurement allows to create an overview of the most frequently tried after penetration through the SSH protocol. Thanks to fact that the level of interaction in the honeypot is high usually only simpler attacks (e.g. bots and script-kiddies) are caught.

4 Distribution of Sensors

Three sensors emulating Windows and one emulating Linux were used.

- The first sensor is located in the server center of the University of Ostrava (Czech Republic) where it is connected to the special VLAN where filtering is not applied. This network is directly connected to the Czech Academic Network CESNET.
- The second sensor is implemented at the virtual private hosting – VPS server in Prague.
- The third sensor is located in Spojena skola in Kysucke Nove Mesto (secondary school in the northwest of Slovak Republic). This network is directly connected to the Slovak Academic Network SANET.

– The sensor focusing to attacks against Linux servers where Kippo honeypot is used is located at the VPS hosting in Prague.

It should be noted that CESNET and SANET academic networks are peered. All honeypots deployed in the study through the area of the Czech Republic and Slovakia as shown in Fig. 2.

Fig. 2. Locations of Dionaea honeypots on the map of the Czech Republic and Slovakia

5   Results

The majority of results in this section is split into two categories in accordance with the OS that is emulated (i.e. Windows and Linux by Dionaea and Kippo honeypot, respectively). One of the results that can be gathered on both types of honeypots is the operating system of an attacker. Unfortunately this data (although gathered) turned out to be unreliable because of high level on "unknown" records due to outdated module for OS recognition in Dionaea. Therefore results from this type of analysis is not presented here in details. The total 99.956 % of attacks against Dionaea honeypots (Windows emulating) came from Windows and almost all remainder was represented by Linux (0.033 %) and SunOS (0.01 %).

5.1   Results for Kippo Honeypot – Linux Emulation

The main 'entry point' to Linux-based systems is usually SSH protocol where attackers try to connect by a guessed password and username (that is often 'root'). One of quite interesting results is listed in Table 1 where the list of the most frequently tried combinations of the user name and password is shown.

The interesting and important result is the overview of activity of IP addresses that is shown in Fig. 3. Also the structure of SSH clients could be interesting as shown in Fig. 4.

Table 1. TOP10 combinations of usernames and passwords tried at Kippo honeypot

Username	Password	Tot. number of attacks
root	123456	215
root	admin	190
root	root	85
root	1qaz2wsx	78
root	cisco123	72
root	abc123	72
root	toor	60
root	1q2w3e	59
root	passw0rd	56
root	1	49

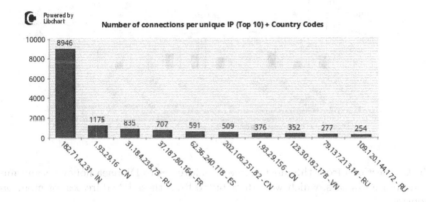

Fig. 3. Most active IP addresses on Kippo including assigned country codes

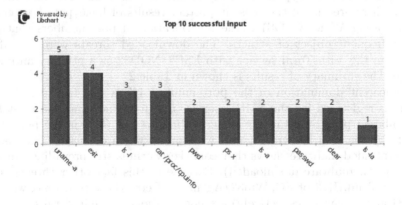

Fig. 4. The most frequently used commands on Kippo

5.2   Results for Dionaea Honeypots – Windows Emulation

One of the most interesting results from Dionaea honeypot is the overview of IP addresses activity throughout the period of measurement. The diagram in Fig. 5 shows the number of all attacks against the honeynet (3 Dionaea honeypots) from 10 the most active IP addresses as well as the number of connection when successful offering and subsequent downloading of a malware piece occurred. As one can see the number of connection towards the honeynet elements is quite high but only small part of attackers are able to offer a malware in a correct way so that it could be subsequently downloaded.The reasons were not investigated in details.

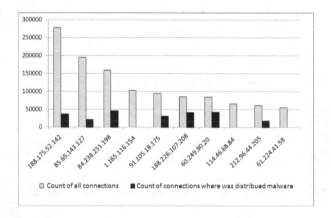

Fig. 5. Connections from the most active IP addresses to Dionaea honeynet and number of malware samples, which were distributed from these IP addresses for monitored time period

The main focus of honeypots is to identify attacks and threats coming from the Internet. Therefore one of the most important results of honeypots (especially those emulating Windows OS) is the identification of port numbers that are the most frequent object of attacks. The downloaded threats were identified according to the VirusTotal service with ESET-NOD32 as a primary antivirus database. The summary of results is shown in Table 2.

The apparent domination of the port number 445 is due to the fact that this port is registered for the SMBD service and it is known as the most frequently abused port by attackers against Windows OS. This is the port used primarily by the well known but still widespread worm conficker. The analysis of downloaded malware shows that conficker worm is still prevailing (almost 99.993 % of all malware downloaded). Because of this fact other threats (e.g. Win32/AutoRun.IRCBot.FC, Win32/Agent.UOT etc. whose occurrence was seldom – 81 and 40 cases, respectively) are not mentioned here in details.

The conficker dominance is despite the fact that this worm is quite old and patches avoiding its distribution is available for virtually every OS both for

Table 2. The most frequently attacked port numbers at Dionaea honeypots

Port number	Number of attacks
445	4279064
80	11806
1433	9220
3306	6818
135	135
21	64
5060	1

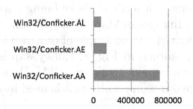

Fig. 6. Statistics of the most popular malware (conficker variants) according the number of connections that distributed malicious code

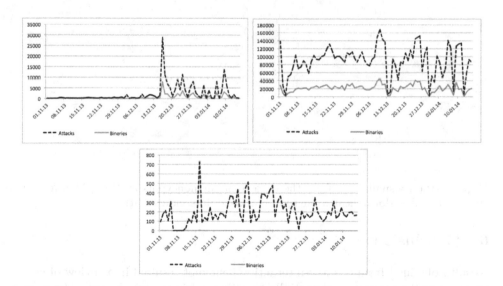

Fig. 7. Statistic outputs of attacks (upper dotted line) and downloaded binaries (lower solid line) for 90-day long period starting Nov 2, 2013, from all 3 Dionaea honeypots as described in Sect. 4 – 1st sensor in the upper left, 2nd in the upper right, 3rd one in the bottom

desktops and servers. On the other hand it should be noted that conficker variants are affluent and new ones are always emerging. The detailed analysis showed total 24 variants of conficker downloaded and another one suspicious to be so. Total 9.3387 million malware pieces were downloaded and only 665 were not identified as conficker (and only 162 pieces were identified positively as other malware. Regarding the occurrence of variants of conficker, the most frequently attacking variants are shown in the diagram in Fig. 6. The day-by-day history of number of attacks and downloaded binaries is shown in Fig. 7.

5.3   Attack Visualization

All attacks against the sensor described above being part of the low-interaction honeynet are visualized using HoneyMap [5] in real-time with places from where attacks were coming indicated. An example of our own experimental data visualized using HoneyMap is shown in Fig. 8. HoneyMap is based on the hpfeeds protocol and its social supplement HpFriends [11]. HoneyMap is a tool that is used here only to visualize our won results as it is used by other bodies (e.g. [12]).

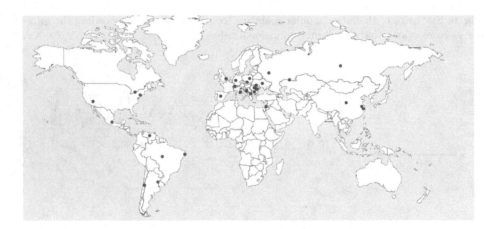

Fig. 8. Attack visualization from the data from our honeypots by HoneyMap in real time. The yellow dot indicates the sensor, the red dots indicate attackers.

6   Conclusions

Results obtained from our low-interaction honeynet resulted in a review of currently spreading threats and partially to activities of attackers as well. The most significant asset obtained is large amount of data in a scale and for that is not available elsewhere. The data show current spreading threats caught by honeypots. On the other hand the interpretation should be careful. For example the high amount of conficker worm attacks should not be interpreted as this is currently the most frequently attacking malware but it is still extremely frequent. This observation witnesses that the update approach of many computers is inappropriate. It indicates the growth of botnets as well. The most important conclusion from our measurements is that is was confirmed that honeypots allow to analyze the behavior of attackers and to reveal trends and changes. Using honeypots researchers are able to find more about techniques and procedures applied by attackers, to map current threats and malware and to apply such information into the improvement of safety tools and rules.

Acknowledgment. The authors thank for financial support for this paper publication to the Student Grant Programme 2014 (SGS) of the University of Ostrava. The authors thank to the secondary school Spojena Skola in Kysucke Nove Mesto and to the Centre for Information Technology of the University of Ostrava for providing a space and connectivity for our research honeypot sensor. Our thank belongs also to Mr. David Vorel, head of Czech Chapter of The HoneyNet Project, and to the staff of CZ.NIC Labs for expert consulting.

References

1. DenyHosts homepage, `http://denyhosts.sourceforge.net/index.html` (online, quoted January 21, 2014)
2. Internet Storm Center (ISC). Dshield.org, `http://www.dshield.org/` (online, quoted January 21, 2014)
3. Unspam Technologies, Inc. Project Honey Pot, `https://www.projecthoneypot.org/` (online, quoted January 21, 2014)
4. The Shadowserver Foundation. Shadowserver, `https://www.shadowserver.org` (online, quoted January 21, 2014)
5. Weingarten, F., Schloesser, M., Gilger, J.: HoneyMap, `https://github.com/fw42/honeymap/` (online, quoted January 21, 2014)
6. Joshi, R.C., Sardana, A.: Honeypots A New Paradigm to Information Security. Science Publishers (2011)
7. Kheirkhah, E., Amin, S.M.P., Sistani, H.A.J., Acharya, H.: An Experimental Study of SSH Attacks by using Honeypot Decoys. Indian Journal of Science and Technology 6(12), 5567–5578 (2013)
8. Sokol, P., Pisarcik, P.: Digital evidence in virtual honeynets based on operating system level virtualization. In: Proceedings of the Security and Protection of Information 2013, Brno. Univ. of Defence, May 22-24, pp. 22–24 (2013)
9. Kippo – SSH honeypot homepage, `http://code.google.com/p/kippo` (online, quoted January 21, 2014)
10. Carnivore.it. Dionaea project homepage, `http://dionaea.carnivore.it` (online, quoted January 21, 2014)
11. Schloesser, M., Gilger, J.: HpFriends homepage, `http://hpfriends.honeycloud.net` (online, quoted January 21, 2014)
12. Deutsche Telecom. SicherheitsTacho, `http://www.sicherheitstacho.eu/` (online, quoted January 21, 2014)

On the Prevention and Detection of Replay Attacks Using a Logic-Based Verification Tool

Anca D. Jurcut, Tom Coffey, and Reiner Dojen

Department of Electronic & Computer Engineering,
University of Limerick, Ireland
{anca.jurcut,tom.coffey,reiner.dojen}@ul.ie

Abstract. This paper is concerned with the design and verification of security protocols. It focuses on how to prevent protocol design weaknesses that are exploitable by intruder replay attacks. The reasons why protocols are vulnerable to replay attack are analysed and based on this analysis a new set of design guidelines to ensure resistance to these attacks is proposed. Further, an empirical study using a verification tool is carried out on a range of protocols that are known to be vulnerable to replay attacks, as well as some amended versions that are known to be free of these weaknesses. The goal of this study is to verify conformance of the protocols to the proposed design guidelines, by establishing that protocols which do not adhere to these guidelines contain weaknesses that are exploitable by replay attacks. Where non-conformance with the design guidelines is established by the verification tool the protocol can be amended to fix the design flaw.

Keywords: security protocols, formal verification, attack detection, design guidelines, replay attacks.

1 Introduction

Security protocols are trusted components in a communication system to protect sensitive information, by providing a variety of services such as confidentiality, authentication and non-repudiation. The design of provably correct security protocols is complex and highly prone to error.

Formal verification of security protocols is concerned with proving that the goals of the protocols are established and demonstrating the presence of any weaknesses that may be exploitable by mountable attacks. Formal verification is an essential part of the design process [1], as it:

- provides a systematic way to detect design flaws,
- identifies the exact cryptographic properties a protocol aims to satisfy,
- identifies protocol assumptions and properties of the environment,
- removes ambiguity in the specifications of the protocol.

However, the absence of formal verification of these protocols can lead to weaknesses and security errors remaining undetected. Many published security

A. Kwiecień, P. Gaj, and P. Stera (Eds.): CN 2014, CCIS 431, pp. 128–137, 2014.

protocols have subsequently been found to contain security weaknesses [2–6], which can be exploited by various attacks on the protocol such as replay and parallel session. The difficulty of designing security protocols that are free of mountable attacks continues today, as highlighted by many instances of replay and parallel session attacks being discovered [3, 5, 7–9].

A replay attack occurs when a message recorded in a previous run of the protocol is replayed by the intruder as a message in the current run. In a parallel session attack the attacker starts new runs of the protocol using knowledge gathered from previous runs. The attacker mixes and matches pieces of sessions running in parallel to achieve advantages, which were not intended by the security protocol designer.

The remainder of this paper is organized as follows: Section 2 presents a new analysis on the reasons why replay attacks succeed against security protocol and a new set of guidelines is proposed to ensure resistance to replay attacks for a broad spectrum of security protocols. In Sections 3 and 4 the effectiveness of the proposed guidelines is evaluated by way of an empirical study on a set of protocols with known vulnerabilities and their amended versions using an automated verification tool. Section 5 concludes the paper.

2 Designing Protocols to be Resistant to Replay Attacks

In this section we establish why replay attacks succeed over a broad spectrum of security protocols and we propose a new set of protocol design guidelines to counter these attacks.

2.1 Analysis on Why Replay Attacks Succeed

Definition 1. *A protocol P is a set of ordered steps $\{S1, S2, \ldots, Sz\}, z \geq 1$, executed in any run of P. A protocol step Sr is defined as: $Sr : A \to B : m$, where A and B are principals involved in P and m is the message transmitted. A is the sender of message m of step Sr and B is the recipient.*

Definition 2. *An initiation step is any step in a protocol, where any principal that is not a TTP (trusted third party) is the sender.*

Definition 3. *A response step Sp is the first step after an initiation step So, where the recipient of Sp is the sender of So.*

Consider two principals A and B involved in a protocol P. In an arbitrary step Sr of the protocol P, A sends a message containing a cryptographic expression $\{x\}k$ to B. If this cryptographic expression does not have at least one component fresh for the recipient B, then the message containing $\{x\}k$ can be replayed. An intruder (I) can mislead the recipient (B) to accept old and possibly compromised data as a component of the cryptographic expression from message containing $\{x\}k$. Further, if message containing $\{x\}k$ is used for authentication, then the intruder can get authenticated by substituting a previously recorded message (from A to B) for $\{x\}k$. In the following cases freshness-based attacks can be mounted:

Case 1. *B* emits fresh data.

Assume principal *B* generates fresh data x_B and emits it at step Sq prior to step Sr. Principal *A* either receives x_B directly at Sq or indirectly at any step prior to Sr. If $\{x\}k$ does not contain x_B, an intruder *I* can replay message *i.Sr* of the first run instead of message *ii.Sr* in the second run. Principal *B* has no way to detect the replay, as $\{x\}k$ does not contain any fresh component recognised by *B*. The attack can be outlined as follows:

$$i.Sq.B \rightarrow C : x_B$$
$$i.Sr.A \rightarrow B : \{x\}k$$
$$ii.Sq.B \rightarrow C : y_B$$
$$ii.Sr.I(A) \rightarrow B : \{x\}k$$

Case 2. Synchronized clock environment.

If the cryptographic expressions $\{x\}k$ does not contain a timestamp, an intruder *I* can replay message *i.Sr* instead of *ii.Sr*. Principal *B* has no way to detect the replay, as the cryptographic expressions $\{x\}k$ of Sr does not contain any fresh component.

Case 3. Exchange of key material.

Assume principal *B* generates fresh data x_B and emits it at step Sq prior to step Sr. Principal *A* either receives x_B directly at Sq or indirectly at any step prior to Sr. The cryptographic expression $\{x,w\}k$ of Sr contains a component w, which is used in the generation of a key. If $\{x,w\}k$ does not contain x_B, an intruder *I* can replay message *i.Sr* instead of *ii.Sr*. *B* has no way to detect the replay, as the cryptographic expression $\{x,w\}k$ of Sr does not contain any component which *B* recognises as being fresh. Thus, *B* considers the replayed *i.Sr* a legitimate message *ii.Sr* and accepts the contained old key material w. The attack can be outlined as follows:

$$i.Sq.B \rightarrow C : x_B$$
$$i.Sr.A \rightarrow B : \{x,w\}k$$
$$ii.Sq.B \rightarrow C : y_B$$
$$ii.Sr.I(A) \rightarrow B : \{x,w\}k$$

2.2   Proposed Guidelines to Ensure Resistance to Replay Attacks

In order to prevent freshness related issues in security protocols that are exploitable by replay attacks, the guidelines described hereafter should be adopted:

(GL1.1) Each cryptographic expression that is part of a response step in a protocol should contain a fresh component generated by the recipient in a previous step of the protocol.

(GL1.2) Under the assumption of synchronized clocks, each cryptographic expression of a protocol should contain a timestamp.

(GL1.3) A component w used in the generation of a key should be sent at least once as part of a cryptographic expression containing at least one component which the recipient recognises as being fresh.

3   A Protocol Verification Tool with Attack Detection

The CDVT/AD verification tool [9, 10] is an automated system developed at the University of Limerick that implements a modal logic of knowledge theory and an attack detection theory. The implemented tool can analyse the evolution of both knowledge and belief during a protocol execution and therefore is useful in addressing issues of both security and trust. Additionally, the verification tool has the capability of detecting protocol weaknesses that can be exploited by replay or parallel session attacks. This attack detection facility incorporates detection rules that are classified into five main categories addressing problems related to: (1) message freshness[1], (2) message symmetries, (3) handshake construction, (4) signed statements and (5) certificates. The resulting automated system, as shown in Fig. 1, enables both attack detection analysis and conventional logic-based protocol verification from a single protocol specification.

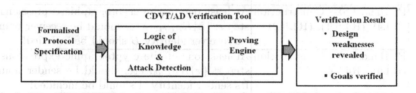

Fig. 1. The CDVT/AD verification tool

The CDVT/AD tool takes an input a text file containing the formalised protocol specification. The specification, written in the language of the tool [9, 10], contains: initial assumptions labelled "An", protocol steps labelled "Sn" and protocol goals labelled "Gn".

The output of the tool provides results for both protocol goal verification and exploitable weaknesses leading to attacks. For the attack detection, the tool establishes whether the prerequisites of any of the implemented detection rules can be derived from the formalised protocol specification. For each derivation found, the tool produces a message revealing weakness leading to possible replay (R) or parallel session attacks (P), as described in Table 1.

4   Empirical Study of Replay Attacks

This study uses the CDVT/AD tool to verify a range of protocols. The protocols chosen for this study incorporate protocols from differing categories such as: authentication, key management, secrecy, etc, with known vulnerability to replay attacks and their published amended versions. The study seeks to identify the reasons why the vulnerable protocols are susceptible to replay attacks. The verification of a key distribution protocol [11] is detailed and this is followed by summary results on the range of verified protocols.

---

[1] This paper only addresses message freshness issues (replay attacks). The other 4 categories relate to parallel session attacks.

Table 1. CDVT/AD Type of Weaknesses

Type	Rule Violation	Described Weakness
A	R: Freshness (R1.1)	Cryptographic expression in step r: $\{x\}k$ does not contain freshness identifier!
B	R: Freshness (R1.2), (R1.3)	Cryptographic expression in step r: $\{x\}k$ is not freshness protected!
C	P: Symmetry (R2.1), (R2.2), (R2.3), (R2.4)	The pair of cryptographic expressions: $\{x\}k1$ from step q and $\{y\}k2$ from step r are symmetric in opposite directions!
D	P: Signed Statement (R3.1), (R3.2), (R3.3)	The signed statement in step r: $\{x\}K_{APriv}$ is not receiver bound! Its receiver identity B should be included!
E	P: Handshake Construction (R4.1.1), (R4.2.1), (R4.3.1)	At least one of the cryptographic expressions in step r: $\{x\}K_{AB}$, etc. should be strong sender bound. Its sender identity A should be included!
F	P: Handshake Construction (R4.4), (R4.5.1), (R4.5.2)	At least one of the cryptographic expressions in step r: $\{x\}K_{APriv}$, etc. should be receiver bound. Its receiver identity B should be included!
G	P: Handshake Construction (R4.6), (R4.7)	At least one of the cryptographic expressions in step r: $\{x\}K_{BPub}$, etc. should be sender bound. Its sender identity A should be included!
H	P: Certificates Construction (R5)	The cryptographic expression in step r: $\{x\}k$ should contain the owner of public key identity A!

4.1   Detection of Replay Attack on Andrew Secure RPC Protocol

The Andrew Secure RPC protocol [12] distributes a fresh shared key between two principals, who already share a key:

1. $A \rightarrow B : A, \{Na\}Kab$
2. $B \rightarrow A : \{succ(Na), Nb\}Kab$
3. $A \rightarrow B : \{succ(Nb)\}Kab$
4. $B \rightarrow A : \{Kab\_1, Nb\_1\}Kab$

Formalisation of Andrew Secure RPC Protocol. Initial assumptions are statements defining what each principal possesses and knows at the beginning of a protocol run. The following specifies the initial assumptions of the Andrew Secure RPC protocol:

```
A1:     A possess at[0] Kab;
A2:     A know at[0] B possess at[0] Kab;
A3:     A possess at[0] Na;
A4:     A know at[0] NOT(Zero possess at[0] Na);
A5:     B possess at[0] Kab;
A6:     B know at[0] A possess at[0] Kab;
A7:     B possess at[0] Nb;
A8:     B know at[0] NOT(Zero possess at[0] Nb);
A9:     B possess at[0] Nb_1;
```

```
A10:    B know at[0] NOT(Zero possess at[0] Nb_1);
A11:    B know at[0] KMaterial(Kab_1);
```

Statements A1–A4 define the initial assumptions for principal A before a protocol run with principal B (i.e. at time t0), as follows: Assumption A1 states that A possesses symmetric key Kab. A2 specifies that A is aware of the fact that principal B possesses the key Kab. A3 specifies that A possesses the nonce Na and assumption A4 states that A knows that nonce Na is fresh for the current run of the protocol. Statements A5–A11 expresses the initial assumptions stating B's possessions and knowledge before the start of the protocol run. A5 states that B possesses symmetric key Kab. A6 specifies that B is aware of the fact that principal A possesses the key Kab. A7 and A9 express the fact that B possesses nonces Nb and Nb_1. A8 and A10 state that B knows that nonces Nb and Nb_1 are fresh for the current run of the protocol. A11 expresses the fact that the component Kab_1 is to be used in the generation of a new key.

The Andrew Secure RPC protocol steps are formalised as follows:

```
S1:    B receivefrom A at[1] A,{Na}Kab;
S2:    A receivefrom B at[2] {(F(Na), Nb)}Kab;
S3:    B receivefrom A at[3] {F(Nb)}Kab;
S4:    A receivefrom B at[4] {(Kab_1, Nb_1)}Kab;
```

The objective of the Andrew Secure RPC protocol is the distribution of a fresh shared key and the mutual authentication of the two principals involved (i.e. the authentication of A to B and of B to A). The formalised goals of the Andrew Secure RPC protocol are as follows:

```
G1:    A possess at[4] Kab_1;
G2:    A know at[4] B possess at[4] Kab_1;
G3:    A know at[2] B send at[2] {(F(Na), Nb)}Kab;
G4:    B possess at[4] Kab_1;
G5:    B know at[4] A possess at[4] Kab_1;
G6:    B know at[3] A send at[3] {F(Nb)}Kab;
G7:    A know at[4] NOT(Zero send at[0] {(Kab_1, Nb_1)}Kab);
```

G1 and G4 state that principals A and B possess at step 4 the new freshly generated key Kab_1. G2 states that A knows at step 4 that B possesses at step 4 the new freshly generated key Kab_1, while G5 states that B knows at step 4 that A possesses at step 4 key Kab_1. G3 states that A knows at step 2 that B is the sender of $\{(F(Na), Nb)\}Kab$, while G6 states that B knows at step 3 that A is the sender of $\{F(Nb)\}Kab$. G7 states that A knows that message in step 4, $\{(Kab\_1, Nb\_1)\}Kab$ has been created during the current protocol run.

Andrew Secure RPC Protocol Verification Results. The results of the automated verification of the Andrew Secure RPC protocol are presented in Fig. 2 and Fig. 3.

The results in Fig. 2 indicate that a number of the security goals (represented by "(3)", "(5)", "(6)" and "(8)") are not satisfied. Investigating the results for failed protocol goals reveals that the protocol suffers from the following weaknesses:

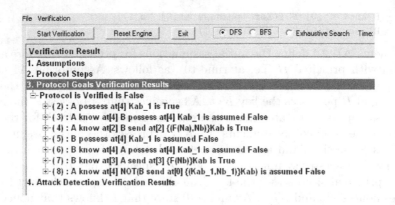

Fig. 2. CDVT/AD Verification results of Andrew Secure RPC protocol

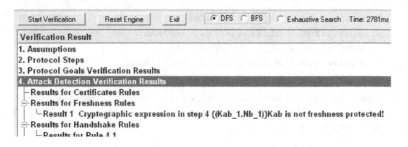

Fig. 3. CDVT/AD Attack Detection results of Andrew Secure RPC protocol

- A's inability to establish that B possesses at step 4 the newly generated key Kab_1 (goal G2), B's inability to establish that A possesses at step 4 Kab_1 (goal G5) and failure of goal G4 indicates that key Kab_1 might be compromised.
- A's inability to establish that message 4 $\{(Kab\_1, Nb\_1)\}Kab$ (goal G7) is fresh indicates that new distributed key, Kab_1 also might not be fresh.

Thus, there is no assurance that key Kab_1 is fresh and is not substituted from a previously recorded message 4.

Additionally, a weakness in the design of Andrew Secure RPC protocol, identifying a replay vulnerability is revealed. It can be seen from Fig. 3 that one of the freshness detection rules of the tool is triggered. The cryptographic expression in step 4 $\{(Kab\_1, Nb\_1)\}Kab$ is not freshness protected. This implies that $\{(Kab\_1, Nb\_1)\}Kab$ does not contain data which receiver A recognises as being fresh.

In summary, this verification reveals that the protocol has weaknesses that are exploitable by repay attacks. Details on how to mount this attack can be found in [11].

Applying the Proposed Guidelines to Amend Andrew Secure RPC.
The proposed design guideline GL1.1 reveals that the cryptographic message
$\{(Kab\_1, Nb\_1)\}Kab$ in step 4 of the Andrew Secure RPC protcol should include
a component which its recipient recognises as fresh. This can be achieved by
including nonce Na, previously generated by principal A in step 1, as follows:

1. $A \to B : A, \{Na\}Kab$
2. $B \to A : \{succ(Na), Nb\}Kab$
3. $A \to B : \{succ(Nb)\}Kab$
4. $B \to A : \{Kab\_1, Nb\_1, Na\}Kab$

Thus, the cryptographic message that contains the new generated key Kab_1
can be identified by principal A as fresh, i.e. as belonging to the current protocol
run. Consequently, any attempt by an attacker to replay message
$\{(Kab\_1, Nb\_1)\}Kab$ of step 4 will fail, as A can identify the replay through
the incorrect value of Na.

4.2 Detection of Replay Attacks – Results Summary

The verification results on the chosen range of protocols using the CDVT/AD
tool are summarised in Table 2, where the second column enumerates previously

Table 2. Empirical Study using CDVT/AD: Replay Attack Vulnerability

Analyzed Protocol	Published Attacks	Triggered Detection Rule	Guideline Violation
NS PK,1978	1981 [13]	R1.1, R1.2[1]	GL1.1, GL1.2
Lowe's fix NS PK,1995	1981 [13]	R1.1, R.2	GL1.1, GL1.2
AS RPC,1989	1990 [11]	R1.2[2]	GL1.2
BAN mod AS RPC,1990 [11]	None	No freshness rules[1]	None
BAN conc.AS RPC,1990 [11]	None	No freshness rules[1]	None
Lowe AS RPC, 1996 [6]	None	No freshness rules	None
SPLICE/AS,1991	1995 [14]	R1.3[1]	GL1.3
HC SPLICE/AS,1995	1995 [14]	R1.3	GL1.3
Kao-Chow Auth.v1,1995	1995 [15], 2008 [16]	R1.3[1]	GL1.3
DLC Kao-Chow v1, 2008[16]	None	No freshness rules[1]	None
TMN,1989	1997 [17]	R1.1, R1.2, R1.3	GL1.1, GL1.2, GL1.3
V2 of TMN,1997	1997 [17]	R1.1, R1.2, R1.3	GL1.1, GL1.2, GL1.3
V3 of TMN,1997	1997 [17]	R1.1, R1.2, R1.3	GL1.1, GL1.2, GL1.3
PKM Auth. IEEE 802.16, 2004	2006 [18]	R1.1, R1.2	GL1.1, GL1.2
CBKM-IC,2008	2008 [19]	R1.1, R1.2[1]	GL1.1, GL1.2
CBKM-RC,2008	2008 [19]	R1.1, R1.2	GL1.1, GL1.2
DZC fix CBKM-IC, 2008 [19]	None	No freshness rules	None
Jin et al. RFID Auth., 2011	2013 [3]	R1.1	GL1.1
LKY Auth, 2005	2013 [9]	R1.1[1]	GL1.1
NKPW mod. LKY, 2007	2013 [9]	R1.1	GL1.1
JDC mod. LKY, 2013, [9]	None	No freshness rules[1]	None

[1]Other attacks also detected

published replay attacks on the analyzed protocol. The third column indicates the rules of the CDVT/AD detection tool thar are triggered revealing the replay attack(s), while the last column presents the proposed design guidelines that are violated.

The study shows that for all the protocols evaluated those with known replay attacks trigger at least one of the freshness rules R1.1–R1.3 and at least one of the proposed design guidelines GL1.1–GL1.3 are violated. Also, none of the freshness rules are triggered for protocols which are known to be secure against replay attacks.

5 Conclusion

This work investigated why weaknesses in the design of security protocols make them vulnerable to mountable attacks by intruders. The paper focused on how to prevent design weaknesses that are exploitable by replay attacks. The reasons why protocols can be vulnerable to replay attacks were analysed and based on this analysis a new set of design guidelines to ensure resistance to these attacks was proposed. An empirical study was conducted using the CDVT/AD verification tool on a range of protocols that are vulnerable to replay attacks as well as their amended versions. The applicability of the proposed design guidelines for security protocol design was demonstrated by example of the Andrew Secure RPC key distribution protocol. The performed verification revealed weaknesses in the authentication scheme that can be exploited by a mountable replay attack. The impact of this attack is that an attacker can replay an old message as the last message in the protocol and thus, misleading principal A into using an old, possibly compromised, session key. It was shown that non-compliance of this protocol with the proposed guidelines resulted in one of the freshness rule of the CDVT/AD tool being triggered. The amended version of the protocol did not trigger any freshness rule of the verification tool, indicating conformance of the protocol with the design guidelines. The empirical study revealed that, for all the protocols evaluated, those with known replay attacks triggered at least one of the freshness rules R1.1–R1.3 and at least one of the proposed design guidelines GL1.1–GL1.3 were violated. Also, none of the freshness rules were triggered for protocols which are known to be secure against replay attacks. The empirical study thus provides confidence that the protocols which are compliant with the design guidelines are resilient against replay attacks.

Acknowledgements. This work was funded by Science Foundation Ireland – Research Frontiers Programme (11/RFP.1/CMS 3340).

References

1. Coffey, T., Dojen, R., Flanagan, T.: Formal verification: An imperative step in the design of security protocols. Computer Networks 43(5), 601–618 (2003)
2. Nam, J., Kim, S., Park, S., Won, D.: Security Analysis of a Nonce-Based User Authentication Scheme Using Smart Cards. IEICE Transactions Fundamentals E90-A(1), 299–302 (2007)
3. Fu, X., Guo, Y.: A Lightweight RFID Mutual Authentication Protocol with Ownership Transfer. In: Wang, R., Xiao, F. (eds.) CWSN 2012. CCIS, vol. 334, pp. 68–74. Springer, Heidelberg (2013)
4. Dojen, R., Pasca, V., Coffey, T.: Impersonation Attacks on a Mobile Security Protocol for End-to-End Communications. In: Schmidt, A.U., Lian, S. (eds.) MobiSec 2009. LNICST, vol. 17, pp. 278–287. Springer, Heidelberg (2009)
5. Dojen, R., Jurcut, A., Coffey, T., Györodi, C.: On Establishing and Fixing a Parallel Session Attack in a Security Protocol. In: Badica, C., Mangioni, G., Carchiolo, V., Burdescu, D.D. (eds.) Intelligent Distributed Computing, Systems and Applications. SCI, vol. 162, pp. 239–244. Springer, Heidelberg (2008)
6. Lowe, G.: Some new attacks upon security protocols. In: Proc. 9th IEEE Computer Security. Foundations Workshop, pp. 162–169 (1996)
7. Hsiang, H.C., Shih, W.K.: Weaknesses and improvements of the Yoon-Ryu-Yoo remote user authentication scheme using smart cards. Computer Communications 32, 649–652 (2009)
8. Wang, D., Ma, C.: Cryptanalysis of a remote user authentication scheme for mobile client-server environment based on ECC. Information Fusion 14, 498–503 (2013)
9. Jurcut, A., Coffey, T., Dojen, R.: Establishing and Fixing Security Protocols Weaknesses using a Logic-based Verification Tool. Journal of Communication 8(11), 795–806 (2013)
10. Coffey, T., Dojen, R., Jurcut, A.: CDVT/AD Verification Tool 2014 Executable, www.dcsl.ul.ie/cdvt-ad-tool
11. Burrows, M., Abadi, M., Needham, R.: A logic of authentication. ACM Transactions on Computer Systems TOCS 8(1), 18–36 (1990)
12. Satyanarayanan, M.: Integrating security in a large distributed system. ACM Transactions on Computer Systems 7(3), 247–280 (1989)
13. Denning, D., Sacco, G.: Timestamps in key distributed protocols. Communication of the ACM 24(8), 533–535 (1981)
14. Clark, J., Jacob, J.: On the security of recent protocols. Information Processing Letters 56, 151–155 (1995)
15. Kao, I., Chow, R.: An efficient and secure authentication protocol using uncertified keys. Operating Systems Review 29(3), 14–21 (1995)
16. Dojen, R., Lasc, I., Coffey, T.: Establishing and Fixing a Freshness Flaw in a Key-Distribution and Authentication Protocol. In: IEEE International Conference on Intelligent Computer Communication and Processing, pp. 185–192 (2008)
17. Lowe, G., Roscoe, A.W.: Using CSP to detect errors in the TMN protocol. Software Engineering 23(10), 659–669 (1997)
18. Xu, S., Huang, C.: Attacks on PKM Protocols of IEEE 802.16 and Its Later Versions. In: Proceedings of 3th International Symposium on Wireless Communication System (ISWCS), pp. 185–189. IEEE Press, Spain (2006)
19. Dojen, R., Zhang, F., Coffey, T.: On the Formal Verification of a Cluster Based Key Management Protocol for Wireless Sensor Networks. In: International Performance Computing and Communications Conference, Texas, USA, pp. 499–506 (2008)

System Network Activity Monitoring
for Malware Threats Detection

Mirosław Skrzewski

Politechnika Śląska, Instytut Informatyki,
Akademicka 16, 44-100 Gliwice, Polska
miroslaw.skrzewski@polsl.pl

Abstract. Monitoring network communication is one of the primary
methods used for years to combat network threats. Recent attacks on
corporations networks shows that classical perimeter centric detection
methods, based on the analysis of signatures, statistical anomalies or
heuristic methods aimed at protection from the outside do not work,
and are easily circumvented by new generations of malware. Increasingly
apparent becomes the need to create additional internal line of defense,
aimed at detecting and blocking what penetrated inside and operates
in a network environment. The paper presents such solution – a new
method for threats detection, based on novel principle – local monitoring
and analysis of the system and application's network activity, detecting
traces of malware operation to the level of process running on the system.

Keywords: outbound traffic monitoring, malware infection detection,
system network activity, multi-level system defense.

1 Introduction

Detecting signs of malware operation by monitoring network communication is
one of the primary methods used for years to combat network threats. Many
IDS / IPS system analyzes in various ways the packet flows in order to identify and block the threats related communications. Recent attacks on systems
of known corporations [1,2] and government [3] institutions shows that classical
outside oriented detection methods, based on the analysis of signatures, statistical anomalies or heuristic methods aimed at protection from the outside do not
work, and are easily circumvented by new generations of malware threats.

Worse, the methods aimed at protecting against new infection of malware
often do not cope well with detection of traces of existing malware infections
[4,5]. Increasingly apparent becomes the need to complement existing methods
of protection with an effective solution to detect traces of malware activities
inside the network of institutions – the need to create additional, the next line
of defense, aimed at detecting and blocking what infiltrated inside and operates
in the network environment.

Emerging new solutions of detection systems [6,5] continues to hold on to the
paradigm of centralization of data collection and threats analysis, ignoring the

A. Kwiecień, P. Gaj, and P. Stera (Eds.): CN 2014, CCIS 431, pp. 138–146, 2014.

possibility of using the data available inside the monitored systems. Focusing exclusively on the analysis of network traffic on the network boundary leads to the processing of large amounts of data from the network and problems with unambiguous classification of the observed effects [7] (detection of symptoms of malware operation).

Many problems related to the amount of irrelevant data processed can be avoided by using the informations identifying programs that generate network traffic in individual systems and to assess them using the communication characteristics (profiles) of individual programs and applications. Required information can be obtained from the records of described in [8] method of system's network activity monitoring, as seen on the level of transport drivers interface (TDI) by tdilog program.

The paper presents new method of threats detection, based on the analysis of network activity of system programs and applications, that enables the detection of traces of malicious activity with an accuracy to the level of process running on the system.

2   Monitoring of System Communication

The operation of a computer system in a network environment always manifests itself in two modes: as a server – passive entity, offering in response to external request the execution of a specific services and as a client – active party initiating the communication operations, sending requests to other systems on the network.

In the classical security solutions the attention is focused mainly on the server side of the communication, protecting access to the system services from network (e.g. by firewall) by carefully controlling access to the system ports on various levels of details. In general, there are no server initiated communication. The client side is considered safe – from the assumption the communication is initiated by the programs installed on the system, by default under the user's control.

This approach gave the model of strong protection of system communication from the outside, with an indulgent treatment of outgoing traffic, by default considered secure. The emergence of the contemporary generation of malware (worms, Trojans, spyware, finally bots) [9,10] has forced change of perspective on the security of the system communication on the network and attempts to cover with monitoring the outbound communication from the system as well.

Implementation of this monitoring encounters certain problems that require a different approach to the rules defining the permitted and prohibited behavior. Connections from outside world come to the well-known port numbers, usually fixed for a specific service – one can easily formulate the access control rules.

Outgoing communication is done from any source ports to any destination – no simple rules exist that determines what output ports are allowed or no for communication. However, administrators can control to which ports on the external systems are they opened, and thereby limit the ability to use certain services by the users.

Some services (like http, mail) are so prevalent that becomes standards necessary for performing everyday tasks, and outside access to the ports of these services are commonly open. Communication on other, untypical ports can easily be (and often is) blocked.

On the other hand, central monitoring systems do not have access to other then source IP information, so they don't know who (what process) is responsible for initiate outbound communication to given service port, and whether it is safe (should be permitted) or not. In most cases such connection is assumed safe and allowed.

This asymmetry of security posture is well known and also used in malicious intents – many malware programs changed their behavior from server to client mode and actively "call home" to check command [11] instead waiting passively to incoming tasks.

3　Network Activity of the System

Monitoring of network communication requires registration of connections being opened by system programs to other systems (active side of the communication), ports opened by supporting them programs (server side) and registration of incoming connections from other systems.

Such registration can be done on communication interfaces of the system at the kernel level. In Windows, there are two levels of interfaces – network-level Network Driver Interface Specification (NDIS) and transport-level Transport Driver Interface (TDI) providing complete information about system connections.

For network monitoring has been used TDI interface. The program tdilog [12] recorded as an event in the log all the data concerning flow of information on the various ports of the system (times of opening and closing the port, establishment and termination of connection, the amount of transferred data, program name and the context from which there has been a communication service performed and the type of event). In newer generations of Windows the TDI interface is marked as deprecated, and therefore traffic monitoring was moved to NDIS interface.

The recorded data allow to conduct analysis of the network activity of individual applications, modules of operating system as well as of incoming requests on the open ports of the system.

3.1　Programs Communication Profiles

By grouping recorded communication events according to the programs involved one can receive a set of types of connectivity, in which participates given program – its model of communication. For the analysis of network threats was defined the concept of the communication profile of the program as a collection of numbers of outbound connections to specific destination ports to which given program establish a connection in a specified period of time. A collection of profiles of active programs creates the profile of the system activity.

To determine the activity profile of the Windows system were recorded network communication of several virtual machines of XP system without installed applications and with installed additional software. Example of network activity profile of Windows XP without installed applications registered within 4 hours of continuous system operation is shown in Table 1.

Table 1. Activity profile of clean XP system

Count	Prot	Dport	Appl-path
3	TCP	139	C:\WINDOWS\system32\lsass.exe
150	TCP	80	C:\WINDOWS\System32\svchost.exe
3	TCP	139	C:\WINDOWS\System32\svchost.exe
99	TCP	443	C:\WINDOWS\System32\svchost.exe
2	TCP	139	C:\WINDOWS\system32\tdilog.exe
2	TCP	139	System
94	UDP	123	C:\ProgramFiles\NetTime\NeTmSvNT.exe
15	UDP	123	C:\ProgramFiles\NetTime\NetTime.exe
11	UDP	137	C:\ProgramFiles\NetTime\NetTime.exe
1	UDP	137	C:\WINDOWS\Explorer.exe
1	UDP	137	C:\WINDOWS\system32\lsass.exe
12	UDP	137	C:\WINDOWS\system32\spoolsv.exe
12	UDP	138	C:\WINDOWS\system32\spoolsv.exe
3881	UDP	53	C:\WINDOWS\system32\svchost.exe
222	UDP	67	C:\WINDOWS\System32\svchost.exe
110	UDP	123	C:\WINDOWS\System32\svchost.exe
180	UDP	137	C:\WINDOWS\System32\svchost.exe
6	UDP	1027	C:\WINDOWS\System32\svchost.exe
53	UDP	1028	C:\WINDOWS\System32\svchost.exe
43	UDP	1029	C:\WINDOWS\System32\svchost.exe
9	UDP	1030	C:\WINDOWS\System32\svchost.exe
333	UDP	1900	C:\WINDOWS\System32\svchost.exe
1672	UDP	137	System
219	UDP	138	System

A lot of Windows programs do not communicate through the network, hence in the network activity profile of the system, there are few program names – svchost.exe, spoolsv.exe, lsass.exe and System, and the communication is done mainly through the ports related to a NetBIOS (137, 138 UDP, 139 TCP). Svchost.exe also communicates via http on ports 80 and 443. The only installed application, time sync program NetTime.exe communicates via port 123, and also uses 137 UDP. The most active programs are System and svchost.exe.

Installing anti-virus program on the system adds a lot of network activity associated with updates (ports 2221 and 2222) and e-mail transmission control (Table 2). Other common applications installed such as Office, Picasa, Firefox does not contribute much to the network activity profile of the system (Table 3).

Table 2. Activity profile of anti-virus program

Count	Prot	Dport	Appl-path
812	TCP	80	C:\ProgramFiles\ESET\ESETNOD32Antivirus\ekrn.exe
8	TCP	110	C:\ProgramFiles\ESET\ESETNOD32Antivirus\ekrn.exe
11	TCP	443	C:\ProgramFiles\ESET\ESETNOD32Antivirus\ekrn.exe
8	TCP	995	C:\ProgramFiles\ESET\ESETNOD32Antivirus\ekrn.exe
12	TCP	2221	C:\ProgramFiles\ESET\ESETNOD32Antivirus\ekrn.exe
239	TCP	2222	C:\ProgramFiles\ESET\ESETNOD32Antivirus\ekrn.exe
22	UDP	137	C:\ProgramFiles\ESET\ESETNOD32Antivirus\ekrn.exe

Table 3. Activity profile of common application programs

Count	Prot	Dport	Appl-path
16	TCP	443	C:\ProgramFiles\MozillaFirefox\firefox.exe
1	UDP	137	C:\ProgramFiles\Intel\Wireless\Bin\S24EvMon.exe
5	UDP	137	C:\ProgramFiles\MicrosoftOffice\OFFICE11\WINWORD.exe
1	UDP	137	C:\ProgramFiles\MicrosoftOffice\Office14\OUTLOOK.exe
2	UDP	137	C:\ProgramFiles\MicrosoftOffice\Office14\POWERPNT.exe
1	UDP	137	C:\ProgramFiles\MotorolaMediaLink\Lite\NServiceEntry.exe
11	UDP	137	C:\ProgramFiles\MozillaFirefox\firefox.exe
1	UDP	137	C:\ProgramFiles\Picasa2\PicasaPhotoViewer.exe
5	UDP	137	C:\ProgramFiles\ThinkPad\ConnectUtilities\AcSvc.exe
1	UDP	137	C:\ProgramFiles\ThinkPad\ConnectUtilities\ACWLIcon.exe
2	UDP	137	C:\ProgramFiles\TrackerSoftware\PDFViewer\PDFXCview.exe
6	UDP	137	C:\WINDOWS\Explorer.exe

All installed programs run from its default locations (C:\Program Files\, C:\WINDOWS\, C:\WINDOWS\system32\) and periodically contact with the pages of manufacturers usually by using the https protocol.

Connections associated with software updates takes place on the well-known to programs destination addresses, and virtually are not accompanied by any communication errors. Communication initiated by the user activity usually affects ports associated with the operation of selected services (http, https, mail).

3.2 Malware Activity in Communication Profiles

The emergence of the malware program changes the network activity profile of the system – there is a new sender contacting with its C&C servers. Operation of malware in the system can cause several types of effects in the communication profile of the system: may appear new processes / programs contacting to known or new destination ports – easily visible malware programs; there might be changes in the behavior of well-known system programs, such as not yet communicating in the network programs begin sending packets (attempts to establish connectivity) or there will be the changes in behavior of known programs in

the communication profile of the system – this corresponds to a situation called
.dll injection – malware code is injected (attached to code of running in RAM
program), changing its behavior.

The problem is to identify all the elements of the activity profile of the malware. During the testing of the honeypot systems and running on them registered
copies of malware has been observed that in the system at the time of infection
are starting several, differently behaving programs.

Often starts a new process communicating on the network, but also appear
malware modules running as part of other components of the Windows which
normally not acting in the network, such as `notepad.exe` or `Explorer.exe`, or
modifying the activity of network programs. Network activity profile of malware
infected test system presents Table 4.

Table 4. Activity profile of malware infected test system

Count	Prot	Dport	Appl-path
92	TCP	80	C:\ProgramFiles\InternetExplorer\iexplore.exe
108	TCP	25	C:\WINDOWS\Explorer.EXE
1	TCP	80	C:\WINDOWS\Explorer.EXE
19	TCP	139	C:\WINDOWS\Explorer.EXE
2	TCP	443	C:\WINDOWS\Explorer.EXE
2	TCP	7081	C:\WINDOWS\Explorer.EXE
148	TCP	8800	C:\WINDOWS\Explorer.EXE
19	TCP	80	C:\WINDOWS\notepad.exe
1	TCP	6777	C:\WINDOWS\notepad.exe
1	TCP	80	C:\WINDOWS\ppdrive32.exe
6349	TCP	445	C:\WINDOWS\ppdrive32.exe
1	TCP	6971	C:\WINDOWS\ppdrive32.exe
19	UDP	137	C:\WINDOWS\Explorer.EXE
1283	UDP	53	C:\WINDOWS\notepad.exe
184	UDP	53	C:\WINDOWS\system32\svchost.exe
16	UDP	67	C:\WINDOWS\System32\svchost.exe
4	UDP	123	C:\WINDOWS\System32\svchost.exe
12	UDP	1900	C:\WINDOWS\System32\svchost.exe

In the network activity profile appeared a new program, probably malware
(`ppdrive32.exe`) and emerged programs that do not support normally network
(`Explorer.exe`, `notepad.exe`) that communicates on non-standard destination
ports.

Detecting such slight changes of system activity may require algorithms, knowing / learning activity profile of "clean" programs and enabling detection of new
elements in the program activity profiles.

Analyzing a number of examples of malware was observed the occurrence
of certain other significant differences between the clean and the infected system profiles, which can be an effective element for classification of suspicious
programs. The program tdilog records as one of the elements of connection

description a class of event, such as CONNECT, DATAGRAM for successful connectivity TCP or UDP, as well as errors (TIMEOUT, RESET, CANCELED, UNREACH, ERR: etc.).

Such errors do not occur practically in activity profile of a "clean" system, but there are quite numerous in the activity profile of the malware, often with regard to non-standard communication destination ports. The Table 5 shows the network activity profile taking into account the class of events (connection errors) for the same as in the Table 4 infected system. And hence the idea of using them as a simple indication of the presence of malware in the system.

Table 5. Activity profile of transmission errors for infected test system

Count	Prot	Dport	State	Appl-path
2	TCP	25	RESET	C:\WINDOWS\Explorer.EXE
2	TCP	8800	TIMEOUT	C:\WINDOWS\Explorer.EXE
15	TCP	8800	RESET	C:\WINDOWS\Explorer.EXE
6	TCP	80	TIMEOUT	C:\WINDOWS\notepad.exe
170	TCP	445	UNREACH	C:\WINDOWS\ppdrive32.exe
1	TCP	445	ERR:c0000207	C:\WINDOWS\ppdrive32.exe
222	TCP	445	RESET	C:\WINDOWS\ppdrive32.exe
5935	TCP	445	TIMEOUT	C:\WINDOWS\ppdrive32.exe
9	TCP	445	CANCELED	C:\WINDOWS\ppdrive32.exe
3	TCP	139	ERR:c000020a	System

3.3   Malware Detection Algorithm

Malware detection algorithm using analysis of system network activity can be formulated as follows:

- We monitor locally all outbound communication from the system (initiated by program or user).
- For the assumed time window we create a profile of system network activity also taking into account the unsuccessful connections.
- The following symptoms can be considered as signs of malware operation in the system:
  - occurrence in the profile of programs known as not communicating in the network,
  - the emergence in the profile of new atypical destination ports for well-known programs,
  - the appearance in the profile of previously not known programs,
  - the occurrence in the profile of numerous connection errors, especially for atypical destination ports.

Some programs intensively operating on the network such as web browsers, instant messaging, peer-to-peer can generate a single failed transmissions for the specified destination ports. The classifications of the program as malware determines the port number and the number of erroneous transmission occurring for this single port.

4   Conclusion

The provided examples of the system activity profiles illustrate basic classification criteria of the algorithm. On standard activity profile (see Table 4) Explorer.exe tries once or twice to contact on ports 80 and 443 (http), and it is satisfactory. In contrast, more than 100 tests for connectivity, to electronic mail (port 25) or to unknown ports 8800, 7081 satisfy the conditions to recognize this behavior as an the action of malware. Similar type of infection exhibit notepad.exe.

In case of ppdrive32.exe attention drown over 6000 connectivity attempts to port 445 (netbios/smb) and on strange port 6971. From the activity profile of transmission errors can be seen that all this attempts were unsuccessful, confirming the Internet worm type activities (port scan 445).

Since the activity profile of the program contains the full path to a running module, it identifies the process linked to the given type of transmission (a copy of the malware running in the system) such as in the case of ppdrive32.exe

The algorithm does not generate false positive errors, while in the case of malware infection of code injection type into the address space of another program it will indicate only the location of the bearer, which exhibits the behavior of malware, so an indication of the algorithm will not be full, as in the case of programs explorer.exe or notepad.exe. The same may apply to malware that may runs under the "cover" of the svchost.

The algorithm is not intended as the front-line tool for system protection from infections. Such tools are available in numerous versions and from time to time fail in performing their duties.

The main goal of the algorithm is to deliver the indication, that something really wrong is going in the system and focus attentions of competent persons on identified symptoms, in order to protect the network environment against losses associated with long-term exposure of valuable information on malicious software.

Delivered information may be used in part to stop the operation of identified copies of malware, using for example kill process or similar tools, or better remove from registry the records which starts the operation of identified malware programs.

References

1. 2013 Data Breach Investigations Report. Verizon,
   http://www.verizonenterprise.com/DBIR/2013/
2. Fortinet 2013 Cybercrime Report. Fortinet,
   http://www.fortinet.com/resource_center/whitepapers/cybercrime_report_
   on_botnets_network_security_strategies.html
3. 2013 Information Security Breaches Survey,
   https://www.gov.uk/government/publications/information-security-
   breaches-survey-2013-technical-report

4. The Demise in Effectiveness of Signature and Heuristic Based Antivirus,
   http://docs.media.bitpipe.com/io_10x/io_102267/item_632588/2013-01-09_
   the_demise_of_signature_based_antivirus_final.pdf
5. Defeating Advanced Persistent Threat Malware. Infoblox,
   http://securematics.com/sites/default/files/secure/default/files/pdfs/
   infoblox-whitepaper-defeating-apt-malware.pdf
6. Piper, S.: Definitive Guide to Next-Generation Threat Protection. CyberEdge
   Group, LLC,
   http://www2.fireeye.com/definitive-guide-next-gen-threats.html
7. Assessing the Effectiveness of Antivirus Solutions, Hacker Intelligence Initiative,
   Monthly Trend Report #14,
   http://www.imperva.com/docs/HII_Assessing_the_Effectiveness_of_
   Antivirus_Solutions.pdf
8. Skrzewski, M.: Analyzing Outbound Network Traffic. In: Kwiecień, A., Gaj, P.,
   Stera, P. (eds.) CN 2011. CCIS, vol. 160, pp. 204–213. Springer, Heidelberg (2011)
9. ENISA Threat Landscape 2013 – Overview of current and emerging cyber-threats,
   https://www.enisa.europa.eu/activities/risk-management/evolving-
   threat-environment/enisa-threat-landscape-2013-overview-of-
   current-and-emerging-cyber-threats
10. IBM X-Force 2013 Mid-Year Trend and Risk Report. IBM,
    http://www-03.ibm.com/security/xforce/downloads.html
11. The Advanced Cyber Attack Landscape. FireEye, Inc.,
    http://www.security-finder.ch/fileadmin/dateien/pdf/studien-
    berichte/fireeye-advanced-cyber- attack-landscape-report.pdf
12. Skrzewski, M.: Monitoring system's network activity for rootkit malware detection.
    In: Kwiecień, A., Gaj, P., Stera, P. (eds.) CN 2013. CCIS, vol. 370, pp. 157–165.
    Springer, Heidelberg (2013)

Measuring, Monitoring, and Analysis of Communication Transactions Performance in Distributed Control System

Marcin Jamro and Dariusz Rzonca

Rzeszow University of Technology, Department of Computer and Control Engineering,
al. Powstancow Warszawy 12, 35-959 Rzeszow, Poland
{mjamro,drzonca}@kia.prz.edu.pl
http://kia.prz.edu.pl

Abstract. Nowadays, distributed control systems often consist of complex software executed on many components connected together. Due to performing crucial tasks in industry, it is important to ensure that communication between devices in distributed control systems works as expected and does not reach nor is dangerously close to the given boundary value. The paper presents a concept of Extended Communication Performance Tests, together with their development and execution. What is more, a way of measuring, monitoring, and analysis of communication in a small distributed control system is proposed. A structure of system, communication tasks, and performance requirements are modeled on three kinds of SysML diagrams, namely Block Definition, Internal Block, and Requirement Diagram. Test cases are automatically generated in the CPTest+ test definition language. A set of dedicated tools allows engineers to monitor communication time parameters in the on-line mode, check constraints, and prepare performance analysis. The concept has been integrated with the CPDev engineering environment designed for programming industrial controllers, as well as small and medium-sized distributed control systems.

Keywords: control systems, communication, performance, testing.

1 Introduction

Control systems frequently perform crucial roles in industry and can be used in several domains to automate processes, speed up production, as well as limit impact coming from human mistakes. While software complexity of such systems is still increasing, they often involve also numerous components connected by communication network, forming a distributed control system (DCS) [1]. Controllers can communicate with external equipment, such as I/O modules, HMI panels, converters, or gateways, for exchanging process data [2]. Typically, industrial fieldbuses [3] and field protocols, defined in the IEC 61158 standard, are used [4]. However, nonstandard solutions are also possible [5,6]. Unfortunately, actual performance of industrial communication systems are often worse than expected based on theoretical models, due to internal behavior of real components [7].

A. Kwiecień, P. Gaj, and P. Stera (Eds.): CN 2014, CCIS 431, pp. 147–156, 2014.
© Springer International Publishing Switzerland 2014

Different models of access to the communication link can be considered, depending on a communication bus and field protocol [2]. An industrial DCS typically uses master-slave, token passing or producer-distributor-consumer models. They define access to the communication link and scenario of data exchanges in a particular case. Such parameters are usually fixed, however, some on-line changes may result in improving the communication flexibility and reliability, as described in [8]. It is worth mentioning that different approaches may be used in horizontal and vertical communications[1]. Thus, testing and performance monitoring of the target industrial system are essential.

It should be stressed that communication performance testing is not the only important way of testing control systems. This problem is addressed by some researchers who indicate that testing in such systems is not systematic [9] and there is still a necessity of automation-supported, systematic, and agile testing approaches [10]. Several methods of checking control system behavior already exist, including using unit tests for testing program organization units [11] from the IEC 61131-3 standard [12], and measuring performance of such units [13]. Other approaches involve the test-driven automation concept using UML and test-driven development paradigm [14], the agile keyword-driven testing method [10], and the deterministic replay debugging solution [15].

The approach of performance testing, proposed in this paper, is dedicated to horizontal communication in small- and medium-sized DCSs using an industrial master-slave protocol (such as Modbus) with fixed scenario of data exchanges. Such a concept allows to model, measure, monitor, and analyze performance results for multiple communication transactions. They can be performed between various devices, such as between a controller and I/O modules, converters, or HMI panels (Fig. 1).

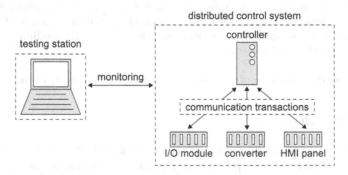

Fig. 1. Monitoring communication transactions in a distributed control system

The prototype version of testing tool, supporting the proposed concept, has been implemented in the CPDev engineering environment [16]. It is a solution

---

[1] Horizontal communication is a process of exchanging data between a controller and I/O modules, while vertical communication regards communication between an engineering station and a controller.

for programming industrial controllers (such as PLCs and PACs), as well as distributed control systems, in languages from the IEC 61131-3 standard. The package supports not only programming, but also modeling, testing, debugging, simulation, as well as configuration of devices existing in the system. It has a few industrial applications, such as in the ship control and automation system developed by Praxis Automation Technology B.V.[2] company (the Netherlands).

The paper is organized as follows. In the second section, the overall concept of the proposed testing approach is described, together with a brief description of development and execution stages. In the next section, a way of modeling the communication tasks and requirements is presented. It contains also a short description of translating diagrams into the CPTest+ code. The process of measurement, monitoring, and analysis of communication performance is covered in two following sections.

2 Overall Concept

The proposed approach is an extension to the concept of Communication Performance Tests, which has been introduced in [17]. Its new version is named Extended Communication Performance Tests, further referred as ECPTs. There are several major improvements in comparison with the previous version, including a significantly more flexible approach to the modeling of communication structure with additional types of SysML diagrams, as well as major extension of monitoring and analyzing capabilities. A structure of tests and models has been redesigned to better suit CPDev requirements, thus some simplifications and limitations have been intentionally introduced. Such an approach might be generalized and adapted for other systems as well. Typically, CPDev is used in small- and medium-sized DCSs where horizontal communication is based on a master-slave protocol and the scenario of data exchanges is fixed, so the proposed approach has been developed with identical assumptions. In this paper, the whole process related to tests development and execution is presented. Such a process consists of a few stages, which are shown in Fig. 2.

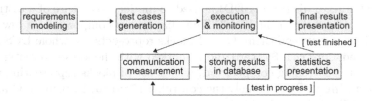

Fig. 2. Extended Communication Performance Tests development and execution

At the beginning, an engineer prepares a model in the SysML [18] graphical modeling language using a few types of diagrams. They allow to model a

---

[2] http://www.praxis-automation.nl/

structure of the distributed control system, communication tasks, as well as their performance requirements. In the next stage, test cases in the dedicated CPTest+ test definition language are generated automatically, based on the model. When tests are prepared, they can be executed. The process is conducted by a dedicated runtime and monitoring environment, which performs a set of operations in a loop, until the test is finished. In each iteration, the system measures time of all communication transactions, stores results in a database, as well as calculates and presents simple real-time statistics. When the test execution is finished, the final version of results is presented to the engineer. Moreover, obtained values may be exported for further analysis by statistical software. All stages of EPCTs development and execution are presented in the following sections of this paper.

To make understanding easier, a simple running example is considered. It is based on DCS consisting of the SMC controller and three SM I/O modules. Such an uncomplicated system has been chosen to simplify explanation of the overall concept and presented models. What is more, the SMC controller is based on CPDev software and has been designed by LUMEL S.A. company[3] in cooperation with Department of Computer and Control Engineering from Rzeszow University of Technology. Authors of this paper are also coauthors of the CPDev engineering environment and the SMC firmware, thus some improvements necessary to implement a prototype version of the testing tool have been possible.

3 Modeling of Communication Tasks and Requirements

The proposed approach of Extended Communication Performance Tests supports the model-based systems engineering (MBSE) concept [19]. It uses SysML (Systems Modeling Language) [18], which is an extension to UML (Unified Modeling Language) [20]. SysML contains nine diagram types that can present requirements, structure and behavior of various system kinds. In the proposed ECPTs concept, the SysML language is used to model a structure of a distributed control system, communication tasks, and their performance requirements. For this purpose, a few kinds of diagrams are used, namely Block Definition Diagram (BDD), Internal Block Diagram (IBD), and Requirement Diagram (REQ).

The first kind of diagram (BDD) is used to model a structure of a distributed control system, as shown in Fig. 3. This diagram is composed of a few blocks and associations between them. One of blocks represents the whole DCS system (Distributed Control System) and is marked with the «*dcs*» stereotype. This block is connected (using composition) with a set of blocks representing various devices operating in DCS, namely the controller (Controller) and I/O modules (AIModule, BIModule, and BOModule). Each block has two stereotypes assigned. The first indicates whether the block represents a resource («*resource*») or other kind of devices («*device*»). The other informs whether the device operates as a master («*master*») or slave («*slave*»). Such stereotypes are defined in the

---

[3] http://www.lumel.com.pl/en/

predefined model library available for engineers creating ECPTs. It is worth mentioning that blocks stereotyped by *«slave»* contain the dedicated compartment, where the number of slave device is shown (number).

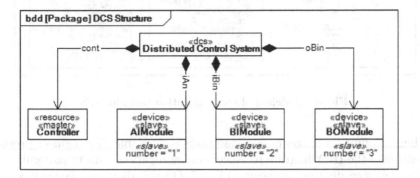

Fig. 3. Modeling of DCS structure

In the next stage of modeling, the engineer creates the BDD diagram to model communication tasks existing in the distributed control system. Such diagram is located in the Communication package and is named Tasks, as shown in Fig. 4. Each communication task is modeled as the block with *«comTask»* stereotype and is specified by a set of data available in the dedicated compartment, namely:

– a function performed by the communication task, such as reading registers;
– an address on the remote device, where data should be stored or read;
– a number of registers that should be written or read;
– an address on the local machine, where data should be stored or read;
– a conversion mode, such as 16 to 8 bits;
– a priority, as one out of three values: low, normal, high;
– a timeout in milliseconds;
– an address on the local machine, where the system stores a value indicating whether the communication transaction is finished successfully;
– an address on the local machine, where the system stores a value indicating whether the communication task is active.

It is worth mentioning that values of these settings differ for various communication tasks. Their proper configuration is crucial, because such data are later used to generate a communication tasks table that specifies which tasks are available in the distributed control system. Based on the example from Fig. 4, the first communication task (CT1) will read 8 registers from 7006 remote address and store values to local memory begin from address 31. The task will perform automatic conversion from little- to big-endian (ABDCtoDCBA), have normal priority, and 100 ms set as a timeout. A value indicating whether a communication transaction is finished successfully will be stored to 30 local address, while a value informing whether the task is active will be available at 24 local address.

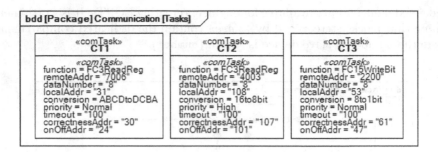

Fig. 4. Modeling of communication tasks in DCS

When a definition of communication tasks is available, the engineer proceeds to creation of the IBD diagram to assign communication tasks to particular connections between devices, as shown in Fig. 5. On this diagram, the engineer uses blocks already defined on BDD diagrams from Fig. 3 and 4. In the example, the engineer takes into account communication between a controller (`cont` of `Controller`) and three input/output modules (`iAn` of `AIModule`, `iBin` of `BIModule`, and `oBin` of `BOModule`) using three communication tasks (`ct1` of `CT1`, `ct2` of `CT2`, and `ct3` of `CT3`), represented by association blocks. The first communication task is assigned to connection between the controller and the input analog module, the second – between the controller and the input binary module, while the other – between the controller and the output binary module.

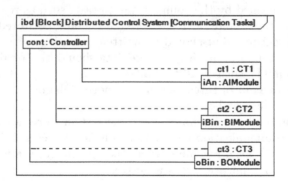

Fig. 5. Modeling of communication tasks assignment

The last stage of modeling involves specification of performance requirements for particular communication tasks, using the Requirement Diagram (Fig. 6). For this purpose, the engineer uses a set of blocks stereotyped by «*performanceRequirement*». Each of them is dedicated to a particular communication task and contains a unique identifier (`id`), a textual description (`text`), as well as an allocation to the block instance representing a particular communication task (*allocatedTo* compartment). Each requirement is verified using a set of

ECPTs, which are presented as «*comPerformanceTest*» stereotypes connected with the «*performanceRequirement*» by the «*verify*» relationship. The test specifies three properties, defined by the «*comPerformanceTest*» stereotype, namely:

- a mode of verification (check) indicating whether correctness of communication transactions is verified (Successful) or not (Finished);
- a mode of test (mode) indicating whether all results or only the average result should not be higher than the boundary value (Always or Average);
- a maximum acceptable result, i.e. the boundary value (time).

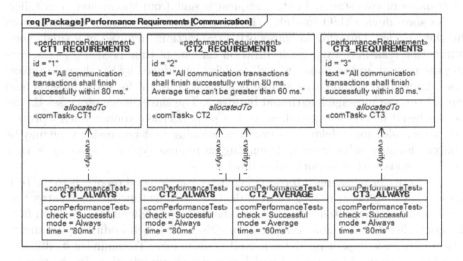

Fig. 6. Modeling of performance requirements for communication tasks

When the process of modeling is finished, a suitable implementation code is generated automatically in the CPTest+ test definition language [11]. The code has the form similar to presented in the preliminary version of CPTs [17], as well as in POU-oriented performance tests [13]. The CPTest+ code generated for the CT1_ALWAYS ECPT (from Fig. 6) is the following:

```
01: DEFINE CT 1
03: DEFINE EXPECTED_TIME T#80ms
04: DEFINE CH_CORR TRUE
05: ASSERT CT_ALWAYS
```

The code above uses two CPTest+ instructions, namely DEFINE and ASSERT. The first is used to configure the testing environment for ECPT execution, by setting a proper communication task identifier (CT), expected time (i.e. boundary value, EXPECTED_TIME), and a value indicating whether correctness should be checked (CH_CORR). The ASSERT instruction is combined with the CT_ALWAYS operator. That makes it possible to start ECPT execution, measure and monitor communication performance, as well as provide data to the engineer.

4 Measurement of Communication Performance

Extended Communication Performance Tests are evaluated by a dedicated tool integrated with the CPDev engineering environment [21]. Such a tool is run on the engineering station connected with the target controller, which is a central node of the distributed control system. While the controller executes control program and communicates with external modules, the CPTest tool collects time parameters and stores them in a local database.

Communication tasks are executed sequentially by the controller. A single execution of a particular task (called transaction) in the master-slave protocol consists of two stages. First, a request is sent from the master (controller) to the slave device (I/O module). Then, the master receives a response from the slave device. Duration of consecutive transactions involved in a single communication task may vary. Such duration consists of time needed for transmitting the request and response, as well as delay introduced by the slave. In typical cases, transmission time on a simple fieldbus is constant for given message and depends on the message length and baudrate. The slave device processes the request when it has been received, as well as prepares the response. Such a process also causes additional delay. It is worth mentioning that processing time may be different for successive frames, depending on request type and slave cycle time (asynchronous with communication link).

The controller measures duration of each transaction and reports it to CPTest. Errors that may occur during communication, such as timeouts or incorrect responses, are also reported. These values should be periodically transmitted from the controller to the engineering station. Obviously, such additional communication may have impact on the achieved performance. To minimize it, different data buses are used for vertical and horizontal communication. In the master-slave protocol, the controller typically works as a master device during horizontal communication, and as a slave in vertical one. Additional factor is introduced in case of ECPTs to reduce impact of vertical data transfers on throughput of horizontal communication. Here, data exchanged with PC are not time critical, thus low priority vertical communication is served only in the controller idle time.

5 Communication Transactions Monitoring and Analysis

As explained in the preceding section, time parameters measured for respective transactions are transferred to the CPTest tool. Performance requirements from the Requirement Diagram (see Sect. 3) are constantly compared with obtained results. If any value exceeds the specified limit, the user is informed. Moreover, all values are stored in the database for the off-line performance analysis. Data can be exported and analyzed in statistical software, such as the R package [22]. Sample results acquired for the system described earlier are shown in Fig. 7.

The statistical boxplot, with typical symbols, is used for data presentation. Here, upper and lower edges of boxes (hinges) indicate the upper and lower quartile of the data set, respectively. The lines inside the boxes mark the median

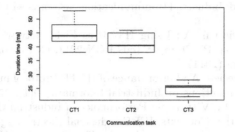

Fig. 7. Communication time measured during tests

value, while vertical whiskers indicate the minimum and maximum value. It is worth mentioning that in the presented example (Fig. 7) tasks CT1 and CT2 need significantly more time than task CT3. It can be explained as a result of different length of transmitted frames – task CT3 uses Modbus FC15 bit function while CT1 and CT2 operate on 16-bit registers (FC3 Modbus function). A slight difference between CT1 and CT2 durations is caused by additional processing delay in analogue input module served by CT1. All communication tasks satisfy performance requirements specified on the REQ diagram (Fig. 6).

6 Conclusion

Responsible and complex operations performed often by distributed control systems require that control software should be thoroughly tested before deployment. Testing should be performed in a systematic and automatic way, as well as should check not only functional requirements, but also non-functional ones.

The approach proposed in this paper allows to test communication time requirements using the Extended Communication Performance Tests. Such a concept facilitates developing an efficient communication subsystem for DCSs, especially small- and medium-sized. According to the methodology, communication tasks and performance requirements are modeled as SysML diagrams during the design stage. Later, the model is translated into the CPTest+ code describing test cases. During execution of ECPTs, time parameters are constantly read, monitored, stored, and presented to the engineer. Then, collected data may be exported for further statistical analysis, calculations, or visual presentation.

References

1. Feng-Li, L., Moyne, W., Tilbury, D.: Network design consideration for distributed control systems. IEEE Trans. on Control Systems Techn. 10(2), 297–307 (2002)
2. Gaj, P., Jasperneite, J., Felser, M.: Computer Communication Within Industrial Distributed Environment – a Survey. IEEE Transactions on Industrial Informatics 9(1), 182–189 (2013)
3. Thomesse, J.P.: Fieldbus Technology in Industrial Automation. Proceedings of the IEEE 93(6), 1073–1101 (2005)

4. IEC 61158 Standard: Industrial Communication Networks – Fieldbus Specifications (2007)
5. Jestratjew, A., Kwiecień, A.: Using HTTP as Field Network Transfer Protocol. In: Kwiecień, A., Gaj, P., Stera, P. (eds.) CN 2011. CCIS, vol. 160, pp. 306–313. Springer, Heidelberg (2011)
6. Jestratjew, A., Kwiecien, A.: Performance of HTTP Protocol in Networked Control Systems. IEEE Transactions on Industrial Informatics 9(1), 271–276 (2013)
7. Seno, L., Tramarin, F., Vitturi, S.: Performance of Industrial Communication Systems: Real Application Contexts. IEEE Industrial Electr. Mag. 6(2), 27–37 (2012)
8. Kwiecień, A., Sidzina, M., Maćkowski, M.: The Concept of Using Multi-protocol Nodes in Real-Time Distributed Systems for Increasing Communication Reliability. In: Kwiecień, A., Gaj, P., Stera, P. (eds.) CN 2013. CCIS, vol. 370, pp. 177–188. Springer, Heidelberg (2013)
9. Kormann, B., Vogel-Heuser, B.: Automated test case generation approach for PLC control software exception handling using fault injection. In: IECON 2011– 37th Annual Conference on IEEE Industrial Electronics Society, pp. 365–372 (2011)
10. Hametner, R., Winkler, D., Zoitl, A.: Agile testing concepts based on keyword-driven testing for industrial automation systems. In: IECON 2012 – 38th Annual Conference on IEEE Industrial Electronics Society, pp. 3727–3732 (2012)
11. Jamro, M., Trybus, B.: Testing Procedure for IEC 61131-3 Control Software. In: 12th IFAC/IEEE International Conference on Programmable Devices and Embedded Systems (PDeS), pp. 192–197 (2013)
12. IEC 61131-3 - Programmable controllers - Part 3: Programming languages (2013)
13. Jamro, M.: Development and Execution of POU-oriented Performance Tests for IEC 61131-3 Control Software. In: Szewczyk, R., Zieliński, C., Kaliczyńska, M. (eds.) Recent Advances in Automation, Robotics and Measuring Techniques. AISC, vol. 267, pp. 91–101. Springer, Heidelberg (2014)
14. Winkler, D., Hametner, R., Biffl, S.: Automation component aspects for efficient unit testing. In: IEEE Conference on Emerging Technologies Factory Automation, ETFA 2009, pp. 1–8 (2009)
15. Prahofer, H., Schatz, R., Wirth, C., Mossenbock, H.: A Comprehensive Solution for Deterministic Replay Debugging of SoftPLC Applications. IEEE Transactions on Industrial Informatics 7(4), 641–651 (2011)
16. Jamro, M., Trybus, B.: An approach to SysML modeling of IEC 61131-3 control software. In: 18th International Conference on Methods and Models in Automation and Robotics (MMAR), pp. 217–222 (2013)
17. Jamro, M., Rzońca, D., Trybus, B.: Communication Performance Tests in Distributed Control Systems. In: Kwiecień, A., Gaj, P., Stera, P. (eds.) CN 2013. CCIS, vol. 370, pp. 200–209. Springer, Heidelberg (2013)
18. OMG: OMG Systems Modeling Language, V1.3 (2012)
19. Friedenthal, S., Moore, A., Steiner, R.: A Practical Guide to SysML. The Systems Modeling Language. Elsevier Inc. (2012)
20. OMG: OMG Unified Modeling Language. Infrastructure, V2.4.1 (2011)
21. Jamro, M., Rzonca, D., Sadolewski, J., Stec, A., Swider, Z., Trybus, B., Trybus, L.: CPDev Engineering Environment for Modeling, Implementation, Testing, and Visualization of Control Software. In: Szewczyk, R., Zieliński, C., Kaliczyńska, M. (eds.) Recent Advances in Automation, Robotics and Measuring Techniques. AISC, vol. 267, pp. 81–90. Springer, Heidelberg (2014)
22. R Core Team: R: A Language and Environment for Statistical Computing. R Foundation for Statistical Computing, Vienna, Austria (2013)

Monitoring of Speech Quality in Full-Mesh Networks

Jan Rozhon, Filip Rezac, Jiri Slachta, and Miroslav Voznak

CESNET z.s.p.o., Zikova 4, 160 00 Prague,
Czech Republic
{rozhon,rezac,slachta}@cesnet.cz, voznak@ieee.org
http://www.cesnet.cz

Abstract. The quality of speech in network environment is a growing concern of all the companies and institutions, that migrated or are planning to migrate their voice communications to IP based technologies. This trend together with the increasing amount of Internet traffic in general poses a great challenge for network administration to be able to configure and manage the network so it can carry the multimedia traffic without the excessive delays, which degrade the quality of speech as it is perceived by the end users. Continuous monitoring of key nodes of internal infrastructure as well as external interconnections is the possible way to increase the quality of multimedia transmissions for most users, because it allows the network administrators to be informed as soon as the problem rises and consequently to change the network routing and queuing policies to bypass the connection experiencing the quality issues. The tool for this type of monitoring and its architecture are described in detail in this paper.

Keywords: network administration, network policing, PESQ, speech quality monitoring.

1 Introduction

Past decade was marked by two important factors from the multimedia communication point of view. The first one, a rapid development of the Voice over IP (VoIP) technology in the business communications was an aspect of the long-term trend of voice and data networks convergence. This trend was driven by the both economic and network management reasons. VoIP technologies, no matter whether commercial or open-source, have been a mean to achieve higher efficiency of intra-organization collaboration by establishing a platform for in given time enterprise level services like conference, joint calendar events, video telephony, call transfer functions, etc. From the economic perspective, the single infrastructure advantage and inherent VoIP features such as performing a least cost routing (LCR) or callback function catalyzed the ongoing process even more, which led to a broad VoIP adoption across all the levels of private and public institutions.

A. Kwiecień, P. Gaj, and P. Stera (Eds.): CN 2014, CCIS 431, pp. 157–166, 2014.

Alongside this fundamental transition between technologies, the Internet multimedia content began to form the most of the Internet content in general. This resulted in heavy burden laid on the back of the network infrastructure, which had to be capable of delivering the content not only reliably and securely, but with given quality parameters as well. Through systematic analysis and upgrades the infrastructure was in many cases made robust enough to handle this load. However, under some special circumstances (device malfunction, traffic peak, etc.) the transmission parameters can temporarily deteriorate. These special occasions can significantly decrease the multimedia service rating by the customers, but can be countered easily as well. To counter the effects of these anomalies many routing and queuing policies can be employed, but to be able to do that efficiently the anomalies need to be identified, which in terms of speech quality (or multimedia quality in general) cannot be achieved directly on network elements, such as routers or switches.

One-time or irregular measurements cannot address this problem, but the continuous monitoring of speech quality can achieve the benefits being sought like already mentioned quick or even immediate notification about the problem or in some cases even the speech quality prediction.

This paper deals with the tool for the aforesaid periodic quality measurements in the full-mesh IP networks, which is designed to use the industry standard for speech quality calculation – PESQ (Perceptual Evaluation of Speech Quality) as well as the state-of-art monitoring and call generation tools available. In this paper the basic idea, functionality and architecture of this system will be presented and the planned future development of the project outlined.

2   State of the Art

The introductory part of this paper mentioned one of the most used algorithms for evaluation of speech quality – PESQ, but this is just one representative of the extensive group of methods to measure and evaluate speech quality. These methods can be sub-divided into two groups according to the approach applied, conversational and listening.

Conversational tests are based on mutual interactive communication between two subjects through the whole transmission chain of the tested communication system. These tests provide the most realistic testing environment but they are very time consuming. Listening tests do not provide such plausibility as conversation tests but are recommended more frequently. According to methods of assessment, speech quality evaluation methodologies can be subdivided as subjective methods and objective methods. To evaluate speech quality, MOS (Mean Opinion Score) scale as defined by the ITU-T recommendation P.800 is applied [1].

In order to avoid misunderstanding and incorrect interpretation of MOS values, ITU-T published ITU-T recommendation P.800.1 [2]. This recommendation defines scales for both subjective and aforementioned objective methods as well as for aforementioned individual conversational and listening tests.

The subjective evaluation methods are based on evaluation by human beings (listeners), i.e. subjects. During the testing, samples are played to a sufficient number of subjects, and their results are subsequently analyzed statistically. Subjects can evaluate the speech quality on a five-degree scale in accordance with the MOS model as defined by ITU-T. The best known representatives of these measurements include methods such as ACR (Absolute Category Rating) or DCR (Degradation Category Rating). Major disadvantages of these methods are high requirements on time, final evaluation being influenced by listener's subjective opinion and most of all impossibility to use them for testing in real time.

The objective evaluation methods substitute the necessity to involve humans in the testing by mathematical computational models or algorithms. Their output is again a MOS value or, depending on the algorithm applied, a different value which can be transferred to a MOS value using a suitable mapping function. The aim of objective methods is to estimate, as precisely as possible, the MOS value which would be obtained by a subjective evaluation involving sufficient number of evaluating subjects. Objective testing's exactness and efficiency is therefore a correlation of results from both subjective and objective measurements. Objective methods can be sub-divided into two groups, Intrusive and Non-intrusive.

The core of intrusive (also referred to as input-to-output) measurements is the comparison of the original sample before releasing it into a transmission chain of a communication system with the output sample, transmitted through the system (degraded).

This type of testing includes, among other, the following methods: PSQM (Perceptual Speech Quality Measurement), PAMS (Perceptual Analysis Measurement System) developed by British Telecommunications, P.OLQA (Perceptual Objective Listening Quality Assessment) and PESQ (Perceptual Evaluation of Speech Quality) [3]. P.OLQA may become the successor of PESQ because avoids weaknesses of the current P.862 model and has an extension towards higher bandwidth audio signals. Yet, PESQ is the most common objective intrusive method. Computational technique applied by this method combines PAMS' robust temporal alignment techniques and the PSQM' exact sensual perception model. Its final version is contained in ITU-T recommendation P.862 [3]. As was stated above, the principle of this intrusive test is based on comparison of original and degraded signals, their mathematical analysis using FFT and interpretation in the cognitive model.

Contrary to intrusive methods which need both the output (degraded) sample and the original sample, non-intrusive methods do not require the original sample. This is why they are more suitable to be applied in real time. Yet, since the original sample is not included, these methods frequently contain far more complex computation models. Examples of these types of measurements frequently use INMD (in-service nonintrusive measurement device) that has access to transmission channels and can collate objective information about calls in progress without disrupting them. These data are further processed using a particular method, with a MOS value as the output. The method defined by ITU-T recommendation P.563 or a more recent computation method E-model defined by ITU-T recommendation G.107 are examples of such measurements [4,5].

As said in the previous paragraphs, there are several methods which can be used for speech quality measurements, from which PESQ and POLQA provide the best results in comparison to subjective measurements. This however comes with the need for the custom call generation and capture so that these intrusive methods can compare the two samples. Better accuracy is thus compensated (in negative way) by the higher bandwidth and computation power requirements. Most of the proposed models (such as this one [6]) therefore rely on ITU-T's E-model, which however has a disadvantage of lower accuracy which needs to be improved [7].

In the system described in this paper, however, the intrusive PESQ measurements will be used, because of the higher accuracy and because of the proposed architecture, that allows for offloading the PESQ calculations to the central server and leaves the individual measuring boxes – the probes – with minimal requirements. In the future development the PESQ computation will be substituted by the model for quality of speech prediction based on the network parameters. This model will be based on the results of PESQ algorithm and conformance of the results from current phase of development and the future one is a reason for adopting intrusive approach as well. The possible model for future development will focus on improving currently published models in this area [8,9].

The PESQ based periodic monitoring tool has not been presented as far as the authors are aware. However, the ideas of mesh networks monitoring were presented in the field of wireless communications [10]. The tool itself together with its philosophy is the main contribution of this paper. The tool which accomplishes accurate results, counters the negative features of the intrusive methods and provides a space for easy and seamless incorporation of models for PESQ results estimation currently being developed.

3   System Architecture

From the previous sections it is clear why and what algorithm was chosen for the speech quality calculation. However, the algorithm itself is just a small piece in a greater picture of the system architecture necessary to provide the full set of functionalities needed for continuous monitoring. To fully understand the system architecture, the necessary background has to be provided.

CESNET as the network of national research interconnects the academic institutions all around the country. Although the main aim of the association is to provide full-featured interconnecting network, the multimedia services are made available to all the member institutions as well. This way the geographically extensive multimedia network of mutually interconnected nodes was created based on common VoIP protocols such as H.323 and SIP. Due to the distances between individual nodes many network elements get involved in the communication. These elements are monitored on the network level, but in some cases, they can influence the quality of speech as it is perceived by end users even without the sign of a problem on the network level. This is due to the specific features and requirements of the speech (or multimedia in general) traffic. Among others the low and stable latencies and small number of lost packets can be named. From the aforementioned two necessary features of the monitoring system are implied. First, to be able to discover and identify the problem on the application level, the application level measurements need to be implemented. Second, the numerous network probes need to be deployed, so that all the key multimedia interconnections can be observed.

The term *probe* itself is very important from the perspective of the system architecture. Due to the fact, that it is expected for the probe to be connected with many other probes in many different locations, the number of function the probe performs must be limited. The reason for this is economic, if the functionalities of the probes were vast, the system resources, it would require, would render the whole system economically inefficient. To counter this, the probes come in form of low-cost hardware computer and the sole purpose of theirs is to initiate, terminate and record the call. When deployed the probe is tapped to the network of the hosting institution. The probe then communicates with other partner probes in the monitored network using standard VoIP protocols, namely SIP for signalling and RTP for media.

Two pieces of software were used for the probe functionalities. The Asterisk PBX for call generation and recording and OpenSSH/Rsync for the security and data transfer. Both programs are open-source and are time-proven with large community of users, which makes them excellent for deployment from the economic and future development point of view.

To make the system fulfill its role, the central element is also required to supplement the probes. This element – a server – provides following features to the system:

- the interface for the user to manage the probes,
- the interface for the probe to register automatically,
- system access control and encryption of data communication,
- PESQ calculation,
- presentation of the results.

The server communicates with the probes using a secure connection based on SSH tunnels. Within these tunnels the information about all the probes, the measuring intervals and probe registration credentials are exchanged. The probe cannot access the server without the SSH tunnel, which can only be established with correct private keys. This mechanism allows control of access for the individual probes and secures the connections between probes and the server. The second role of the server is to calculate the MOS score using the PESQ algorithm. To be able to handle the number of PESQ calculations, the server needs to be of high performance and should therefore be implemented as the virtual computer with high-end hardware resource parameters. Virtualization offers two key aspects – the existing infrastructure can be used and the server backup can easily be created and deployed in case of hardware malfunction. The server also provides the means for the result visualisation and probe management to the system administrator.

For the server to be able to perform the abovementioned tasks, the well-tested and stable open-source software was used. Upon Debian linux, OpenSSH and Zabbix are deployed. While the former provides the security layer, the latter forms the user-system interface to present the results to the administrator and to allow him to manage the probes and to control the monitoring process. Both the programs share the active community of users and developers, which makes them safe for long term deployment even in critical systems.

The entire system architecture in the example deployment is depicted in the Fig. 1.

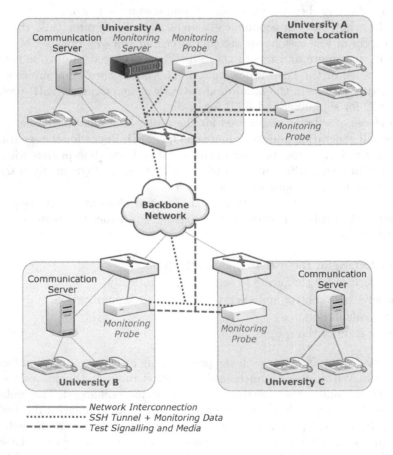

Fig. 1. The system architecture in the example environment with monitoring data (dotted) and test signalling and media (dashed) inteconnections

4   Monitoring Algorithm

The functionality of the system, the architecture of which was described in a previous section, can be viewed as fully modular. Each module has its role and offers its functionality to other modules to create a compact and efficient whole. The list of system modules is comprised of:

- provisioning module,
- communication and security module,
- interface module,
- test module.

4.1   Provisioning Module

This module's task is to make the probe operational with as little effort of the administrator as possible. This module can fully automate the addition of new probes to the system, but it requires the access to the local DHCP (Dynamic Host Control Protocol) service, where the administrator needs to enter the records the probe requires to work properly. First of all, it is the probe's IP address and hostname, which are used to communicate with the rest of the network. Then it is the server's IP address or its resolvable domain name to allow the probe to communicate with the server and the last item (apart from standard ones, e.g. default gateway, etc.) is the path from which the probe can download its private key. The last item has to be kept secret and the communication must be secured using HTTPS protocol and restricted only for intended hosts. With all these information in place the probe then can register itself to the server without the user interaction, which is useful in large deployment scenarios.

4.2   Communication and Security Module

The tasks of this module were partly discussed in the system architecture, because they are closely connected with both the architecture and the functionality. This module provides the secure communication layer for the system based on public keys infrastructure, which is also used to provide access control, so that only authorized probes can join the system and fetch the sensitive data about the other probes. This module is only used as the mean for the instruction transfer from the server to probes or to send recorded sound samples from the probe to the server. It is not used for the test-connected communication.

4.3   Interface Module

The transfer of network information from the server to the probes uses this module, which is built upon the Zabbix monitoring server. By using the custom made tools and standard Zabbix communication application interface the direct link from the probes to network information database of the Zabbix server is created.

The probe then reads the complete list of the probes, which is then used for testing. The interface module's second role is to allow the administrator to easily manage the network, the individual probes, the intervals between individual test rounds, etc. For this communication, a standard Zabbix web interface is used together with visualization tool, that allows to render a map of probes and the tested interconnections, so that the results of measurements are easily accessible and understandable.

4.4 Test Module

This test performs the key functionality of the system – the speech quality testing itself. It is based on open-source software PBX (Public Exchange) Asterisk, which provides the features necessary to establish, record and terminate the call. Due to the intended full-mesh network topology the even call distribution becomes an issue. To prevent some probes to be overloaded while others being idle the call distribution function has been implemented to this module. This function is based on the mathematical combinations and evenly, or at least almost evenly distributes the call generation orders. Since this function is implemented in the Test module, which is only present on the probes, each probe generates the same list of call orders. Due to the natural both-way communication, it is only necessary to perform a call from probe A to probe B to get the information about speech quality in both directions.

Together with the information about the partner probes, server sends the parameters of the test as well. The most important parameter carries the information about the time period between individual test rounds. The testing is performed in individual rounds and each round comprises of calls among all the probes in the database. The call length is set to one minute to provide long enough samples for the analysis. All the test calls initiated by one probe are performed at once, which in case of precise time synchronization across the network would result in all calls to be performed at once. Due to relatively low bandwidth of a single call it is very difficult to reach measurable network load in the intended environment of backbone networks. When the call (or test round) is finished, the recorded wav files are then transferred to the server for analysis. The test rounds are then performed periodically and their frequency depends on the time interval set by the administrator of the network. This way the compromise between network load and responsiveness of the system can be defined thus making the system usable in vast number of different network environments.

5   Current Deployment and Future Development

The system itself is in the final stage of the development, which means it only operates in the testing environment of Technical University of Ostrava, where the simulated network is being monitored. In this environment the system is operational for more than one month without significant issue, therefore it is ready for the deployment in the networks of the member institutions of the CESNET association.

The future development is focused on maintaining the high accuracy of the PESQ algorithm in comparison to subjective speech quality tests while decreasing the load of the network caused by the system operation. The way to achieve this goal, which is being investigated, is the development of the mathematical model based on the neural networks, which would approximate the results of PESQ calculation according to the network statistics that are commonly obtainable such as network delay, jitter or packet loss. Several papers have been published on this topic and each of the approaches makes the approximation more accurate. The possible usage of adaptive neuro-fuzzy inference system is a way to achieve higher accuracy while preserving the benefits of the mathematical model approach.

6 Conclusion

In the limited space of this paper the idea and proposition of the system for continuous monitoring of speech quality was presented. This monitoring system introduces application layer monitoring of the speech-oriented multimedia infrastructure, which makes it possible to detect the irregularities and anomalies of the network on the application layer, thus allowing for faster and more appropriate response to the situation in the form of routing and queuing policy changes. This system is based on PESQ algorithm to provide the high accuracy in comparison to subjective methods for speech quality measurement. This algorithm works in conjunction with broadly used open-source applications such as Asterisk PBX and Zabbix.

The system is intended for the non-commercial operation in the network of the academic institutions, but thanks to its robust design it can easily be used in any environment, for example to allow the private organizations to monitor the stability of their VPN connections. The full-mesh operation of the system is a possibility but not the necessity, becuse the probes can be instructed to test only certain interconnections. This is done by splitting the probes into the separate groups in the monitoring system, which results in a subset of the probes to be propagated from the monitoring system to the individual probes, thus allowing to monitor only certain interconnections.

The contribution of the paper is an integration of speech quality evaluation into common monitoring system. Our design of monitoring probes was implemented and verified in testbed. This testbed was created in the experimental environment with the custom defined traffic and distortion generation. During the one month period of testing the visible correlation between network impairments and resulting speech quality was observed, thus confirming the expected behaviour of the system.

The future development will be marked with an increased effort for transition from the usage of PESQ algorithm to the neural network based model for speech quality prediction to further decrease the resources needed by the system and to increase the responsiveness of the system as the whole.

Acknowledgement. This research has been supported by the Ministry of Education of the Czech Republic within the project LM2010005.

References

1. ITU-T P.800 – Recommendation P.800 of the International Telecommunication Union: Methods for subjective determination of transmission quality. ITU-T (1996)
2. ITU-T P.800.1 - Recommendation P.800.1 of the International Telecommunication Union: Mean Opinion Score (MOS) terminology. ITU-T (July 2006)
3. ITU-T P.862 – Recommendation P.862 of the International Telecommunication Union: Perceptual evaluation of speech quality (PESQ) - An objective Method for end-to-end speech quality assessment of narrow-band telephone networks and speech codecs. ITU-T (February 2001)
4. ITU-T G.107 - Recommendation G.107 of the International Telecommunication Union: The E-model: A computational model for use in transmission planning, Geneva (2009)
5. Voznak, M., Tomes, M., Vaclavikova, Z., Halas, M.: E-model Improvement for Speech Quality Evaluation Including Codecs Tandeming. In: Advances in Data Networks, pp. 119–124. Communications, Computers, Faro (2010)
6. Conway, A.E.: A Passive Method for Monitoring Voice-over-IP Call Quality with ITU-T Objective Speech Quality Measurement Methods. In: IEEE Conference on Communications, pp. 2583–2586. ICC (2002)
7. Kovac, A., Halas, M.: E-model MOS estimate precision improvement and modelling of jitter effects. Advances in Electrical and Electronic Engineering 10, 276–281 (2012)
8. Fu, Q., Yi, K., Sun, M.: Speech quality objective assessment using neural network. In: 2000 IEEE International Conference on Acoustics, Speech, and Signal Processing, pp. 1511–1514. IEEE, Piscataway (2000)
9. Al-Akhras, M., Zedan, H., John, R., Almomani, I.: Non-intrusive speech quality prediction in VoIP networks using a neural network approach. Neurocomputing 72(10), 2595–2608 (2009)
10. Morais, A., Cavalli, A.: A quality of experience based approach for wireless mesh networks. In: Masip-Bruin, X., Verchere, D., Tsaoussidis, V., Yannuzzi, M. (eds.) WWIC 2011. LNCS, vol. 6649, pp. 162–173. Springer, Heidelberg (2011)

Time Domain Estimation of Mobile Radio Channels for OFDM Transmission

Grzegorz Dziwoki[1], Jacek Izydorczyk[1], and Marcin Szebeszczyk[2]

[1] Silesian University of Technology, Institute of Electronics,
Gliwice, Poland
{grzegorz.dziwoki,jacek.izydorczyk}@polsl.pl
[2] Silesian University of Technology,
Automatix sp. z o.o, Katowice
m.szebeszczyk@automatix.com.pl

Abstract. Time domain synchronous OFDM (TDS-OFDM) is a kind of orthogonal multicarrier transmission, where a guard period between consecutive parts of useful information is filled with a pseudorandom noise that acts as a training sequence. This sequence can be used both for synchronization and channel estimation. The paper presents a new estimation procedure for mobile channel recovery, that is based on the time domain training sequence and utilizes the compressive sensing approach. The proposed solution does not require any additional help from any deliberately deployed training information in the frequency domain (pilot tones). The method is experimentally explored in a simulated doubly selective sparse transmission environment. The quality metrics, obtained for an uncoded OFDM transmission system with implementation of the proposed estimation method, are compared to the ones that were obtained in case of ideal channel state information as well as the time-frequency training method.

Keywords: mobile channels, OFDM modulation, time domain estimation, compressive sensing.

1 Introduction

Time Domain Synchronous Orthogonal Frequency Division Multiplex (TDS-OFDM) transmission belongs to the broad family of orthogonal multicarrier modulation schemes that use IFFT/FFT as the key processing method. The specific feature that distinguishes it from the others OFDM techniques is a pseudorandom noise sequence with good autocorrelation property inserted between consecutive information symbols [1]. Although this modulation type is less popular than the cyclic prefix OFDM (CP-OFDM) [2–4], the one has been successfully applied in the Chinese digital television network [5].

The guard time is part and parcel of any OFDM transmission. It is a period of time, that splits the transmitted information conveyed by the consecutive OFDM symbols into independent data parts, provided that its duration is longer than the delay spread of the channel impulse response. Fulfillment of this condition

A. Kwiecień, P. Gaj, and P. Stera (Eds.): CN 2014, CCIS 431, pp. 167–176, 2014.

eliminates inter-block interference (IBI) and is obligatory for effective transmission. In the case of TDS-OFDM modulation, content of the guard period is known to the receiver, so it can be used as a training sequence for synchronization and channel estimation. Because majority of the initialization process may be based on the time-domain training in this case, an additional support form the frequency domain pilot (training) signals may be reduced as compared to CP-OFDM systems. This better utilization of the frequency spectrum for data transmission improves spectral efficiency about 10 % [6].

In the paper, the main attention is directed at channel estimation problem in TDS-OFDM transmission system, so synchronization will be assumed as perfect. Characteristics of the considered channels are doubly selective, i.e. they are dynamic both in the time and frequency domains. From practical point of view, relation between the Doppler frequency f_D and the duration of the OFDM symbol T usually meets condition [7]

$$f_D T \leq 0.1 . \tag{1}$$

For example, for the carrier frequency of 500 MHz and the relative speed of transmission's participants of 150 $\frac{km}{h}$, the OFDM symbol should span less than 1.4 ms. For comparison purposes, this symbol duration can corresponds to a 8K mode OFDM symbol (8192 samples of data plus 2048 samples of the guard time) of the digital terrestrial television system DVB-T2.

Beside the time-frequency variability, sparsity in the time domain is another important channel feature adopted here. It means that only a few propagation paths between transmitter and receiver have significant impact on transmission. The scientific reports show that the sparsity phenomenon is inherent feature of many wireless environments [8]. For instance, the COST 207 TU-6 propagation model consists of only six resolvable paths.

Simulation analysis of a new channel estimation procedure, that uses only pseudorandom training sequence of the TDS-OFDM symbol, even in the case of doubly selective channels, is the main contribution of this paper. The remainder of this paper is organized as follows. Section 2 presents the selected approaches to channel estimation for TDS-OFDM systems. Their short description acts as background for presentation of the new one in Sect. 3. The method performance is evaluated in Sect. 4 and conclusion are in Sect. 5.

2   Channel Estimation Essentials for TDS-OFDM

Concatenation of the information data sequence x_i of length N and the pseudorandom noise training sequence c_i of length M forms $N + M$ samples of the i-th TDS-OFDM symbol in the time domain. The vector x_i is IFFT transform of N complex numbers X_i coming from a QAM signal constellation. The two parts of the OFDM symbol are independent each other as opposed to the CP-OFDM symbol, where a copy of M samples from the end of x_i is transmitted during the guard period. Preservation of the subcarriers orthogonality precludes immediate use of N-point FFT even on properly synchronized received signals until the mutual channel interference between $y_i = x_i * h_i$ and $d_i = c_i * h_i$ are eliminated and the cyclic property reconstruction of y_i is made. Symbol "$*$" is the convolution operator and h_i denotes a channel impulse response. This overlapping phenomenon is presented in Fig. 1, where every beginning part of the received data sequence y_i as well as the received training sequence d_i is distorted by the tail of the preceding component. The only way to obtain the "pure" y_i for further Fourier transformation is to eliminate d_i and d_{i+1} from the received signal. This processing step is feasible only if the channel impulse responses h_i and h_{i+1} are known to the receiver.

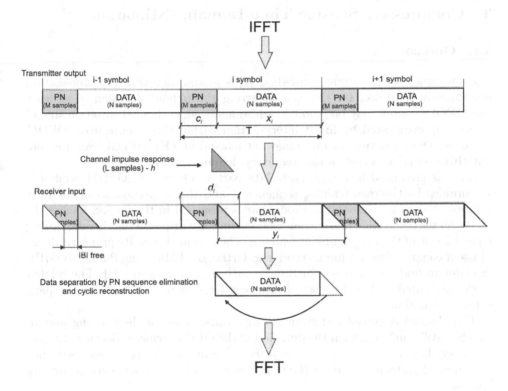

Fig. 1. Time domain processing steps in TDS-OFDM transmission

The method of channel impulse response estimation in TDS-OFDM system, that only employs the pseudorandom training sequence, is called iterative cancellation algorithm [1]. The first supposition about a current channel characteristic comes from the linear extrapolation of the channel coefficients, that were estimated during the previous two OFDM symbols. Thus, channel variability over one symbol duration should be small enough, that is typical for slow fading frequency selective channels only. Next adjustments of the recovered channel parameters are carried out in iterative manner using sequentially estimated x_i (X_i) and d_i vectors.

For better tracking of the channel time variability, the time domain training sequence can be supported by pilots in the frequency domain as proposed in Time-Frequency Training OFDM (TFT-OFDM) transmission scheme [7]. The number of pilots depends on complexity (sparsity) of the channel impulse response and its dynamic. The channel estimation method distinguishes two separate stages. The first one uses the time-domain training sequence only for paths delay estimation. The second one, that is carried out in the frequency domain, determines the paths gains for the delays selected before.

3   Compressive Sensing Time Domain Estimation

3.1   Outline

Main weakness of the iterative method [1] is assumption that the propagation environment is static, i.e. changes occurring throughout duration of at least one OFDM symbol may be ignored. Additionally, slow channel variation should be well approximated by linear interpolation within three consecutive OFDM symbols. Those restrictions are somewhat relaxed in TFT-OFDM transmission, but this is done at a cost of the frequency-domain training [7].

As it is presented in Fig. 1, each data part x_i of the i-th OFDM symbol is surrounded by the own training sequence c_i and the sequence c_{i+1} of the next symbol. A channel recovery method, that is proposed in this paper, focuses directly on the estimation of a current channel impulse response that is valid during transmission of the c_{i+1} training sequence. The assumed sparsity property allows to use a compressive sensing method, e.g. Orthogonal Matching Pursuit (OMP), in order to find the elements of the channel impulse response [9]. The related works presented in [10, 11] use the compressive sensing approach for the path delays estimation only.

The channel is considered static during transmission of the training period, which is still much less than the duration of the OFDM symbol. Because the previous symbol, i.e. $i - 1$, has been already recovered, there is assumed that the channel state information (CSI) in the time of the c_i sequence is already known.

Consequently, having at receiver's disposal two estimates of the channel impulse response – valid right before and after the data part of the OFDM symbol – and taking into account the principle expressed in (1), the channel variation within data part of the symbol may be linearly approximated.

Next subsection presents a mathematical description of the proposed procedure.

3.2   Algorithm Details

The time domain interference between the i-th channel distorted data samples y_i and the next channel distorted training sequence d_{i+1} at the receiver's input, that occurs during the $i+1$-th guard period, can be expressed in matrix notation as follows:

$$
\begin{bmatrix}
y_{i,N-1} \\
y_{i,N} + d_{i+1,0} \\
y_{i,N+1} + d_{i+1,1} \\
\vdots \\
y_{i,N+L-1} + d_{i+1,L-1} \\
d_{i+1,L} \\
\vdots \\
d_{i+1,M-1}
\end{bmatrix}
=
\begin{bmatrix}
x_{i,N-1} & x_{i,N-2} & \cdots & x_{i,N-L} \\
c_{i+1,0} & x_{i,N-1} & \cdots & x_{i,N-L+1} \\
c_{i+1,1} & c_{i+1,0} & \cdots & x_{i,N-L+2} \\
\vdots & \ddots & \ddots & \ddots \\
c_{i+1,L-1} & c_{i+1,L-2} & \cdots & x_{i,N-1} \\
c_{i+1,L} & c_{i+1,L-1} & \cdots & c_{i+1,0} \\
\vdots & \ddots & \ddots & \ddots \\
c_{i+1,M-1} & c_{i+1,M-2} & \cdots & c_{i,M-L}
\end{bmatrix}
\begin{bmatrix}
h_{i+1,0} \\
h_{i+1,1} \\
\vdots \\
h_{i+1,L-1}
\end{bmatrix}
, \quad (2)
$$

where the matrix in the right side of the Equation (2) is called the measurement matrix according to terminology introduced in the compressive sensing theory [9].

Because the sought channel impulse response is considered sparse, the number of the active paths S in the channel is significantly less than its delay spread L. The Orthogonal Matching Pursuit (OMP) algorithm works iteratively in two steps. The first step is to determine a single delay of a path and appends it to the delays estimated in the previous iterations. The second step is to estimate of the gains for uncovered paths using the LS (Least Squares) method. The number of iteration equals to the maximum number of resolvable paths in the channel. Two modes of sparse estimations are proposed:

- *direct* (dir) – all $M - L$ received samples in the IBI free region of the guard period, i.e. $[d_{i+1,L}, d_{i+1,L+1}, \cdots, d_{i+1,M-1}]^T$ are directly used by the OMP. It can be done because the corresponding submatrix of the measurement matrix in the Equation (2) is completely known to the receiver (it consists of the training sequence only). Estimation of the sparse vector representing the propagations paths is correct with high probability (about 99 %) if the following inequality is met [9]:

$$
M - L > 2S \ln L , \tag{3}
$$

- *iterative* (iter) – it applies when the condition (3) is false, i.e. the IBI free region is too short. The OMP algorithm is run for all received samples of the guard period but the end part of data sequence x_i must be estimated first. It is done in an iterative manner using last results of channel estimation (h_i) as the first approximation of the current channel impulse response h_{i+1}. Then, the first estimate of the x_i is calculated as follows:

$$
\hat{x}_i^1 = \mathrm{IFFT}\left\{ \mathcal{Q}\left[\frac{\mathrm{FFT}\{\tilde{y}_i\}}{\mathrm{FFT}\{h_i\}} \right] \right\} , \tag{4}
$$

where $\tilde{y}_i$ is the received data sequence of length N that is cyclically reshaped according to the scheme outlined in Fig. 1. The $\mathcal{Q}[\cdot]$ refers to a hard decision operator that implements the minimum distance rule to find the likely values of the transmitted complex number in each OFDM subchannel. When the first estimate $\hat{x}_i^1$ is already known, the OMP begins to compute new values of the channel impulse response h_{i+1}.

The next estimates of x_i (the iterative mode) as well as final one (the direct and iterative modes), are calculated assuming the linear interpolation of the respective gains in the channel impulse response within the i-th data sequence. The linear variation of the channel gain within data sequence for the k-th path is calculated as follows:

$$h(k)_{i,n}^t = \frac{h(k)_{i+1}^t - h(k)_i}{N} n + h(k)_i \quad \text{for} \quad n = 0 \ldots N - 1 \tag{5}$$

where n is the discrete time index and t is the current iteration number. The straightforward LS (Least Squares) estimate of the transmitted data is:

$$\hat{x}_i^t = \text{IFFT} \left\{ \mathcal{Q} \left[\mathbf{G_i^{-1}} \text{FFT} \{ \tilde{y}_i \} \right] \right\}, \tag{6}$$

where $\mathbf{G_i}$ is the frequency domain channel matrix for the i-th OFDM symbol. The elements of $\mathbf{G_i}$, due to inter-channel interference produced by the channel time variability, are generally non zero out of the diagonal. Any $g_{p,q}$ element of the $\mathbf{G_i}$ matrix is calculated according to equation [7]:

$$g_{p,q} = \sum_k \left(\frac{1}{N} \sum_{n=0}^{N-1} h(k)_{i,n} e^{-j2\pi \frac{p-q}{N} n} \right) e^{-j2\pi q \Delta_k}, \tag{7}$$

where both p and $q \in < 0 \ldots N - 1 >$, and Δ_k means the delay value of the k-th path in the channel impulse response.

4   Simulations

The simulation environment designed for analysis of the proposed method assumes as follows:

- the transmitted symbol of TDS-OFDM modulation consists of 1024 data samples and 64 or 128 samples of the pseudorandon noise training sequence in the guard period. No subcarriers are used as pilot tones. The 16-QAM constellation is used for data coding in the subchannels. The total number of the OFDM symbol transmitted during one simulation cycle is 300 (300x1024=307200 QAM data symbols for SER estimation);
- the model of the time and frequency selective wireless channel for terrestrial propagation in an urban area (COST 207) is considered in the investigations. It consists of six active path with the delay spread about 5 μs. According to the sampling frequency about 9 MHz (DVB-T), the delays of the active paths can be approximated by discrete time indices $-$ [0,1,4,14,21,45];

- two cases of the channel time selectivity was considered – $f_d T = [0.1, 0.02]$;
- no additional error correction coding is used in order for a better understanding of the underlying capabilities of the proposed method;
- signal to noise ratio varies from 5 dB to 25 dB.

Both modes of the proposed method was explored regardless of the guard period duration. In fact, the direct one (marked as "dir" on the graphs) is intended for relatively long guard periods, where the condition (3) holds for the IBI-free area. This mode is designed so that it would not be necessary to use previous decoded data (full independence of reception of the consecutive OFDM symbols). But it does not preclude to support the estimation procedure with that information. On the opposite side, when the delay spread of the channel impulse response and the guard period are comparable in the time, there is advised to use the iterative mode.

Each aforementioned mode is explored for another two cases, that are related to operation principles of the OMP algorithm. The basic research assumes that both the paths delays and gains are estimated by the OMP. But, the delay evaluation (the first processing step in the OMP) may be incorrect in high noisy environment, so in order to assess the influence of this inaccuracy on the final result, the operation of the OMP was evaluated in case of perfect information about the channel path delays.

Figures 2 and 3 present estimated Symbol Error Rate (SER) where "Symbol" in the metric name refers to the element transmitted in every subchannel. The SER expresses an average value over the one simulation cycle. The solid line without any markers on it, that is on every graph in Fig. 2 and 3 represents the absolute minimum of SER for the given instance of the channel. The perfect knowledge of the impulse response (perfect CSI) in every time sample was assumed to draw those lines. Because the simulation procedure reflects a deterministic approach, the level of the minimum may vary according to current

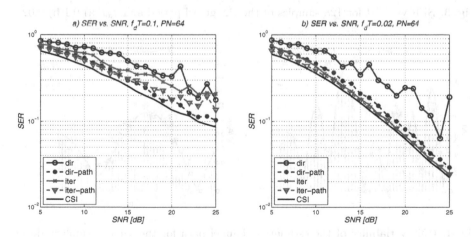

Fig. 2. SER vs SNR for 64 samples of the PN guard period and $f_d T$ a) 0.1 b) 0.02

channel realization. The lines with "path" suffix in the names refer to the simulation with the prior information about the correct path delays.

The iterative mode characterizes good estimation accuracy in case of relatively slow channel variability (Fig. 2b and 3b) regardless of the guard time duration. The risk of an initial fault estimation of the path delay is compensated by iterative procedure. It is seen especially for the short guard period, where the precise delay estimation has clear impact on SER, if the direct mode is used.

The high channel variability and the "short" guard time (64 samples) destroy effectiveness of the methods if the exact values of the paths delays are not known in advance (Fig. 2). The poor results of the iterative mode may come from error propagation in the estimated x_i data sequence. The error correction coding of x_i is probably the simplest way to improve the estimation accuracy. It should be noted that for the "long" guard time (128 samples) the iterative mode is no longer necessary regardless of the channel variability.

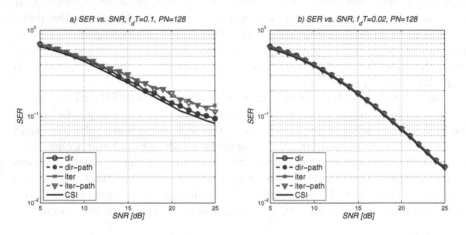

Fig. 3. SER vs SNR for 128 samples of the PN guard period and $f_d T$ a) 0.1 b) 0.02

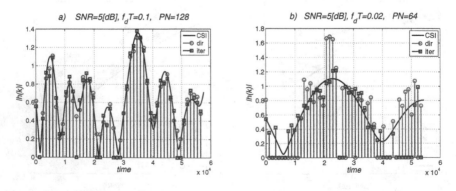

Fig. 4. Gain variability of the estimated channel path for the same exemplary delay and different channel properties and the guard period duration

To sum up, if the IBI free region in the guard time is large enough to find the parameters of the channel paths with the help of the OMP algorithm, then it is possible to use the direct mode. In the opposite situation, the iterative one with error correction should be used and better quality of the prior path delay estimation should be ensured.

If the obtained SER values are in the close vicinity of the one for the perfect CSI, the estimated paths coefficients should roughly approximate the perfect ones. It can be noticed in Fig. 4, that depicts the absolute gains values of the one of the estimated path (bars on the graphs) visible against the background of ideal channel path coefficients (solid line).

In TFT-OFDM system [7], the pilot tones are used for initial channel estimation. This algorithm was reconstructed here for comparison purposes. Fig. 5 presents that better initial performance (lower SER) achieves the proposed compressive sensing time domain method. The TFT one gets the worse results regardless of amount of pilots allocated in the OFDM symbol and type of impulse response interpolation within symbol duration.

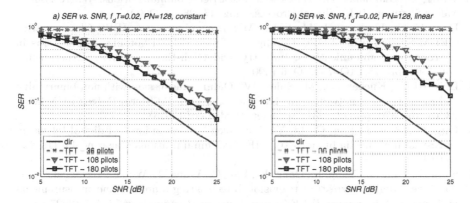

Fig. 5. Compressed sensing time domain estimation vs. time-frequency training for a) constant interpolation b) linear interpolation of the paths gain

5   Conclusions

Utilization of some subcarriers as the pilot tones for efficient estimation of the transmission channel is the standard approach used in the most OFDM systems. The high estimation quality is substantial for the transmission throughput, but undoubtedly, every pilot irrecoverable consumes some part of the system capacity on its own. Therefore, the pilot reduction is a serious challenge especially in case of the mobile radio environment, that usually requires more training information for the channel recovery, because of its dynamic properties both in the time and frequency domains. The guard time, as inherent part of every OFDM symbol, is another natural way for training delivery. It is successfully implemented in TDS-OFDM system by the pseudorandom noise sequence. Nonetheless, a support from pilot tones is still taken into consideration.

The paper presents the new channel estimation method that is based on the time domain training completely. No pilot tones is required to do so. The estimated channel is assumed sparse, which is confirmed in many practical situations. The latter property allows to use the compressed sensing method for channel paths acquisition. The simulation results with the OMP algorithm demonstrate fine estimation accuracy for mobile channels, even better than for the time frequency training method implemented in TFT-OFDM. The method was investigated for the fully implemented frequency domain channel matrix. Next research plans cover the issues of an improvement of the paths delay estimation and system analysis with error correction codes.

References

1. Wang, J., Yang, Z.X., Pan, C.Y., Song, J., Yang, L.: Iterative padding subtraction of the pn sequence for the tds-ofdm over broadcast channels. IEEE Transactions on Consumer Electronics 51, 1148–1152 (2005)
2. Wang, Z., Giannakis, G.: Wireless Multicarrier Communications: Where Fourier Meets Shannon. IEEE Signal Process. Mag. 47, 28–48 (2000)
3. Dziwoki, G., Kucharczyk, M., Sulek, W.: Transmission over UWB Channels with OFDM System using LDPC Coding. In: Proc. SPIE, Photonics Applications in Astronomy, Communications, Industry, and High-Energy Physics Experiments, vol. 7502, pp. 75021Q–75021Q–6 (2009)
4. Dziwoki, G., Kucharczyk, M., Sulek, W.: Ofdm transmission with non-binary ldpc coding in wireless networks. In: Kwiecień, A., Gaj, P., Stera, P. (eds.) CN 2013. CCIS, vol. 370, pp. 222–231. Springer, Heidelberg (2013)
5. Ong, C., Song, J., Pan, C., Li, Y.: Technology and standards of digital television terrestrial multimedia broadcasting. IEEE Communications Magazine 48, 119–127 (2010)
6. Song, J., Yang, Z., Yang, L., Gong, K., Pan, C., Wang, J., Wu, Y.: Technical review on chinese digital terrestrial television broadcasting standard and measurements on some working modes. IEEE Transactions on Broadcasting 53, 1–7 (2007)
7. Dai, L., Wang, Z., Yang, Z.: Time-frequency training ofdm with high spectral efficiency and reliable performance in high speed environments. IEEE Journal on Selected Areas in Communications 30, 695–707 (2012)
8. Duarte, M., Eldar, Y.: Structured compressed sensing: From theory to applications. IEEE Transactions on Signal Processing 59, 4053–4085 (2011)
9. Tropp, J., Gilbert, A.: Signal recovery from random measurements via orthogonal matching pursuit. IEEE Transactions on Information Theory 53, 4655–4666 (2007)
10. Pan, C., Dai, L.: Time domain synchronous ofdm based on compressive sensing: A new perspective. In: IEEE Global Communications Conference (GLOBECOM), pp. 4880–4885 (December 2012)
11. Dai, L., Wang, Z., Yang, Z.: Compressive sensing based time domain synchronous ofdm transmission for vehicular communications. IEEE Journal on Selected Areas in Communications 31, 460–469 (2013)

Influence of Feature Dimensionality and Model Complexity on Speaker Verification Performance

Adam Dustor, Piotr Kłosowski, and Jacek Izydorczyk

Silesian University of Technology, Institute of Electronics
Akademicka 16, 44-100 Gliwice, Poland
{adam.dustor,piotr.klosowski,jacek.izydorczyk}@polsl.pl

Abstract. This paper provides description of a text dependent speaker recognition system based on vector quantization approach. The scope of this paper is to check influence of feature dimensionality and the complexity of the speaker model on verification process. Provided results show that MFCC features yield the lowest possible verification errors among all tested parameters. Although dimensionality of feature vectors is important, there is no need to increase it above some level as the improvement in verification performance is relatively low and computational complexity increases. Far more important than dimensionality is complexity of the speaker model.

Keywords: biometrics, security, speaker verification, voice identification, feature extraction.

1 Introduction

Division of Telecommunication, a part of the Institute of Electronics and Faculty of Automatic Control, Electronics and Computer Science of the Silesian University of Technology, for many years specializes in speech and speaker recognition [1–4]. One of the results of conducted research is presented in this paper which is devoted to speaker verification.

Speaker recognition is the process of automatically recognizing who is speaking by analysis speaker-specific information included in spoken utterances. This process encompasses identification and verification. The purpose of speaker identification is to determine the identity of an individual from a sample of his or her voice and it can be divided into two main categories, i.e. closed-set and open-set. In a closed-set identification there is an assumption that only registered speakers have an access to the system which makes a decision 1 from K, where K is the number of previously registered speakers. In an open-set identification there is no such an assumption so the identification system has to determine whether the testing utterance comes from a registered speaker or not and if yes it should determine his or her identity. The purpose of speaker verification is to decide whether a speaker is whom he claims to be. Most of the applications in which voice is used to confirm the identity claim of a speaker are classified as speaker verification. Speaker recognition systems can also be divided

A. Kwiecień, P. Gaj, and P. Stera (Eds.): CN 2014, CCIS 431, pp. 177–186, 2014.

into text-dependent and text-independent. In text-dependent mode the speaker has to provide the same utterance for training and testing, whereas in text-independent systems there are no such constraints. The text-dependent systems are usually based on template matching techniques in which the time axes of an input speech sample and each reference template are aligned and the similarity between them is accumulated from the beginning to the end of the utterance. Because these systems can directly exploit voice individuality associated with each phoneme or syllable, they usually achieve higher recognition performance than text-independent systems [5].

The paper is organized in the following way. At first fundamentals of speaker verification are discussed, next feature parameters and construction of speaker model are presented. At last achieved speaker verification results for the given dimensionality and complexity of the speaker model are shown.

2   Speaker Recognition

Basic structure of speaker verification system is shown in Fig. 1. Speech signal is cut into short fragments, which usually last for 20–30 ms known as speech frames. Feature extraction is responsible for extracting from each frame a set of parameters known as feature vectors. Extracted sequence of vectors is then compared to speaker model (in verification) or speaker models (in identification) by pattern matching. The purpose of pattern matching is to measure similarity between test utterance and speaker model. In identification an unknown speaker is identified as the speaker whose model best matches the test utterance. In verification the similarity between input test sequence and claimed model must be good enough to accept the speaker as whom he claims to be. As a result, verification requires choosing decision threshold. If computed distance is less than this threshold, a decision can be made that the speaker is whom he claims to be. How to find an optimum value of this threshold still remains a problem for scientist [6]. Another very desired property of this threshold is its independence of a speaker, which means that there is one threshold for all speakers. Since these problems are not solved satisfactorily, they still remain very important research issues, apart from problem of finding the best set of speech parameters, which must be studied further to make an improvement in speaker recognition technology.

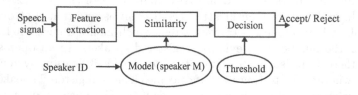

Fig. 1. Speaker verification scheme

3 Feature Parameters

One of the most important procedures in speaker recognition is feature extraction. The extracted parameters should have high speaker discrimination power, high interspeaker variability and low intraspeaker variability. Only such parameters guarantee very good speaker recognition results. Although there are a lot of techniques for extracting speaker specific information from the speech signal, probably the most important are features based on frequency spectrum of the speech as linear prediction coefficients LPC and parameters derived from them like LPC cepstrum known as LPCC fetaures.

3.1 LPC Parameters

Calculation of these parameters is based on the linear model for speech production shown in Fig. 2, where the glottal pulse, vocal tract and radiation are individually modeled as linear filters.

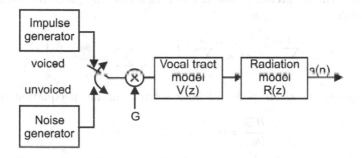

Fig. 2. The linear model of speech production

The source is either a random sequence for unvoiced sounds or a quasi-periodic impulse sequence for voiced sounds. The gain factor G controls the intensity of the excitation. The vocal tract is modeled by transfer function $V(z)$ whereas the radiation model $R(z)$ describes the air pressure at the lips. Combining these parts of the vocal tract yields an all-pole transfer function

$$H(z) = G(z)V(z)R(z) = \frac{G}{A(z)} \quad , \tag{1}$$

$$H(z) = \frac{G}{A(z)} = \frac{G}{1 - \sum\limits_{k=1}^{p} a_k z^{-k}} \quad , \tag{2}$$

where p is the prediction order and a_k are predictor coefficients. The LPC model of the speech signal specifies that a speech sample $s(n)$ can be represented as a linear sum of the p previous samples plus an excitation term

$$s(n) = \sum_{k=1}^{p} a_k s(n-k) + Gu(n) \ . \tag{3}$$

As in speech applications the excitation term is usually unknown it is ignored and the LPC approximation of $s(n)$ depends only on the past output samples. Unfortunately some speaker specific information is included in the excitation term (e.g. fundamental frequency) which affects on the performance of the LPC based speaker recognition systems. Since vocal-tract changes its configuration over time, in order to model it, the predictor coefficients a_k must be computed adaptively over short intervals (10 ms to 30 ms) called frames during which time-invariance is assumed. There are two standard methods of solving for the predictor coefficients: autocorrelation and covariance method. Both of them are based on minimizing the mean-square value E of the prediction error $e(n)$ which is the difference between the actual and the predicted value of the speech sample

$$E = \sum_{n=0}^{N-1+p} e^2(n) = \sum_{n=0}^{N-1+p} \left[s(n) - \sum_{k=1}^{p} a_k s(n-k) \right]^2 . \tag{4}$$

The a_k parameters can be found after solving the linear equations resulting from

$$\frac{\partial E}{\partial a_i} = 0 \ , \quad i = 1, 2, \ldots, p \ . \tag{5}$$

Assuming that speech samples outside the frame of interest are zero and defining the autocorrelation function as

$$r(\tau) = \sum_{i=0}^{N-1-\tau} s(i)s(i+\tau) \ , \tag{6}$$

where N is the number of samples in a frame, the autocorrelation method yields the Yule-Walker equations given by [7]

$$\begin{bmatrix} r(0) & r(1) & r(2) & \ldots r(p-1) \\ r(1) & r(0) & r(1) & \ldots r(p-2) \\ r(2) & r(1) & r(0) & \ldots r(p-3) \\ \vdots & \vdots & \vdots & \ddots & \vdots \\ r(p-1) & r(p-2) & r(p-3) & \ldots & r(0) \end{bmatrix} \begin{bmatrix} a_1 \\ a_2 \\ a_3 \\ \vdots \\ a_p \end{bmatrix} = \begin{bmatrix} r(1) \\ r(2) \\ r(3) \\ \vdots \\ r(p) \end{bmatrix} . \tag{7}$$

Since the matrix is Toeplitz, a computationally efficient algorithm known as the Levinson-Durbin recursion can be used to find the predictor coefficients. During this iterative procedure also other useful features known as reflection coefficients RC are found.

3.2 Cepstral Parameters

Cepstrum, defined as the inverse Fourier transform of the log of the signal spectrum, is an important spectral representation for speech and speaker recognition. It can be calculated from LPC coefficients or from the filter-bank spectrum. In the first case it is known as the LPC based cepstral coefficients LPCC. In the latter case as a mel frequency cepstral coefficients MFCC.

LPCC parameters can be calculated from the transfer function of the vocal tract $H(z)$ in (1) which requires calculating poles of the $H(z)$, or more computationally efficient recursion formula is used [7]

$$c_{lp}(n) = \begin{cases} a_n + \sum_{k=0}^{n-1} \frac{k}{n} a_{n-k} c_{lp}(k), & 1 \le n \le p , \\ \sum_{k=n-p}^{n-1} \frac{k}{n} a_{n-k} c_{lp}(k), & n > p . \end{cases} \tag{8}$$

The LPC based cepstrum has many interesting properties. It is causal for the minimum phase $H(z)$ and of infinite duration. As the cepstrum represents the log of the signal spectrum, signals represented as the cascade of two effects which are products in the spectral domain are additive in the cepstral domain. This property of separability of pitch excitation and vocal tract is considered as one of the reasons that cepstral parameters are more effective for speaker recognition than other representations of speech signal. Another interesting property is the fact that $c_{lp}(n)$ decays as fast as $1/n$ as n approaches $+\infty$ so the feature vector consists of the finite number, most significant components $c_{lp}(1)$ to $c_{lp}(x)$, where $x \approx 1.5p$.

MFCC parameters are based on the nonlinear human perception of the frequency of sounds. They can be computed as follows: window the signal, take the FFT, take the magnitude, take the log, warp the frequency according to the mel scale and finally take the inverse FFT. Mel warping transforms the frequency scale to place less emphasis on high frequencies [8].

4 Pattern Recognition

Since speaker recognition is based on similarity calculation between test utterance and the reference model, it is obvious that the problem of construction of the good model is crucial. The simplest approach to this problem but also the most computationally demanding during recognition process is to store in a memory all feature vectors extracted from the speech during training of the system. As speech is a very redundant signal, it can be easily seen that for text-independent recognition, which requires providing a lot of training speech to find a speaker model, it consists of thousands of multidimensional vectors. Computational efficiency of such model is very low. In order to compute distance between such model and vectors extracted from the test speech it is necessary to find for each test vector the most similar vector, known as a nearest neighbor NN, from the model.

Another method used for representing speaker in a speaker recognition system is based on vector quantization VQ. Speaker is represented as a set of several (less than 100) vectors that possibly in the best way represent speaker. This set of vectors is called a codebook. In this case during recognition each test vector is compared with its nearest neighbour from the codebook and the overall distance for the whole test utterance is computed. Calculation of normalized distance D for M frames of speech is given by

$$D = \frac{1}{M} \sum_{i=1}^{M} \min(d(x_i, c_q)) \quad 1 \leq q \leq L \ , \tag{9}$$

where x_i is a test vector and c_q a code vector from a codebook of size L. As it can be seen for M frames and L code vectors its necessary to calculate ML distances. The most often used measure of similarity is an Euclidean distance

$$d(x_i, c_q) = \sum_{k=1}^{p} (x_i(k) - c_q(k))^2 \tag{10}$$

where p is a dimension of a vector. VQ method is faster than NN technique, but unfortunately requires to find a codebook for each speaker, whereas in NN method a model consists just of all vectors. How to find the best codebook for speaker from a lot of training data? To solve this problem a kind of clustering technique is required, which can find a small set of the best representative vectors of a speaker. One of applied algorithms are k-means and Linde Buzo Gray procedure.

K-means algorithm is an iterative procedure and consists of four major steps. At first arbitrarily choose L vectors from the training data, next for each training vector find its NN from the current codebook, which corresponds to partitioning vector space into L distinct regions. The third step requires updating the code vectors using the centroid of the training vectors assigned to them and the last step – repeat steps 2 and 3 until some converge criterion is satisfied. The converge criterion is usually an average quantization error expressed in the same way as in (9) with an exception that x_i is a training vector.

Although k-means training method works well, it is even better to design a codebook in steps by using a splitting procedure, which leads to LBG algorithm. It starts with one cluster, which is the centroid of all training vectors, and then the code vector is split into two, $c_0 + \delta$ and $c_0 - \delta$, where δ is a small perturbation vector. With these two clusters k-means procedure is run. After the averaged distortion reaches steady level, the codebook is split again and the new codebook is trained with k-means method. This splitting is repeated until the desired codebook size is reached.

5   Speaker Verification in Matlab

All research was done on Polish database ROBOT [9]. This database consists of 2 CD with 1 GB of speech data. The speech utterances were collected from

30 speakers of both sex in a several time-separated sessions to catch intraspeaker variability. Main specifications of ROBOT are the following: sampling frequency 22 kHz, language – Polish, quantization 16 bit, file format ".wav", lack of files compression, recording environment – quiet, each file is preceded and followed by the silence. Recorded utterances consist of the words belonging to three dictionaries (L1, L2, and L3). Words in L1 and L3 are numbers from 0 to 9 and 10 to 99 respectively. Dictionary L2 consists only from commands used in robot control (start, stop, left, right, up, down, drop, catch, angle). These dictionaries were used to construct seven different sets of utterances Z1...Z7.

During training and testing of the speaker verification system the same signal processing procedure was used. Speech files, before feature extraction, were processed to remove silence. Voice activity detection was based on the energy of the signal. Next signal was preemphasized with a standard parameter of

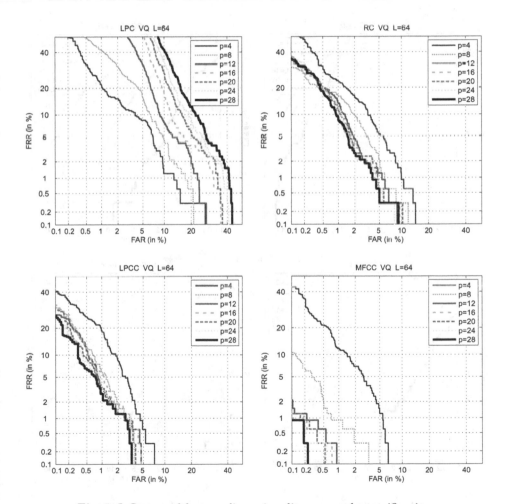

Fig. 3. Influence of feature dimensionality on speaker verification

$\alpha = 0.95$ and segmented into 10 ms frames every 5 ms. Hamming windowing was applied. For each frame LPC analysis was applied to obtain LPC and RC coefficients. LPC parameters were then transformed into LPCC coefficients using Equation (8). From each frame MFCC parameters were also computed. All utterances from Z3 set were used to obtain model of each speaker. Each model was constructed from approximately 90 s of speech after silence removing. Text dependent speaker verification was implemented. All test utterances were from Z4 (combination of numbers from Z3 set). Each speaker provided 11 test sequences of approximately 5 s each. As a result there were 9900 verification trials – 330 valid trails (30*11) and 9570 impostor trials (30*11*29) for each combination of dimensionality of the feature vector and the size of the speaker model.

In order to check the influence of dimensionality, testing was done for the prediction order p equal to 4, 8, 12, 16, 20, 24, and 28. Actual dimensionality

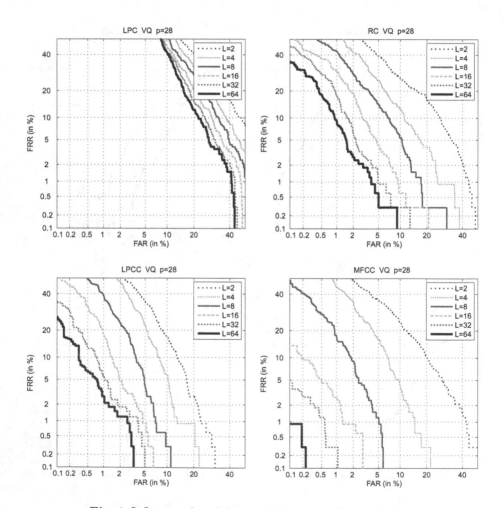

Fig. 4. Influence of model complexity on speaker verification

for the LPC and RC feature vectors were equal to p but for cepstral parameters (LPCC and MFCC) was $1.5p$ (6 to 42). LBG procedure was applied to obtain codebooks for each speaker. In order to check the influence of model size L, testing was done for the codebooks consisting of 2, 4, 8, 16, 32 and 64 code vectors.

Verification performance was characterized in terms of the two error measures, namely the false acceptance rate FAR and false rejection rate FRR. These measures correspond to the probability of acceptance an impostor as a valid user and the probability of rejection of a valid user. Changing the decision level, DET curves which show dependence between FRR and FAR can be plotted. Another very useful performance measure is an equal error rate EER which corresponds to error rate achieved for the decision threshold for which FRR=FAR. In other words EER is just given by the intersection point of the main diagonal of DET plot with DET curves.

Achieved results for the most complex model ($L = 64$ code vectors per speaker) as a function of dimensionality of the feature vectors p were shown in Fig. 3. Achieved results for the highest dimensionality ($p = 28$ parameters extracted from each segment of the utterance) as a function of model complexity L were shown in Fig. 4. The best achieved results of speaker verification were shown in Fig. 5.

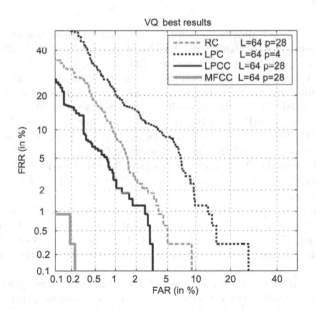

Fig. 5. The best achieved results for the speaker verification

6   Conclusion

The lowest FRR and FAR errors were achieved for the MFCC parameters (Fig. 5) For the best combination of dimensionality and complexity of the model ($p = 28$

and $L = 64$) achieved $EER = 0.27\%$ is definitely better than $EER = 1.55\%$ for LPCC, $EER = 2.42\%$ for RC and $EER = 6.43\%$ for LPC features. Such low error rate indicates that for limited number of speakers and high quality of speech, speaker verification may be implemented as an additional level in security systems or which is the final goal of this research may be implemented in a mobile phone. What is interesting is the fact that there is no need to increase dimensionality of feature vectors above some level. From the Fig. 3 optimum value of p may be estimated as between 12 and 20 for the MFCC and LPCC. Much more important than dimensionality is complexity of the model which was shown in the Fig. 4. Error rates are highly dependent on the number of code vectors per speaker model. Summarizing if enough learning data is available, the more code vectors per speaker the better.

Acknowledgment. This work was supported by The National Centre for Research and Development (www.ncbir.gov.pl) under Grant number POIG.01.03.01-24-107/12 *(Innovative speaker recognition methodology for communications network safety)*.

References

1. Kłosowski, P.: Speech processing application based on phonetics and phonology of the polish language. In: Kwiecień, A., Gaj, P., Stera, P. (eds.) CN 2010. CCIS, vol. 79, pp. 236–244. Springer, Heidelberg (2010)
2. Dustor, A.: Voice verification based on nonlinear Ho-Kashyap classifier. In: International Conference on Computational Technologies in Electrical and Electronics Engineering, SIBIRCON 2008, Novosibirsk, pp. 296–300 (2008)
3. Dustor, A.: Speaker verification based on fuzzy classifier. In: Cyran, K.A., Kozielski, S., Peters, J.F., Stańczyk, U., Wakulicz-Deja, A. (eds.) Man-Machine Interactions. AISC, vol. 59, pp. 389–397. Springer, Heidelberg (2009)
4. Dustor, A., Kłosowski, P.: Biometric voice identification based on fuzzy kernel classifier. In: Kwiecień, A., Gaj, P., Stera, P. (eds.) CN 2013. CCIS, vol. 370, pp. 456–465. Springer, Heidelberg (2013)
5. Beigi, H.: Fundamentals of speaker recognition. Springer, New York (2011)
6. Togneri, R., Pullella, D.: An overview of speaker identification: Accuracy and robustness issues. IEEE Circuits and Systems Magazine 11(2), 23–61 (2011)
7. Rabiner, L.R., Juang, B.H.: Fundamentals of speech recognition. Prentice Hall (1993)
8. Fazel, A., Chakrabartty, S.: An overview of statistical pattern recognition techniques for speaker verification. IEEE Circuits and Systems Magazine 11(2), 62–81 (2011)
9. Adamczyk, B., Adamczyk, K., Trawiński, K.: Zasób mowy ROBOT. Biuletyn Instytutu Automatyki i Robotyki WAT 12, 179–192 (2000)

Generation of Entangled Qudits States with XY-Like Dynamics in 1D Qudits Spins

Marek Sawerwain[1] and Joanna Wiśniewska[2]

[1] Institute of Control & Computation Engineering,
University of Zielona Góra, Licealna 9, Zielona Góra 65-417, Poland
M.Sawerwain@issi.uz.zgora.pl
[2] Institute of Information Systems, Faculty of Cybernetics,
Military University of Technology, Kaliskiego 2, 00-908 Warsaw, Poland
jwisniewska@wat.edu.pl

Abstract. At present, the spin chains are discussed as the solution which may be used in future to construct quantum communication networks. In this paper the definition of XY-type Hamiltonian is presented. The Hamiltonian realizes the transfer of information in a one-dimensional spin chain, which is built of generalised units of quantum information – so-called qudits. The main presented example in the paper shows that the proposed Hamiltonian's definition allows generating the entanglement between boundary elements of the chain. The potentiality of entanglement generation is a very important feature of future quantum communication networks. The paper also contains a numerical example for analyzing the level of entanglement with use of CCNR criterion.

Keywords: quantum information transfer, qudits chains, creation of entangled states, numerical simulations.

1 Introduction

At present, it is impossible to point out the technology which will be used in future to transfer quantum data. However, the current research in the quantum computing field [1, 2] allow to suppose that spin chains and quantum entanglement [3] may be very significant solutions for realizing quantum transfer.

The problem of state transfer (ST) was analysed i.a. in [4], but the presented solution allows the degradation of information during the ST process, so it is not so-called perfect transfer. However, in [5], the schemes of spin chains for quantum data transfer were shown and the described solution, in terms of Fidelity measure, is a perfect state transfer (PST). Further information about PST – not only in the case of qubits – can be found in [6–8].

The spin chains may be used to realize the quantum data transfer, but also to:

- initialize the state of single qubit/qudit,
- initialize the state of whole quantum system,

A. Kwiecień, P. Gaj, and P. Stera (Eds.): CN 2014, CCIS 431, pp. 187–196, 2014.

– amplify the signal during the process of quantum measurement,
– generate the entanglement phenomenon between chosen chain's elements.

The main matter of this paper is to verify, if proposed in [9] Hamiltonian definition for realizing XY type dynamic in qudit chain is suitable to generate the entanglement for chosen chain's initial states. It should be noted that exists others techniques to create and generate quantum entanglement [10–14].

The paper was organized in the following way: in the first Sect. 2 the generalized information on XY-type qudits' dynamics and remark concerning perfect transfer's features were presented. The Section 3 is dedicated to transfer's process which results with the entanglement of qudits from chain's both ends. The numerical tests of transfer's correctness for qubit and qutrit chains were shortly discussed in the Sect. 4. The conclusions are presented in the Sect. 5.

2   Definition of Hamiltonian for Spin Chain with Qudits

In this section a Hamiltonian H^{XY_d} will be defined (firstly presented in [9]). This Hamiltonian will be used to realize the perfect transfer of quantum information in qudits chains for entanglement creation between chosen points of a chain.

In the definition of the XY-like Hamiltonian for qudits' chain, given below, the Lie algebra's generator for a group $SU(d)$ was used, where $d \geq 2$ is to define a set of operators responsible for transfer dynamics. For clarity, the construction procedure of $SU(d)$ generators will be recalled – in the first step a set of projectors is defined:

$$(P^{k,j})_{v,\mu} = |k\rangle\langle j| = \delta_{v,j}\delta_{\mu,k} , \quad 1 \leq v , \quad \mu \leq d . \tag{1}$$

The first suite of $d(d-1)$ operators from the group $SU(d)$ is specified as

$$\Theta^{k,j} = P^{k,j} + P^{j,k} , \quad \beta^{k,j} = -i(P^{k,j} - P^{j,k}) , \tag{2}$$

and $1 \leq k < j \leq d$.

The remaining $(d-1)$ generators are defined in the following way

$$\eta^{r,r} = \sqrt{\frac{2}{r(r+1)}}\left[\left(\sum_{j=1}^{r} P^{j,j}\right) - rP^{r+1,r+1}\right] , \tag{3}$$

and $1 \leq r \leq (d-1)$. Finally, the $d^2 - 1$ operators belonging to the $SU(d)$ group can be obtained.

The above construction of $SU(d)$ generator is used to create the XY-like Hamiltonian for qudits. Assuming that each qudit has the same freedom level $d \geq 2$ (the qudit is defined similar to qubit, however a computational base for qudits is expressed with d orthonormal vectors – in case of qubits, the base contains two orthonormal vectors):

$$H^{XY_d} = \sum_{(i,i+1)\in\mathcal{L}(G)} \frac{J_i}{2}\left(\Theta^{k,j}_{(i)}\Theta^{k,j}_{(i+1)} + \beta^{k,j}_{(i)}\beta^{k,j}_{(i+1)}\right) , \tag{4}$$

where J_i is defined as follows: $J_i = \frac{\sqrt{i(N-i)}}{2}$ for $1 \le k < j < d$ and $\Theta^{k,j}_{(i)}$, $\beta^{k,j}_{(i)}$ are $SU(d)$ group operators defined by (2) applied to the (i)-th and $(i+1)$-th qudit. The Hamiltonian (4) will be also called the transfer Hamiltonian.

The state transfers which were studied in [4] use Hamiltonians H which have the following property

$$\left[H, \sum_{i=1}^{N} Z_{(i)} \right] = 0 , \qquad (5)$$

where Z represents the sign gate for qubits. This means that spins are preserved and dynamics generated by H is divided into series of subspaces denoted by the number of qubit in state $|1\rangle$ – see in [7]. In the case discussed here, it is not hard to show that

$$\left[H^{XY_d}, \sum_{i=1}^{N} \eta^{r,r}_{(i)} \right] = 0 , \qquad (6)$$

for $1 \le r \le (d-1)$, so the Equation (6) generalizes the situation mentioned in the Equation (5) – preserving spins and separating dynamics into subspaces.

To specify the features of PST from a point a to a point b, where to these points respectively correspond states $|a\rangle$ and $|b\rangle$, the initial state will be expressed as:

$$|\Psi(t_0)\rangle = |\psi_a 00 \ldots 00_b\rangle = \alpha_0 |0_a 00 \ldots 00_b\rangle + \alpha_1 |1_a 00 \ldots 00_b\rangle +$$
$$\alpha_2 |2_a 00 \ldots 00_b\rangle + \ldots + \alpha_{(d-1)} |(d-1)_a 00 \ldots 00_b\rangle . \qquad (7)$$

The mentioned case of transfer applies to sending an information through the path of length l, what can be presented as a scheme at Fig. 1.

$|\psi_0\rangle \qquad |\psi_1\rangle \qquad |\psi_2\rangle \qquad |\psi_{n-2}\rangle \qquad |\psi_{n-1}\rangle$

$||$ $\qquad\qquad\qquad\qquad\qquad\qquad\qquad\qquad\qquad ||$

$|a\rangle$ $\qquad\qquad\qquad\qquad\qquad\qquad\qquad\qquad |b\rangle$

A one-dimensional chain for transferring state of a single qudit

Fig. 1. The realization of quantum information transfer in a spin chain for a single qudit – the case with path length l

Generally, the system's evolution for N qudits (with freedom level d), taking time t and starting from initial state $|a\rangle$, may be expressed as:

$$|\Psi(t)\rangle = e^{-iH^{XY_d}t}|a\rangle = \sum_{j=1}^{d^N} \beta_j(t)|j\rangle \qquad (8)$$

with a complex coefficients $\beta_j(t)$, and $\left(\sum_{j=1}^{n} |\beta_j(t)|^2 \right) = 1$.

The state described in (8) possesses a perfect transfer feature from a point a to a point b in a time t when:

$$\left| \langle b | e^{-i H^{XY_d} t} | a \rangle \right| = 1 \ . \tag{9}$$

The goal of this paper is not proving the correctness of PST for the Hamiltonian H^{XY_d}. Naturally, it would require to describe a spectrum of Hamiltonian H^{XY_d}. For short chains, when $l = 2$ or $l = 3$, the spectral decomposition may be calculated directly. Only the example for $l = 2$ and for the case when H^{XY_d} describes the dynamics of qubit $(d = 2)$ will be given (naturally, calculation given below reproduces a result given in works [4, 6]).

For the transfer of two qubits the following Hamiltonian and transfer's operator may be used:

$$H^{XY_2} = \begin{pmatrix} 0 & 0 & 0 & 0 \\ 0 & 0 & 1 & 0 \\ 0 & 1 & 0 & 0 \\ 0 & 0 & 0 & 0 \end{pmatrix} , \quad e^{-i H^{XY_2} t} = \begin{pmatrix} 0 & 0 & 0 & 0 \\ 0 & \cos(t) & -i \sin(t) & 0 \\ 0 & -i \sin(t) & \cos(t) & 0 \\ 0 & 0 & 0 & 0 \end{pmatrix} \tag{10}$$

then the spectral decomposition is:

$$e^{-i H^{XY_2} t} = \frac{2 e^{it(\lambda_0)}}{4} \begin{pmatrix} 0 & 0 & 0 & 0 \\ 0 & 1 & 1 & 0 \\ 0 & 1 & 1 & 0 \\ 0 & 0 & 0 & 0 \end{pmatrix} + \frac{2 e^{it\lambda_1}}{4} \begin{pmatrix} 0 & 0 & 0 & 0 \\ 0 & 0 & 0 & 0 \\ 0 & 0 & 0 & 0 \\ 0 & 0 & 0 & 2 \end{pmatrix} +$$

$$\frac{2 e^{it\lambda_2}}{4} \begin{pmatrix} 2 & 0 & 0 & 0 \\ 0 & 0 & 0 & 0 \\ 0 & 0 & 0 & 0 \\ 0 & 0 & 0 & 0 \end{pmatrix} + \frac{2 e^{i\lambda_3}}{4} \begin{pmatrix} 0 & 0 & 0 & 0 \\ 0 & 1 & -1 & 0 \\ 0 & -1 & 1 & 0 \\ 0 & 0 & 0 & 0 \end{pmatrix} , \tag{11}$$

where the eigenvalues are: $\lambda_0 = -1$, $\lambda_1 = 0$, $\lambda_2 = 0$, $\lambda_3 = 1$. The above equation may be simplified to:

$$e^{-i H^{XY_2} t} = \begin{pmatrix} 1 & 0 & 0 & 0 \\ 0 & \frac{e^{-it}}{2} + \frac{e^{it}}{2} & \frac{e^{-it}}{2} - \frac{e^{it}}{2} & 0 \\ 0 & \frac{e^{-it}}{2} - \frac{e^{it}}{2} & \frac{e^{-it}}{2} + \frac{e^{it}}{2} & 0 \\ 0 & 0 & 0 & 1 \end{pmatrix} = \begin{pmatrix} 0 & 0 & 0 & 0 \\ 0 & \cos(t) & -i \sin(t) & 0 \\ 0 & -i \sin(t) & \cos(t) & 0 \\ 0 & 0 & 0 & 0 \end{pmatrix} . \tag{12}$$

If two-qubit chain with initial state has a form: $|\Psi\rangle = |\psi 0\rangle$, where $|\psi\rangle = \alpha|0\rangle + \beta|1\rangle$, then:

$$e^{-i H^{XY_2} t} |\Psi\rangle = \begin{pmatrix} \alpha \\ -i\beta \sin(t) \\ \beta \cos(t) \\ 0 \end{pmatrix} \tag{13}$$

and for the value $t = \frac{\pi}{2}$ the realized transfer is a perfect transfer.

Similarly, for two-qutrit chain we obtain:

$$e^{-iH^{XY_3}t}|\Psi\rangle = \begin{pmatrix} \alpha \\ -i\beta\sin(t) \\ -i\gamma\sin(t) \\ \beta\cos(t) \\ 0 \\ 0 \\ \gamma\cos(t) \\ 0 \\ 0 \end{pmatrix} \tag{14}$$

and again for the value $t = \frac{\pi}{2}$ the realized transfer is a perfect transfer.

Remark 1. In the same way the features of PST in a two-qudit chain can be shown. The spectrum values -1, 0, 1 stay the same, but the their multiplicity will be changed – the multiset of eigenvalues producing the spectrum may be expressed as:

$$\sigma_{H^{XY_d}}^{l=2} = \{-1_{(d-1)}, 0_{(d^2-2(d-1))}, 1_{(d-1)}\} , \tag{15}$$

where the lower index contains the value of multiplicity for specified eigenvalue.

Remark 2. For longer chains the spectrum of H^{XY_d} may be calculated with use of adequate graphs. Just like for qubit chains – where co-spectral graphs for Hamiltonian dynamics may be proposed – it is also possible to use well-known spectrum of Johnson's graph [15] (Johnson's graph $J(n,k)$ is a graph where vertices are given by k-subsets $(1, \ldots, n)$ and two vertices are connected when their intersection has $k - 1$ elements):

$$\lambda_j(J(N,k)) = k(N-k) - j(N+1-j) , \quad j = 0, 1, 2, \ldots, k . \tag{16}$$

There can be the biggest eigenvalue of H^{XY_d} pointed out for the path of length l, using $J(N,k)$:

$$\lambda_{\max}\left(H^{XY_d}\right) = \lambda_j\left(J\left(l, \left\lfloor \frac{l}{2} \right\rfloor\right)\right) . \tag{17}$$

Then the ascending ordered sequence of eigenvalues will be denoted as $\sigma_{H^{XY_d}}$ for the path of length l:

$$\sigma_{H^{XY_d}} = \begin{cases} [-\lambda_{\max}, -\lambda_{\max}+1, \ldots, 0, \ldots, \lambda_{\max}-1, \lambda_{\max}] & \text{when } l \text{ is odd,} \\ [-\lambda_{\max}, -\lambda_{\max}+2, \ldots, 0, \ldots, \lambda_{\max}-2, \lambda_{\max}] & \text{when } l \text{ is even.} \end{cases} \tag{18}$$

3   Entanglement Creation

In the process of entanglement generation the initial state of N qudit chain is defined as follows:

$$|\Psi\rangle = |+\rangle|0^{N-2}\rangle|+\rangle . \tag{19}$$

The definition of a plus state $|+\rangle$ for qubit chains is represented by a superposition of states $|0\rangle$ and $|1\rangle$:

$$|+\rangle = \frac{|0\rangle + |1\rangle}{\sqrt{2}} \ , \tag{20}$$

what is an equivalent operation to using Hadamard gate on state $|0\rangle$, i.e. $H|0\rangle = |+\rangle$. The definition of Hadamard gate for qudits is:

$$H|j\rangle = \frac{1}{\sqrt{d}} \sum_{k=0}^{d-1} \omega^{j \cdot k} |k\rangle \ . \tag{21}$$

where the symbol ω represents the k-th primitive root of unity, defined as

$$\omega_k^d = \cos\left(\frac{2k\pi}{d}\right) + \mathrm{i}\sin\left(\frac{2k\pi}{d}\right) = e^{\frac{2\pi \mathrm{i}k}{d}} \ , \quad k = 0, 1, 2, \ldots, d-1 \tag{22}$$

and d is a degree of the root, and i is a complex unit.

The above definition allows to express generalized plus state using Hadamard gate on the state $|0_d\rangle$ for qudits – with any freedom level d – in the following way:

$$|+\rangle = H|0_d\rangle = \frac{|0\rangle + |1\rangle + \ldots + |d-1\rangle}{\sqrt{d}} \ . \tag{23}$$

The process of entanglement generation may be presented at Fig. 2, where the initial chain's state is shown and after the transfer process the first and the last chain's elements are entangled.

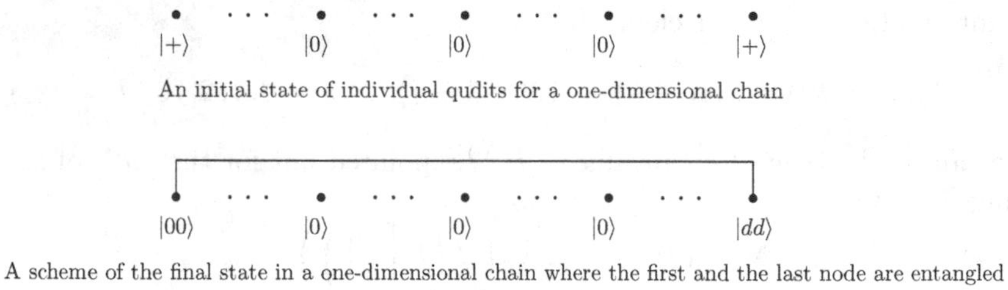

An initial state of individual qudits for a one-dimensional chain

A scheme of the final state in a one-dimensional chain where the first and the last node are entangled

Fig. 2. After the realisation of transfer protocol, the entangled state was created between the first and the last chain's node

Remark 3. It should be noted that the obtained quantum state is considered to be entangled and it may be expressed as a graph state:

$$|\Psi_e\rangle = C_Z|++\rangle \tag{24}$$

where $|+\rangle$ is defined as in Equation (23), and C_Z represents controlled sign gate. For qubit case this gate is formulated as $C_Z = I \oplus \sigma_z$, where $\oplus$ is a direct sum operator. For qudit case it is possible to create several version of sign gate.

This is direct consequence of a form the $\eta^{r,r}$ operator from $SU(d)$ generator, because forms of $\eta^{r,r}$ operator can be regarded as direct counterparts of σ_z in qubit case.

According to the above definitions, for the initial state – given by (19) – the final state in a qubit chain is determined by the number of chain's elements.

$$e^{-iH^{XY_2}t}|\Psi\rangle = \frac{1}{2}\Big(|00\rangle \pm i|01\rangle \pm i|10\rangle + |11\rangle\Big) , \quad \text{for } N = 4, 6, 8, \dots ,$$

$$e^{-iH^{XY_2}t}|\Psi\rangle = \frac{1}{2}\Big(|00\rangle \pm |01\rangle \pm |10\rangle - |11\rangle\Big) , \quad \text{for } N = 5, 7, 9, \dots . \quad (25)$$

It is so because the number of chain's elements tells how many σ_x and σ_y operators will be used to describe the system's dynamic and σ_y operators have the direct influence on the probability amplitudes. The given description corresponds to the first and the last qubit – other $N-2$ qubits were ignored with use of the partial trace operation.

In the case of generating entanglement for qutrit chain (d=3), the final state also depends on a chain's length (represented by N). These are some short chains' examples with a given number of inner elements:

$$e^{-iH^{XY_3}t}|\Psi\rangle = \frac{1}{3}\Big(|00\rangle + i|01\rangle + i|02\rangle + i|10\rangle + |11\rangle + |12\rangle$$
$$+ i|20\rangle + |21\rangle + |22\rangle\Big) , \quad \text{for } N = 4 , \quad (26)$$

$$e^{-iH^{XY_3}t}|\Psi\rangle = \frac{1}{3}\Big(|00\rangle + |01\rangle + |02\rangle + |10\rangle - |11\rangle - |12\rangle$$
$$+ |20\rangle - |21\rangle - |22\rangle\Big) , \quad \text{for } N = 5 , \quad (27)$$

$$e^{-iH^{XY_3}t}|\Psi\rangle = \frac{1}{3}\Big(|00\rangle - i|01\rangle - i|02\rangle - i|10\rangle + |11\rangle + |12\rangle$$
$$- i|20\rangle + |21\rangle + |22\rangle\Big) , \quad \text{for } N = 6 , \quad (28)$$

The obtained state is based on states: $|00\rangle, |11\rangle, \dots, |(d-1)(d-1)\rangle$. Therefore the chain's state may be expressed as:

$$e^{-iH^{XY_d}t}|\Psi\rangle = \frac{1}{3}\Big(|00\rangle \pm (\dots) \pm |11\rangle \pm (\dots) \pm |(d-1)(d-1)\rangle\Big) . \quad (29)$$

where the elements $(\dots)$ stand for additional states, which probability amplitudes depend on the chain's length.

4   Tracing of Entanglement with CCNR Criterion

During the process of information transfer with use of Hamiltonian H^{XY_d} – where the initial state is given by (19) – the entanglement level should rise

to the maximal value in the middle stage of the process and then decrease to
the level corresponding to the maximally entangled state:

$$|\Psi\rangle = \frac{1}{3}\left(|00\rangle \pm |11\rangle \pm |22\rangle \pm \ldots \pm |(d-1)(d-1)\rangle\right) . \qquad (30)$$

To track the entanglement's behaviour during the transfer process the CCNR
criterion [16] may be used. Generally, the criterion itself may be defined by
the fact that if the matrix ρ_{AB} of a bipartite $m \times n$ system is separable, then:

$$||\rho_{AB}^R|| = \sum_{i=1}^{q} \sigma_i\left(\rho_{AB}^R\right) \leq 1 , \qquad (31)$$

where σ_i represents a singular value of ρ_{AB}^R from Singlar Value Decomposition
(SVD) and $q = \min(m^2, n^2)$. The additional advantage of CCNR criterion is
that it also may be used for qudits.

Figure 3 shows the entanglement level obtained by the use of CCNR criterion.
According to the expectations, the entanglement level rises in the middle stage of
the process and then stabilizes at the level corresponding to the entanglement of
outer chain's elements. The second case depicted at Fig. 4 refers to the situation
where the CCNR was used to obtain level of entanglement only to the first and
the last element of a qudit chain, the others elements was removed with the
partial trace operation.

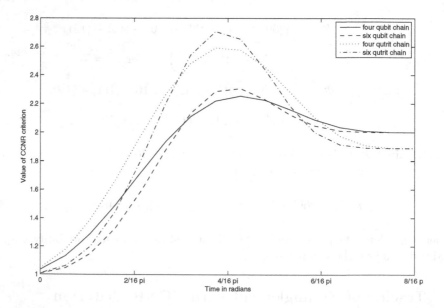

Fig. 3. The values of entanglement level for different chains, obtained by the use of
CCNR criterion on the whole chain

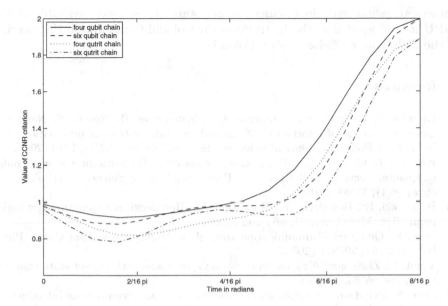

Fig. 4. The values of entanglement level for different chains, obtained by the use of CCNR criterion on the first and the last element of chain, other qubits/qudits are traced out

5 Conclusions

The presented procedure of entanglement generation between the first and the last chain's elements seems to be very important feature, which designates defined Hamiltonian to support information transfer in qudit chains. The generated entanglement allows – using the teleportation protocol – to directly transfer additional information between outer chain's elements. This shows the possibility of information transfer in a spin chain not only with use of XY-type dynamics.

The generated entanglement, written in a graph state form, allows further research on dynamics of presented Hamiltonian in bigger structures like grids and graphs with qudit nodes. One of the problems for further work with Hamiltonian H^{XY_d} is a quantum information routing problem in a qudit networks. Another important problem is also an influence of noise and the issue of decoherence – the paper [17] shows an exemplary scenarios of noise impact into quantum channels and shows the acceptable levels of noise at computation process. These problems are worth of further study in the context of entangled state creation with transfer state protocols with qudits.

Acknowledgments. We would like to thank for useful discussions with the *Q-INFO* group at the Institute of Control and Computation Engineering (ISSI) of the University of Zielona Góra, Poland. We would like also to thank to anonymous referees for useful comments on the preliminary version of this paper. The numerical results were done using the hardware and software available at the "GPU μ-Lab" located at the Institute of Control and Computation Engineering of the University of Zielona Góra, Poland.

References

1. Klamka, J., Węgrzyn, S., Znamirowski, L., Winiarczyk, R., Nowak, S.: Nano and quantum systems of informatics. Nano and quantum systems of informatics. Bulletin of the Polish Academy of Sciences. Technical Sciences 52(1), 1–10 (2004)
2. Klamka, J., Gawron, P., Miszczak, J., Winiarczyk, R.: Structural programming in quantum octave. Bulletin of the Polish Academy of Sciences. Technical Sciences 58(1), 77–88 (2010)
3. Horodecki, R., Horodecki, P., Horodecki, M., Horodecki, K.: Quantum Entanglement. Rev. Mod. Phys. 81, 865–942 (2009)
4. Bose, S.: Quantum Communication through an Unmodulated Spin Chain. Phys. Rev. Lett. 91, 207901 (2003)
5. Vinet, L., Zhedanov, A.: How to construct spin chains with perfect state transfer. Phys. Rev. A 85, 12323 (2012)
6. Bose, S.: Quantum Communication through Spin Chain Dynamics: an Introductory Overview. Contemporary Physics 48, 13–30 (2007)
7. Kay, A.: A Review of Perfect State Transfer and its Application as a Constructive Tool. International Journal of Quantum Information 8(4), 641–676 (2010)
8. Wu, L.A., Miranowicz, A., Wang, X., Liu, Y., Nori, F.: Perfect function transfer and interference effects in interacting boson lattices. Phys. Rev. A 80, 012332 (2009)
9. Sawerwain, M., Gielerak, R.: Transfer of quantum continuous variable and qudit states in quantum networks. In: Kwiecień, A., Gaj, P., Stera, P. (eds.) CN 2012. CCIS, vol. 291, pp. 63–72. Springer, Heidelberg (2012)
10. Leoński, W., Miranowicz, A.: Kerr nonlinear coupler and entanglement. J. Opt. B: Quantum Semiclass. Opt. 6, S37–S42 (2004)
11. Miranowicz, A., Leoński, W.: Dissipation in systems of linear and nonlinear quantum scissors. J. Opt. B: Quantum Semiclass. Opt. 6, S43–S46 (2004)
12. Leoński, W., Kowalewska-Kudłaszyk, A.: Quantum scissors – finite-dimensional states engineering. Progress in Optics 56, 131–185 (2011)
13. Kowalewska-Kudłaszyk, A., Leoński, W., Peřina Jr., J.: Generalized Bell states generation in a parametrically excited nonlinear couple. Phys. Scr. T147, 014016 (2012)
14. Bartkowiak, M., Miranowicz, A., Wang, X., Liu, Y., Leoński, W., Nori, F.: Sudden vanishing and reappearance of nonclassical effects: General occurrence of finite-time decays and periodic vanishings of nonclassicality and entanglement witnesses. Phys. Rev. A 83, 053814 (2011)
15. van Dam, E.R., Haemers, W.H.: Spectral Characterizations of Some Distance-Regular Graphs. J. Algebraic Combin. 15, 189–202 (2003)
16. Rudolph, O.: Some properties of the computable cross-norm criterion for separability. Phys. Rev. A 67, 32312 (2003)
17. Gawron, P., Klamka, J., Winiarczyk, R.: Noise Effects in the Quantum Search Algorithm from the Viewpoint of Computational Complexity. Int. J. Appl. Math. Comput. Sci. 22(2), 493–499 (2012)

Effective Noise Estimation for Secure Quantum Direct Communication over Imperfect Channels

Piotr Zawadzki

Institute of Electronics,
Silesian University of Technology,
Akademicka 16, 44-100 Gliwice, Poland
piotr.zawadzki@polsl.pl

Abstract. In quantum communication actions of an eavesdropper are perceived by legitimate parties as some additional noise so correct estimation of the noise level is vital element of any protocol. The Ping-Pong protocol, one of the most widely recognized applications of entanglement for quantum direct communication, has been formulated with perfect quantum channels in mind and noise estimation is not addressed in its design. The control mode of the protocol's seminal version cannot be used for this purpose as it is insensitive to phase flips of a signal qubit which at the same time contribute to errors observed in message mode. As a result, failures of the control mode are not related to errors in the message mode and, in consequence, in noisy environments it is possible to mount undetectable attacks which provide non-zero information gain to the eavesdropper. The control mode improvement which permits correct estimation of errors occurring in message mode is proposed and analyzed. The proposed modification explores the fact that local measurements in mutually unbiased bases reveal the coherence loss between distant components of the entangled system.

Keywords: Ping-Pong protocol, quantum direct communication, quantum noise.

1 Introduction

Quantum Direct Communication (QDC) protocols, which constitute a significant conceptual progress compared to Quantum Key Distribution (QKD) as they do not require, at least in theory, to cooperate with classic algorithms and their security solely results from the laws of quantum mechanics [1]. The Ping-Pong protocol [2] is one of the most widely recognized applications of entanglement for QDC. The seminal version of the protocol [3] and many of its modifications to higher dimensional systems [4,5] have been proven to be asymptotically secure in lossless channels [3,6,7]. The theoretical success of the protocol has been closely followed by the experimental implementation and the proof on concept installation has been realized in the laboratory [8].

The protocol operates in message mode or control mode. The former is used for secure communication while the latter serves for eavesdropping detection.

A. Kwiecień, P. Gaj, and P. Stera (Eds.): CN 2014, CCIS 431, pp. 197–204, 2014.
© Springer International Publishing Switzerland 2014

However, the mentioned above analyses have been carried out under assumption of perfect quantum channel. Unfortunately, the existence of the noise is rather a rule than an exception in quantum world and its exclusion from the analysis seems to be an oversimplification. If losses are permitted then eavesdropper can construct a quantum circuit which provides him non-zero information but remains undetectable by the protocol's control mode [9]. The problem of protocol adaptation to its operation in such scenario has been addressed in [10,11], where its cooperation with properly designed classic layer has been proposed. The security of the aforementioned layer heavily depends on quantum bit error rate ($QBER$) perceived by communicating parties in message mode. It has been shown therein that noise level estimation is the key factor in provision of high security margin and the protocol can be almost as secure as in perfect channels as long as quantum bit error rate ($QBER$) in message mode is determined reliably. However, the methods of $QBER$ estimation have not been considered in [10,11]. Some step in this direction has been taken in [12] where the protocol description has been ported to formalism adequate for its analysis in noisy environments. As an exemplification of the method usefulness it has been shown that control mode proposed in seminal version of the protocol is unable to detect phase flips of the signal qubit. As a result, the protocol is in some sense self contradictory. Information is encoded as phase flip of a signal qubit in message mode, but control mode is unable to detect errors of this kind. Similarly, control mode detects bit flip errors but message mode is immune to such errors. In effect, there is no correlation between probability of errors observed in control mode and the ones perceived in message mode.

An improvement of the control mode, which overcomes the above mentioned difficulties, is introduced and analyzed in the following paper. It is based on techniques exploiting features of mutually unbiased bases [13] and it remedies deficiencies of the seminal version of the protocol. The control mode of the modified version reliably determines $QBER$ observed in message mode so it provides means for estimation of protocol's security margin in noisy environment. In this paper only the aspect of correct $QBER$ estimation is discussed and the detailed analysis of the modified protocol resistance to already known attacks is left for another study.

2   Ping-Pong Protocol Operation over Noisy Channel

In Ping-Pong protocol, the communication process is started by Bob, the recipient of information, who prepares an EPR pair. Without loss of generality it may be assumed that it has the form

$$|\phi^+\rangle = (|0_B\rangle|0_A\rangle + |1_B\rangle|1_A\rangle)/\sqrt{2} \ . \tag{1}$$

Next Bob sends a signal qubit A to Alice. This qubit on its way can be influenced by two factors: quantum noise because of channel imperfection and/or malicious activities of Eve who may entangle it with the system controlled by herself.

However, from the perspective of legitimate parties, the actions of Eve are indistinguishable from the environmental noise. The density matrix of the system just before signal qubit enters environment controlled by Alice reads

$$\rho_{BA}^{(1)} = \mathcal{N}_{BA}\, \rho_{BA}^{(0)} \ , \tag{2}$$

where operator $\mathcal{N}_{BA}$ describes the action of a noisy channel. At that point of protocol execution Alice can select control or message mode. In the latter case Alice encodes classic bit μ applying ($\mu = 1$) or not ($\mu = 0$) operator Z_A to the possessed qubit. Due to entanglement this local operation has global effects. If channel is perfect then the logical "1" is encoded as the state

$$|\phi^-\rangle = (|0_B\rangle|0_A\rangle - |1_B\rangle|1_A\rangle)/\sqrt{2} \ . \tag{3}$$

In imperfect channel the system state after encoding is given by

$$\rho_{\mu\, BA}^{(2)} = (\mathcal{I}_B \otimes Z_A^\mu)\, \rho_{BA}^{(1)} \left(\mathcal{I}_B \otimes (Z_A^\mu)^\dagger\right) \ . \tag{4}$$

The qubit A is sent back to Bob after encoding operation. It is again influenced by the noise and/or Eve

$$\rho_{\mu\, BA}^{(3)} = \mathcal{N}_{BA}\, \rho_{\mu\, BA}^{(2)} \ , \tag{5}$$

and Bob's task is to distinguish between the states $\rho_0{}_{BA}^{(3)}$ and $\rho_1{}_{BA}^{(3)}$. The bit error rate is equal to [14]

$$QBER = \frac{1}{2}\left(1 - \frac{1}{2}\mathrm{Tr}\left(\left|\rho_0{}_{BA}^{(3)} - \rho_1{}_{BA}^{(3)}\right|\right)\right) \tag{6}$$

when he applies the minimum error discrimination as the reception strategy. As the Eve actions are indistinguishable from noise, the reliable estimation of $QBER$ before the transfer of any sensitive information is the only way to verify that protocol meets security requirements [10,11]. The control mode is supposed to serve this purpose. In this mode Alice and Bob perform local control measurements on the possessed qubits and check the correlation imposed by the initial state (1). It will be shown further that control mode proposed in the protocol's seminal version fails to complete this task. To exemplify the failure let us consider its operation over dephasing and bit flip channels. The dephasing channel for the single qubit is described by the Kraus operators

$$E_0 = \begin{bmatrix} 1 & 0 \\ 0 & \sqrt{r} \end{bmatrix} \ , \quad E_1 = \begin{bmatrix} 0 & 0 \\ 0 & \sqrt{1-r} \end{bmatrix} \tag{7}$$

where r describes channel reliability i.e. $r = 1$ corresponds to perfect channel. Similarly, the bit flip channel can be described as operators

$$E_0 = \sqrt{r}\mathcal{I} \ , \quad E_1 = \sqrt{1-r}\mathcal{X} \tag{8}$$

and for $r = 1$ it is just a perfect channel while for $r = 0$ the signal qubit is always flipped. However, the quantum channel in the Ping-Pong protocol, is composed of two qubits, from which only the travel qubit is affected by the noise. In such situation the Kraus operators can be obtained by simple extension [12,13]

$$E_{BAk} = \mathcal{I}_B \otimes E_k \tag{9}$$

and

$$\mathcal{N}_{BA}\,\rho_{BA} = \sum_k E_{BAk}\,\rho_{BA}\,E_{BAk}^\dagger \ . \tag{10}$$

The von Neumann control measurements in computational basis are described by the projectors $M_{\alpha,A} = \mathcal{I}_B \otimes |z_{\alpha,A}\rangle\langle z_{\alpha,A}|$, $\alpha = 0,1$ and $|z_\alpha\rangle$ is an eigenvector of the $\mathcal{Z}$ operator.

Alice and Bob perform control measurements in computational basis and the probability of error in control mode should be related to error rate in message mode. Probability that Alice finds qubit under investigation in state $|z_\alpha\rangle$ (measures ± 1) is given by

$$p_A\,(\alpha) = \mathrm{Tr}\left(\rho_{BA}^{(1)} M_{\alpha,A}\right) \ . \tag{11}$$

After measurement the state of the whole system collapses to

$$\sigma_{\alpha BA}^{(1)} = \frac{M_{\alpha,A}\,\rho_{BA}^{(1)}\,M_{\alpha,A}}{\mathrm{Tr}\left(\rho_{BA}^{(1)}\,M_{\alpha,A}\right)} \ . \tag{12}$$

Subsequently Bob measures his qubit in computational basis using projectors $M_{\beta,B} = |z_{\beta,B}\rangle\langle z_{\beta,B}| \otimes \mathcal{I}_A$, $\beta = 0,1$. Probability that Bob finds his qubit in state $|z_\beta\rangle$ provided that Alice has found his qubit in state $|z_\alpha\rangle$ is given by

$$p_{B|A}\,(\beta|\alpha) = \mathrm{Tr}\left(\sigma_{\alpha BA}^{(1)} M_{\beta,B}\right) \ . \tag{13}$$

It follows that errors in control mode appear with probability

$$P_{EC} = p_{B|A}\,(1|0)\,p_A\,(0) + p_{B|A}\,(0|1)\,p_A\,(1) \ . \tag{14}$$

Probability of error in control mode P_{EC} and bit error rate $QBER$ in message mode as a functions of reliability r for dephasing and bit flip channels are shown on Fig. 1. It follows that control mode is insensitive to dephasing errors but exactly that kind of noise induces errors in message mode. The opposite is also true for bitflip channel – message mode is immune to that kind of noise while the probability of error in control mode is non zero. Such behavior is a direct consequence of control mode definition as it checks only for bit flip errors. On the other hand, information is encoded as the change of the relative phase of the EPR pair components (see Equation (3)). As a result, seminal version of the Ping-Pong protocol suffers some kind of idiosyncrasy because control mode detects only errors which are irrelevant to information transfer in message mode. In the next section it will be shown that simple change in control mode operation can significantly improve protocol features and remedy its incompatible sensitivity to noise.

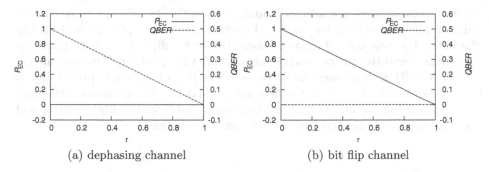

(a) dephasing channel   (b) bit flip channel

Fig. 1. Probability of error in control mode (P_{EC}) and bit error rate ($QBER$) in message mode as a functions of reliability r for dephasing (on the left) and bit flip (on the right) channels

3   Error Rate Estimation with Improved Control Mode

Information transfer in message mode is based on non locality of entanglement. Alice by the local application of the $\mathcal{Z}_A$ operator to the possessed qubit changes the relative phase of the EPR pair components. Such change can be detected only by the collective measurement on both qubits, thus the signal qubit has to be sent back to Bob for information decoding. In fact, this feature is the foundation of the Ping-Pong protocol security — an eavesdropper cannot draw any conclusions about Alice's operation by observing the signal qubit only. But the control mode assumes local measurements of Alice and Bob. How can those measurements reveal a coherence loss between components of an EPR pair? Although it is impossible for the single copy of the quantum state, the coherence change can be statistically verified by the series of measurements on different copies of the quantum state. In particular, the following modification of the control mode is proposed. Instead of using a computational basis only formed by the eigenvectors of $\mathcal{Z}$ operator, let the Alice and Bob execute control measurements in two mutually unbiased bases spanned by the eigenvectors of $\mathcal{X}$ and $\mathcal{Y}$ operators and let the control cycles executed in those bases be selected with equal probability. Those bases are related to the computational basis with the relations

$$|x_k\rangle = \left(|z_0\rangle + (-1)^k |z_1\rangle\right)/\sqrt{2} \ , \tag{15}$$

$$|y_k\rangle = \left(|z_0\rangle - (-1)^k i |z_1\rangle\right)/\sqrt{2} \ . \tag{16}$$

The Alice's and Bob's operators of control measurements will take the form

$$M_{\alpha,A}^{(X)} = \mathcal{I}_B \otimes |x_{\alpha,A}\rangle\langle x_{\alpha,A}| \ , \quad M_{\alpha,A}^{(Y)} = \mathcal{I}_B \otimes |y_{\alpha,A}\rangle\langle y_{\alpha,A}| \ , \tag{17}$$

$$M_{\beta,B}^{(X)} = |x_{\beta,B}\rangle\langle x_{\beta,B}| \otimes \mathcal{I}_A \ , \quad M_{\beta,B}^{(Y)} = |y_{\beta,B}\rangle\langle y_{\beta,B}| \otimes \mathcal{I}_A \ . \tag{18}$$

In each of the bases corresponding probability of error can be found as in (14) and the aggregate probability of error in control mode is given by

$$P_{EC} = (P_{EC}^{(X)} + P_{EC}^{(Y)})/2 \tag{19}$$

because control bases are selected equally frequently. The Figure 2 presents results of P_{EC} calculation for improved control mode for dephasing and bit flip channels. The new control mode is insensitive to bit flips and detects phase errors, thus is sensitive to errors which also disturb the operation of message mode. Similarly $QBER$ in message mode (6) as a function of aggregate probability of error in control mode (19) for the dephasing and depolarizing quantum channels is shown on Fig. 3. The curves are indistinguishable as the both control and message mode are now sensitive to the same kind of noise.

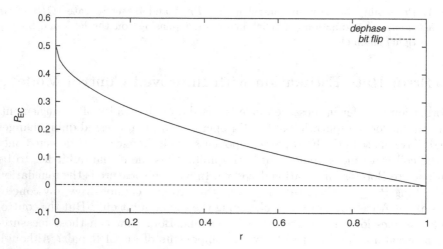

Fig. 2. Probability of error in the improved control mode (P_{EC} from (19)) as a function of reliability r for dephasing and bit flip channels

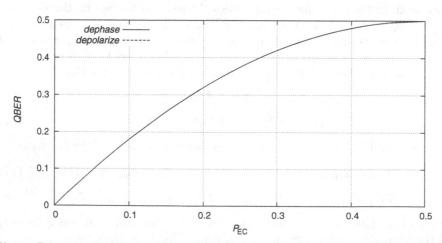

Fig. 3. Bit error rate in message mode as a function of probability of error in improved control mode for the dephasing and depolarizing channels

4   Conclusion

The purpose of control mode in the Ping-Pong protocol is twofold. It is used to verify whether eavesdropping Eve is on the line and that way implicitly transfer authentication from the classic control channel to the quantum one. In noisy environments it should also correctly estimate the expected $QBER$ in the message mode to verify if the communication conditions meet security requirements. The control mode of the protocol's seminal version tests the occurrence of only one kind of errors which are also incompatible with the way the information is coded in the message mode. As a result, such defined control mode fails to achieve both tasks – it is possible to mount undetectable attack [9,15] which returns nonzero information to the eavesdropper and it is impossible to correlate errors in control and message modes.

It is shown that inclusion of mutually unbiased bases in control mode permits correct $QBER$ estimation in message mode. Such modification also allows on detection of attacks considered so far as undetectable, but the detailed discussion of this topic is left for the separate text. The version of control mode proposed in Sect. 3 detects only dephasing errors. If legitimate parties are interested in detection of errors of any kind then they should also use a computational basis in control mode. However, the estimation of the expected $QBER$ in message mode should take into account tests performed only in $\mathcal{X}$ and $\mathcal{Y}$ bases. It follows that the control mode based on mutually unbiased bases is capable to fulfill requirements imposed by secure communication demands.

References

1. Long, G.L., Deng, F.G., Wang, C., Li, X.H., Wen, K., Wang, W.Y.: Quantum secure direct communication and deterministic secure quantum communication. Front. Phys. China 2(3), 251–272 (2007)
2. Boström, K., Felbinger, T.: Deterministic secure direct communication using entanglement. Phys. Rev. Lett. 89(18), 187902 (2002)
3. Boström, K., Felbinger, T.: On the security of the ping-pong protocol. Phys. Lett. A 372(22), 3953–3956 (2008)
4. Vasiliu, E.V.: Non-coherent attack on the Ping-Pong protocol with completely entangled pairs of qutrits. Quantum Inf. Process. 10, 189–202 (2011)
5. Zawadzki, P.: Security of Ping-Pong protocol based on pairs of completely entangled qudits. Quantum Inf. Process. 11(6), 1419–1430 (2012)
6. Jahanshahi, S., Bahrampour, A., Zandi, M.H.: Security enhanced direct quantum communication with higher bit-rate. Int. J. Quantum Inf. 11(2), 1350020 (2013)
7. Jahanshahi, S., Bahrampour, A., Zandi, M.H.: Three-particle deterministic secure and high bit-rate direct quantum communication protocol. Quantum Inf. Process. 12(7), 2441–2451 (2013)
8. Ostermeyer, M., Walenta, N.: On the implementation of a deterministic secure coding protocol using polarization entangled photons. Opt. Commun. 281(17), 4540–4544 (2008)
9. Wójcik, A.: Eavesdropping on the Ping-Pong quantum communication protocol. Phys. Rev. Lett. 90(15), 157901 (2003)

10. Zawadzki, P.: Improving security of the Ping-Pong protocol. Quantum Inf. Process. 12(1), 149–155 (2013)
11. Zawadzki, P.: The Ping-Pong protocol with a prior privacy amplification. Int. J. Quantum Inf. 10(3), 1250032 (2012)
12. Zawadzki, P.: An analysis of the ping-pong protocol operation in a noisy quantum channel. In: Kwiecień, A., Gaj, P., Stera, P. (eds.) CN 2013. CCIS, vol. 370, pp. 354–362. Springer, Heidelberg (2013)
13. Zawadzki, P., Puchała, Z., Miszczak, J.: Increasing the security of the Ping-Pong protocol by using many mutually unbiased bases. Quantum Inf. Process. 12(1), 569–575 (2013)
14. Fuchs, C.A., van de Graaf, J.: Cryptographic distinguishability measures for quantum-mechanical states. IEEE Trans. Inform. Theor. (4), 1216–1227 (1999)
15. Pavičić, M.: In quantum direct communication an undetectable eavesdropper can always tell ψ from ϕ Bell states in the message mode. Phys. Rev. A 87, 042326 (2013)

Ping-Pong Protocol Strengthening against Pavičić's Attack

Piotr Zawadzki

Institute of Electronics,
Silesian University of Technology,
Akademicka 16, 44-100 Gliwice, Poland
Piotr.Zawadzki@polsl.pl

Abstract. A quantum circuit providing an undetectable eavesdropping of information encoded by bit flip operations in Ping-Pong protocol has been recently proposed by Pavičić [Phys. Rev. A, vol. 87, pp. 042326, 2013]. A modification of the protocol's control mode is proposed. The introduced improvement remedies deficiencies of the protocol seminal version and permits Pavičić's attack detection with overwhelming probability. The improved version is also immune to famous Wójcik's attack [Phys. Rev. Lett., vol. 90, pp. 157901, 2003]. As a result, the Ping-Pong protocol asymptotic security is restored both in perfect and lossy quantum channels.

Keywords: Ping-Pong protocol, quantum direct communication, quantum cryptography.

1 Introduction

Division of Telecommunication, a part of the Institute of Electronics and Faculty of Automatic Control, Electronics and Computer Science of Silesian University of Technology, for many years has been conducting research on advanced fields of telecommunication engineering [1–5]. Recently, it has been involved in investigation of quantum features of physical systems which can enhance our abilities to compute and communicate. Non-locality and entanglement, the most prominent signatures of non-classicality, form a foundation for developing a new computation technology which exceeds classical limits [6–9] and construction of novel cryptographic tools [10]. Quantum Key Distribution (QKD) protocols provide distribution of unconditionally secure random keys [10] which are subsequently used to protect classical telecommunication links with methods known from classic cryptography. In contrary, Quantum Direct Communication (QDC) protocols do not require prior key agreement and their security directly results from the laws of quantum mechanics [11].

The so called Ping-Pong [12] protocol has attracted a lot of attention as it is asymptotically secure in lossless channels [13]. The protocol has been recently reformulated to higher dimensional systems [14, 15] and security characteristics throughly analyzed [16, 17]. Also a proper joint usage of Ping-Pong protocol

A. Kwiecień, P. Gaj, and P. Stera (Eds.): CN 2014, CCIS 431, pp. 205–212, 2014.

with primitives well known from classic cryptography can further improve its security [18, 19]. The theoretical success of the protocol has been closely followed by the experimental implementation and the proof on concept installation has been realized in the laboratory [20]. The Ping-Pong protocol, similarly to other QDC protocols, operates in two modes: a message mode is designed for information transfer and a control mode is used for eavesdropping detection. However, the situation looks worse in noisy environments when legitimate users tolerate some level of transmission errors and/or losses. If that level is too high compared to the quality of the channel, then an eavesdropper can peek some fraction of signal particles hiding himself behind accepted Quantum Bit Error Rate (QBER) threshold using quantum circuit proposed by Wójcik [21]. Moreover, Pavičić proposed a quantum circuit capable to eavesdrop in undetectable manner also in lossless channels the information encoded via bit flipping [22]. In such situation quantum advantage of superdense coding is suppressed [23] and legitimate parties have to resort to phase flip encoding.

An improvement of the control mode is introduced and analyzed in the following paper. The proposal remedies diffidences of the seminal version which were exploited in the Wójcik's and Pavičić's attacks. Similar techniques for Ping-Pong protocol security improvement have been also proposed in [15, 24] but the attacks based on exploiting vacuum state properties were not addressed therein. The proposed countermeasures are also much simpler than the ones proposed in [17].

The paper is organized as follows. In Section 2 we review basic concepts of Pavičić's attack. In Section 3 a modification of the Ping-Pong protocol is proposed. It is also explained why it permits detection of the aforementioned attacks. Some general remarks are subject of the conclusion from Section 4.

2   Pavičić's Attack Summary

The Pavičić's attack applies to the Ping-Pong protocol variant in which communicating parties take the quantum advantage of superdense coding and transmit two classic bits per qubit transfer. The communication process is started by Bob, the recipient of information, who prepares two maximally entangled qubits. Without loss of generality it may be assumed that this is an Einstein-Podolsky-Rosen (EPR) pair

$$|\psi^-\rangle = (|0_t\rangle|1_h\rangle - |1_t\rangle|0_h\rangle)/\sqrt{2} \ . \tag{1}$$

One of the qubits, denoted as "home", is kept confidential, while the the second one, named the "travel", is sent to Alice via unprotected quantum channel. Alice randomly selects message or control mode. In message mode Alice sends information to Bob, while in control mode communicating parties check for the presence of the eavesdropper. In a former case she encodes a pair of classic bits by the application of one of Pauli operators $\mathcal{Z}$, $\mathcal{X}$, $i\mathcal{Y} = \mathcal{Z}\mathcal{X}$ or identity $\mathcal{I}$ to the received travel qubit. The entanglement of qubits causes that Alice's local operation causes non local effects. The state composed from the home and travel qubits is transformed into another EPR pair

$$\mathcal{Z}|\psi^-\rangle = |\psi^+\rangle = (|0_t\rangle|1_h\rangle + |1_t\rangle|0_h\rangle)/\sqrt{2} \ ,$$
$$\mathcal{X}|\psi^-\rangle = |\phi^-\rangle = (|0_t\rangle|0_h\rangle - |1_t\rangle|1_h\rangle)/\sqrt{2} \ , \qquad (2)$$
$$\mathcal{ZX}|\psi^-\rangle = |\phi^+\rangle = (|0_t\rangle|0_h\rangle + |1_t\rangle|1_h\rangle)/\sqrt{2}$$

or left unchanged. In general, any set of four unitary transformations can be used for encoding as long as they are build around two complementary operators to fully explore the quantum advantage of superdense coding [23]. Next, the travel qubit is sent back to Bob, who performs collective measurement on both qubits. There exists one-to-one correspondence between encoded bits and the state detected by Bob, so Bob can decode information sent by Alice.

Alice switches to the control mode in some randomly selected protocol cycles. In this mode she measures the received travel qubit and the fact of switching is announced via public authenticated classic channel. This is equivalent to the assumption that classic information is available to Eve, but she cannot control its content. Bob subsequently measures the home qubit and reveals the value of measurement outcome. The result of Bob's measurement is fully determined by the value obtained by Alice because of the fragile entanglement of the EPR pair. Any deviation from that correlation indicates the presence of Eve.

Pavičić's has proposed the usage of quantum circuit shown on Fig. 1 for Ping-Pong protocol eavesdropping [22]. The circuit operates on three registers. Register "t" denotes travel qubit on its way forth and back between Alice and Bob. Registers "x" and "y" represent ancilla system controlled by Eve. The action of the circuit is described by the equation

$$\mathcal{Q} = (\mathcal{I}_t \otimes \mathcal{H}_x \otimes \mathcal{H}_y)\,(\mathcal{CNOT}_{tx} \otimes \mathcal{I}_y)\,(\mathcal{CNOT}_{ty} \otimes \mathcal{I}_x) \times$$
$$\times (\mathcal{T}_t \otimes \mathcal{PBS}_{xy})\,(\mathcal{CNOT}_{tx} \otimes \mathcal{I}_y)\,(\mathcal{CNOT}_{ty} \otimes \mathcal{I}_x) \ . \qquad (3)$$

The schematic illustration of the eavesdropping scenario is shown on Fig. 2. Please note that circuit is unitary thus $\mathcal{Q}^\dagger = \mathcal{Q}^{-1}$.

Let us consider in more details the action of the circuit under consideration. Initially Alice and Bob are decoupled from the ancilla. Eve leaves the register "x" empty and sets register "y" to $|0_y\rangle$. Thus the initial state of entire system (protocol's qubits plus ancilla) reads

$$|q_0\rangle = |\psi_{ht}^-\rangle|v_x\rangle|0_y\rangle \qquad (4)$$

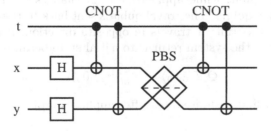

Fig. 1. Circuit proposed by Pavičić in [22] for Ping-Pong protocol eavesdropping

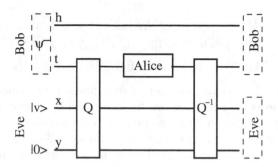

Fig. 2. Schematic illustration of the eavesdropping scenario (Q denotes device from Fig. 1)

where $|v\rangle$ denotes vacuum state. Taking into account (3) one can easily verify that system state when travel qubit arrives at Alice end is described by the expression

$$|q_A\rangle = Q|q_0\rangle = |0_h\rangle|1_t\rangle\frac{|1_x\rangle|v_y\rangle + |v_x\rangle|0_y\rangle}{2} - |1_h\rangle|0_t\rangle\frac{|0_x\rangle|v_y\rangle + |v_x\rangle|1_y\rangle}{2} \ . \quad (5)$$

The special attention is required in analysis of the Polarization Beam Splitter (PBS). It is assumed that state $|1\rangle$ is reflected and $|0\rangle$ transmitted

$$\begin{aligned}
\mathcal{PBS}_{xy}|v_x\rangle|0_y\rangle &= |0_x\rangle|v_y\rangle \ , \\
\mathcal{PBS}_{xy}|v_x\rangle|1_y\rangle &= |v_x\rangle|1_y\rangle \ , \\
\mathcal{PBS}_{xy}|0_x\rangle|v_y\rangle &= |v_x\rangle|0_y\rangle \ , \\
\mathcal{PBS}_{xy}|1_x\rangle|v_y\rangle &= |1_x\rangle|v_y\rangle \ .
\end{aligned} \quad (6)$$

It is equivalent to the assumption that $|1\rangle$ is encoded as photons with vertical polarization and $|0\rangle$ as horizontally polarized ones.

If Alice selects control mode then she measures travel qubit in computational basis. It is clear that (5) preserves anticorrelation required for successful protocol execution. As a result, Alice and Bob do not detect the presence of Eve's circuit.

If Alice decides to encode information then she does nothing or she applies one of the encoding operations (2). Application of $\mathcal{X}$ is equivalent to bit flipping of the travel qubit, while application of $\mathcal{Z}$ changes the relative phase of the travel and home qubits. The travel qubit is sent back to Bob. It enters again Eve's circuit, but this time it travels in opposite direction. If Alice applies no transformation then the system returns to initial state because $Q^\dagger\mathcal{I}Q = \mathcal{I}$

$$Q^\dagger\mathcal{I}Q|q_0\rangle = |\psi_{ht}^-\rangle|v_x\rangle|0_y\rangle \ . \quad (7)$$

Similarly, one can find system states after application of phase flip or bit flip operations

$$Q^\dagger\mathcal{Z}Q|q_0\rangle = |\psi_{ht}^+\rangle|v_x\rangle|0_y\rangle \ , \quad (8)$$

$$\mathcal{Q}^\dagger \mathcal{X} \mathcal{Q} |q_0\rangle = |\phi_{ht}^-\rangle |0_x\rangle |v_y\rangle \ , \tag{9}$$

$$\mathcal{Q}^\dagger \mathcal{Z} \mathcal{X} \mathcal{Q} |q_0\rangle = |\phi_{ht}^+\rangle |0_x\rangle |v_y\rangle \ . \tag{10}$$

It follows that Eve observes click in "x" register when $\mathcal{X}$ has been used in encoding or click in "y" register otherwise. Moreover, in each case, travel and home qubits are decoupled from Eve's ancilla. As a consequence, the presence of circuit is not detected by the control mode and introduces no errors in message mode.

3   Control Mode Improvement

A control mode improvement which removes above mentioned deficiency is presented in this section. Let us closely analyze expression (5) on system's global state at Alice end. It is clear that classic correlation of measurements' outcomes is preserved. But pairs $|0_t\rangle|1_h\rangle$ and $|1_t\rangle|0_h\rangle$ has lost their coherence as a result of Eve eavesdropping. Fortunately, coherence preservation can be checked with local measurements executed in mutually unbiased bases. Let us take into account dual basis

$$|+\rangle = \frac{|0\rangle + |1\rangle}{\sqrt{2}} \ , \quad |-\rangle = \frac{|0\rangle - |1\rangle}{\sqrt{2}} \ . \tag{11}$$

We have

$$|0_h\rangle|1_t\rangle = \frac{1}{2}\left[|+_h\rangle|+_t\rangle - |+_h\rangle|-_t\rangle + |-_h\rangle|+_t\rangle - |-_h\rangle|-_t\rangle\right] \ , \tag{12}$$

$$|1_h\rangle|0_t\rangle = \frac{1}{2}\left[|+_h\rangle|+_t\rangle + |+_h\rangle|-_t\rangle - |-_h\rangle|+_t\rangle - |-_h\rangle|-_t\rangle\right] \ , \tag{13}$$

and

$$|\psi^-\rangle = \left(|0_t\rangle|1_h\rangle - |1_t\rangle|0_h\rangle\right)/\sqrt{2} = \left(|-_t\rangle|+_h\rangle - |+_t\rangle|-_h\rangle\right)/\sqrt{2} \ . \tag{14}$$

It follows that anticorrelation between Alice's and Bob's measurements outcomes is preserved also in case of control mode executed in dual basis as long there is no eavesdropping. Let us investigate state (5) behavior under basis change

$$
\begin{aligned}
|q_A\rangle = \ &|+_h\rangle|+_t\rangle \frac{|1_x\rangle|v_y\rangle + |v_x\rangle|0_y\rangle - |0_x\rangle|v_y\rangle - |v_x\rangle|1_y\rangle}{4} - \\
&- |+_h\rangle|-_t\rangle \frac{|1_x\rangle|v_y\rangle + |v_x\rangle|0_y\rangle + |0_x\rangle|v_y\rangle + |v_x\rangle|1_y\rangle}{4} + \\
&+ |-_h\rangle|+_t\rangle \frac{|1_x\rangle|v_y\rangle + |v_x\rangle|0_y\rangle + |0_x\rangle|v_y\rangle + |v_x\rangle|1_y\rangle}{4} - \\
&- |-_h\rangle|-_t\rangle \frac{|1_x\rangle|v_y\rangle + |v_x\rangle|0_y\rangle - |0_x\rangle|v_y\rangle - |v_x\rangle|1_y\rangle}{4} \ .
\end{aligned}
\tag{15}
$$

It is clear that control measurement of travel qubit resulting in value "+1" (projection $|+_t\rangle\langle +_t|$) will induce home qubit collapse to states $|\pm_h\rangle$ with equal probability. In effect, Alice and Bob observe 50 % error rate in control mode executed in dual basis when travel qubit passes through eavesdropping circuit. Thus to detect eavesdropping Alice should interleave control measurements executed in computational basis with the ones performed in the dual basis. As a result, the presence of eavesdropping circuit is detected with 25 % probability, what restores Ping-Pong protocol asymptotic security. The single cycle of the modified version of the Ping-Pong protocol is described by the following steps.

1. Bob creates an EPR pair $|\psi^-\rangle$ and sends one of its qubits to Alice.
2. Alice with some probability selects control mode or otherwise continues in message mode. If control mode is selected she goes to point 5.
3. Alice encodes information using one of unitary transformations (2) and then she sends signal qubit back to Bob.
4. Bob makes collective measurement on both qubits to identify encoding operation. This finishes message mode cycle.
5. Alice randomly selects computational or dual basis and measures received qubit in the selected basis.
6. She announces to Bob that control mode is activated and measurement basis.
7. Bob measures home qubit in a basis imposed by Alice and returns the outcome to Alice.
8. Alice checks for anticorrelation. This finishes control mode cycle.

The measurement basis selection in point 5 is the only postulated modification compared to the seminal version of the protocol.

The proposed improvement also permits detection of Wójcik's attack [21]. In this attack eavesdropper is able to detect phase flip encoding at the price of losses observed by Alice and Bob. Expression (16) describes the system state (legitimate qubits plus ancilla) when the travel qubit reaches Alice and Eve's eavesdropping circuit is enabled [21, Equation (4)]

$$|q_A\rangle = \frac{1}{2}|0_h\rangle|v_t\rangle|1_x\rangle|0_y\rangle + \frac{1}{2}|1_h\rangle|v_t\rangle|0_x\rangle|1_y\rangle +$$
$$+ \frac{1}{2}|0_h\rangle|1_t\rangle|1_x\rangle|v_y\rangle + \frac{1}{2}|1_h\rangle|0_t\rangle|0_x\rangle|v_y\rangle \ . \tag{16}$$

The first two terms are responsible for losses observed by Alice, while the last two preserve the correlation of outcomes of control measurements although home and travel qubits are coupled with the ancilla registers. However, after transformation to dual basis the above expression takes the form

$$|q_A\rangle = \frac{1}{2\sqrt{2}}\left(|+_h\rangle + |-_h\rangle\right)|v_t\rangle|1_x\rangle|0_y\rangle + \frac{1}{2\sqrt{2}}\left(|+_h\rangle - |-_h\rangle\right)|v_t\rangle|0_x\rangle|1_y\rangle +$$
$$+ \frac{1}{4}|+_h\rangle|+_t\rangle\left(|1_x\rangle|v_y\rangle + |0_x\rangle|v_y\rangle\right) +$$
$$+ \frac{1}{4}|+_h\rangle|-_t\rangle\left(-|1_x\rangle|v_y\rangle + |0_x\rangle|v_y\rangle\right) +$$
$$+ \frac{1}{4}|-_h\rangle|+_t\rangle\left(|1_x\rangle|v_y\rangle - |0_x\rangle|v_y\rangle\right) +$$
$$+ \frac{1}{4}|-_h\rangle|-_t\rangle\left(-|1_x\rangle|v_y\rangle - |0_x\rangle|v_y\rangle\right) \ . \tag{17}$$

It follows, that the home qubit collapses to states $|+_h\rangle$ and $|-_h\rangle$ with equal probability independent on the outcome of the measurement in dual basis of a nonempty travel qubit. Thus similarly to the previous case, Alice and Bob will observe 50 % error rate in control modes executed in dual basis.

4 Conclusion

It has been shown that simple modification of the control mode make the Ping-Pong protocol immune to attacks which have been so far recognized as undetectable. Those attacks exploited the fact, that control mode in seminal version of the protocol does not detect coherence loss between travel and home qubits. Fortunately, decoherence can be detected when local projective measurements of entangled subsystems are conducted in mutually unbiased bases. The proposed modification is a direct consequence of this observation.

References

1. Kłosowski, P.: Speech processing application based on phonetics and phonology of the Polish language. In: Kwiecień, A., Gaj, P., Stera, P. (eds.) CN 2010. CCIS, vol. 79, pp. 236–244. Springer, Heidelberg (2010)
2. Kucharczyk, M.: Blind signatures in electronic voting systems. In: Kwiecień, A., Gaj, P., Stera, P. (eds.) CN 2010. CCIS, vol. 79, pp. 349–358. Springer, Heidelberg (2010)
3. Sułek, W.: Pipeline processing in low-density parity-check codes hardware decoder. B. Pol. Acad. Sci.-Tech. 59(2), 149–155 (2011)
4. Dustor, A., Kłosowski, P.: Biometric voice identification based on fuzzy kernel classifier. In: Kwiecień, A., Gaj, P., Stera, P. (eds.) CN 2013. CCIS, vol. 370, pp. 456–465. Springer, Heidelberg (2013)
5. Dziwoki, G., Kucharczyk, M., Sulek, W.: OFDM transmission with non-binary LDPC coding in wireless networks. In: Kwiecień, A., Gaj, P., Stera, P. (eds.) CN 2013. CCIS, vol. 370, pp. 222–231. Springer, Heidelberg (2013)
6. Shor, P.W.: Polynomial-time algorithms for prime factorization and discrete logarithms on a quantum computer. SIAM J. Comput. 26(5), 1484–1509 (1997)
7. Zawadzki, P.: A numerical simulation of quantum factorization success probability. In: Tkacz, E., Kapczynski, A. (eds.) Internet – Technical Development and Applications. AISC, vol. 64, pp. 223–231. Springer, Heidelberg (2009)
8. Zawadzki, P.: A fine estimate of quantum factorization success probability. Int. J. Quantum Inf. 8(8), 1233–1238 (2010)
9. Izydorczyk, J., Izydorczyk, M.: Microprocessor scaling: What limits will hold? IEEE Computer 43(8), 20–26 (2010)
10. Gisin, N., Ribordy, G., Tittel, W., Zbinden, H.: Quantum cryptography. Rev. Mod. Phys. 74, 145–195 (2002)
11. Long, G.L., Deng, F.G., Wang, C., Li, X.H., Wen, K., Wang, W.Y.: Quantum secure direct communication and deterministic secure quantum communication. Front. Phys. China 2(3), 251–272 (2007)
12. Boström, K., Felbinger, T.: Deterministic secure direct communication using entanglement. Phys. Rev. Lett. 89(18), 187902 (2002)
13. Boström, K., Felbinger, T.: On the security of the ping-pong protocol. Phys. Lett. A 372(22), 3953–3956 (2008)
14. Vasiliu, E.V.: Non-coherent attack on the Ping-Pong protocol with completely entangled pairs of qutrits. Quantum Inf. Process. 10, 189–202 (2011)
15. Zawadzki, P.: Security of Ping-Pong protocol based on pairs of completely entangled qudits. Quantum Inf. Process. 11(6), 1419–1430 (2012)

16. Jahanshahi, S., Bahrampour, A., Zandi, M.H.: Security enhanced direct quantum communication with higher bit-rate. Int. J. Quantum Inf. 11(2), 1350020 (2013)
17. Jahanshahi, S., Bahrampour, A., Zandi, M.H.: Three-particle deterministic secure and high bit-rate direct quantum communication protocol. Quantum Inf. Process. 12(7), 2441–2451 (2013)
18. Zawadzki, P.: Improving security of the Ping-Pong protocol. Quantum Inf. Process. 12(1), 149–155 (2013)
19. Zawadzki, P.: The Ping-Pong protocol with a prior privacy amplification. Int. J. Quantum Inf. 10(3), 1250032 (2012)
20. Ostermeyer, M., Walenta, N.: On the implementation of a deterministic secure coding protocol using polarization entangled photons. Opt. Commun. 281(17), 4540–4544 (2008)
21. Wójcik, A.: Eavesdropping on the Ping-Pong quantum communication protocol. Phys. Rev. Lett. 90(15), 157901 (2003)
22. Pavičić, M.: In quantum direct communication an undetectable eavesdropper can always tell ψ from ϕ Bell states in the message mode. Phys. Rev. A 87, 042326 (2013)
23. Coles, P.J.: Role of complementarity in superdense coding. Phys. Rev. A 88, 062317 (2013)
24. Zawadzki, P., Puchała, Z., Miszczak, J.: Increasing the security of the Ping-Pong protocol by using many mutually unbiased bases. Quantum Inf. Process. 12(1), 569–575 (2013)

A Numerical Comparison of Diffusion and Fluid-Flow Approximations Used in Modelling Transient States of TCP/IP Networks

Tomasz Nycz[1], Monika Nycz[1], and Tadeusz Czachórski[2]

[1] Institute of Informatics, Silesian University of Technology
Akademicka 16, 44-100 Gliwice, Poland
{tomasz.nycz,monika.nycz}@polsl.pl
[2] Institute of Theoretical and Applied Informatics, Polish Academy of Sciences
Bałtycka 5, 44-100 Gliwice, Poland
tadek@iitis.gliwice.pl

Abstract. The paper presents a comparison of two approaches to queuing models used in performance evaluation of computer networks: diffusion approximation and fluid-flow approximation. Both methods are well known and are used alternatively in similar problems but the differences between their results and the errors they introduce when applied to TCP driven time dependent flows were not sufficiently investigated.

Keywords: diffusion approximation, fluid-flow approximation, computer networks, TCP flows.

1 Introduction

Time dependent flows are typical to computer network transmissions. Therefore queuing models used in network performance evaluation should master transient state analysis. Queues observed at IP routers are dynamically changing and an evaluation of any algorithm of traffic control should involve a study of this dynamics which influences waiting times (hence transmission times) as well as loss probabilities. The both factors are essential for transmission quality of service. However, classical analytical models existing in queueing theory are able to perform the transient analysis only in the simplest cases such as single M/M/1 and M/M/1/N stations (following Kendal's notation) and even there the transient state solutions are quite complex. Moreover, the results refer to transient states but it is assumed that the model parameters, the input rate in particular, are constant. We may develop more complex Markovian models where exponential distributions of interarrival and service times are replaced by linear combinations of exponentially distributed phases, e.g. by Cox distribution. There are tools able to match a Cox distribution to any empirical histogram, e.g. [1]. This way we may adapt Markovian models to flows and to packet sizes observed in real networks but the number of states, i.e. the number of differential equations (Chapman-Kolmogorov equations) to be solved numerically in

A. Kwiecień, P. Gaj, and P. Stera (Eds.): CN 2014, CCIS 431, pp. 213–222, 2014.

the resulting Markovian model grows rapidly. If we include in the model also transmission rules, e.g. a Markovian model of TCP congestion window, soon we attain the reasonable computational limits. We are using this approach and develop a numerical tool (packet OLYMP) [2] which is able to solve large models having hundreds of millions of states but we feel that this approach, although exact from mathematical point of view, flexible and general, is not suitable for typical to Internet complex topologies and numerous flows. It is too complex because, like simulations, it is modelling events on the level of packets. Therefore we investigate here two approximations which are modelling networks behaviour on a flow level. There are known and being used but their built-in errors that may be visible in performance evaluation of TCP/IP network nodes are not sufficiently investigated.

2   Fluid-Flow Approximation

In this method, already adapted to model Internet transmissions [3–5], only the mean values of flow changes are considered and for this reason it should introduce larger errors than the diffusion approach which is a second order approximation.

The fluid approximation uses first-order ordinary linear differential equations to determine the dynamics of the average length of node queues and the dynamics of TCP congestion windows in a modelled network. The changes of a queue length at a station j, $dq_j(t)/dt$, Equation (1), are defined as the intensity of the input stream, i.e. the sum of all flows $i = 1, \ldots, K$ traversing a particular node, minus the constant intensity of output flow C_j, i.e. the number of packets sent further in a time unit:

$$\frac{dq_j(t)}{dt} = \sum_{i=1}^{K} \frac{W_i(t)}{R_i(\boldsymbol{q}(t))} - 1(q_v(t) > 0) \ C_j \ . \tag{1}$$

A router allows reception of traffic from K TCP flows $(K \leqslant N)$, where N is the entire number of flows in the network. Each flow i $(i = 1, \ldots, N)$ is determined by its time varying congestion window size W_i. The window size, Equation (2), increases by one at each RTT (round trip time) in the absence of a packet loss and decreases by half after every packet loss occurring in nodes on the flow path. The amount of loss for the entire TCP connection is defined as flow throughput intensity multiplied by total drop probability – the probability which specifies that the loss occurs in nodes on the route. It is based on a matrix $\boldsymbol{B}$ that stores drop probabilities in each router in all flows in the network.

$$\frac{dW_i(t)}{dt} = \frac{1}{R_i(\boldsymbol{q}(t))} - \frac{W_i(t)}{2} \cdot \frac{W_i(t-\tau)}{R_i(\boldsymbol{q}(t-\tau))} \cdot \left(1 - \prod_{j \in V_i}(1 - B_{ij})\right) \ . \tag{2}$$

The values B_{ij} give drop probability *ploss* at node j for packets of connection i; V_i is the set of nodes belonging to this connection, and $q(t)$ is the vector of queues at these nodes. Delays R_i in the above formulas determine the time needed for the information on congestion and packet loss to propagate through the network back to the sender of a flow i, it consists of queue delays at all nodes j, defined as $q_j(t)/C_j$ along this connection and the propagation delay Tp_i:

$$R_i(q(t)) = \sum_{j \in V_i}^{M} \frac{q_j(t)}{C_j} + Tp_i \ . \tag{3}$$

The drop probability $ploss_j(x_j)$, Equation (4), in a single node is determined according to RED mechanism, e.g. [3] as a function on the moving average queue length $x_j(t)$ which is the sum of current queue $q_j(t)$ taken with a weight parameter w and previous average queue taken with $(1 - w)$ weight parameter,

$$ploss_j(x_j) = \begin{cases} 0 \ , & 0 \leqslant x_j < t_{\min_j} \\ \dfrac{x_j - t_{\min_j}}{t_{\max_j} - t_{\min_j}} p_{\max_j} \ , & t_{\min_j} \leqslant x_j < t_{\max_j} \\ 1 \ , & t_{\max_j} \leqslant x_j \ . \end{cases} \tag{4}$$

We used already fluid flow approximation to investigate other than RED algorithms in router queues, [6–8].

3   Diffusion Approximation

In this approach, proposed by [9], the queue length distributions are given by a diffusion process, the density function $f(x, t; x_0)$ of which is defined by Equation (5):

$$\frac{\partial f(x, t; x_0)}{\partial t} = \frac{\alpha}{2} \frac{\partial^2 f(x, t; x_0)}{\partial x^2} - \beta \frac{\partial f(x, t; x_0)}{\partial x} \ . \tag{5}$$

The diffusion parameters α and β reflect the characteristics of input stream, service time distribution, and queueing discipline.

In a single node with FIFO queue where $A(x)$ is the distribution of interarrival times, $B(x)$ is the distribution of service time, the parameters are selected as, [9]

$$\beta = \lambda - \mu \ , \quad \alpha = \sigma_A^2 \lambda^3 + \sigma_B^2 \mu^3 = C_A^2 \lambda + C_B^2 \mu \ ,$$

where mean values are denoted $E[A] = 1/\lambda$, $E[B] = 1/\mu$, the variances are $Var[A] = \sigma_A^2$, $Var[B] = \sigma_B^2$, and squared coefficients of variation are $C_A^2 = \sigma_A^2 \lambda^2$, $C_B^2 = \sigma_B^2 \mu^2$.

Diffusion equation is supplemented with balance equations for probabilities $p_0(t)$ and $p_N(t)$ of the process being at the barriers $x = 0$ (queue is empty) and

$x = N$ (queue is full), the process leaves these states by jumps from $x = 0$ to $x = 1$ and from $x = N$ to $x = N - 1$:

$$\frac{\partial f(x, t; x_0)}{\partial t} = \frac{\alpha}{2} \frac{\partial^2 f(x, t; x_0)}{\partial x^2} - \beta \frac{\partial f(x, t; x_0)}{\partial x} +$$
$$+ \lambda_0 p_0(t) \delta(x - 1) + \lambda_N p_N(t) \delta(x - N + 1) ,$$
$$\frac{dp_0(t)}{dt} = \lim_{x \to 0} \left[\frac{\alpha}{2} \frac{\partial f(x, t; x_0)}{\partial x} - \beta f(x, t; x_0) \right] - \lambda_0 p_0(t) ,$$
$$\frac{dp_N(t)}{dt} = \lim_{x \to N} \left[-\frac{\alpha}{2} \frac{\partial f(x, t; x_0)}{\partial x} + \beta f(x, t; x_0) \right] - \lambda_N p_N(t) , \qquad (6)$$

where $\delta(x)$ is Dirac delta function.

The applied below solution of these equations is based on the representation of the density function $f(x, t; x_0)$ of the diffusion process with barriers and jumps by a superposition of the density functions $\phi(x, t; x_0)$ of diffusion processes with absorbing barriers at $x = 0$ and $x = N$, as described in detail in [10]. The model equations are subject to Laplace transform. The transform of the density function $f(x, t; x_0)$ is determined analytically and then its original is computed numerically using Stehfest algorithm [11]. This approach is valid when transient states are considered in a model with constant parameters. If the parameters are changing with time, e.g. in presence of time-dependent input flow, the time axis is divided into small intervals inside which the parameters are considered constant (hence they are piecewise constant in the model) and at each interval the solution is applied with the initial conditions taken from the solution obtained at the end of precedent interval.

This approach gives satisfactory numerical results and does not increase the errors of the method, e.g. [12]. The change of the parameters may reflect with great flexibility the changes of flows due to traffic control mechanisms, e.g. introduced by TCP an RED algorithms.

For example, in the analysis of the bottleneck node we may modify the diffusion equations in the following way. In the diffusion model we cannot distinguish the moments of packet arrivals when the moving average – which is the basis to determine the probability of dropping a packet in RED mechanism – should be computed, but we may compute it in intervals corresponding to the current mean interarrival time, $\Delta = 1/\lambda(t)$. The traffic intensity λ remains fixed within an interval. When λ changes, also the length $1/\lambda$ of the interval is changed.

At the beginning of an interval i we compute the mean value of the queue: $E[N_i] = 1 + \sum_{n=1}^{N} p_i(n)n$, the weighted mean

$$x_i = (1 - w)x_{i-1}p_i(0) + [1 - p_i(0)] [(1 - w)x_{i-1} + wE[N_i]]$$

and then we determine on this basis, using the drop function $ploss(x)$ of RED, the probability of deleting the packet to announce the congestion.

We know the queue distribution $f(x, t; x_0)$ at the beginning of an interval. This distribution also gives us the distribution of the distribution $r(x, t)$ of response time of this queue. This gives us means to compute current RTT inside the considered connection, $RTT = E[N_i] \cdot 1/\mu + Tp$, where Tp represents all other delays outside the bottleneck router. Following it, we determine the changes of traffic sent by the source and applied with delay of current RTT :

$$\lambda_{\text{new}} = [1 - ploss(x)] (\lambda_{\text{old}} + \Delta\lambda) + \frac{ploss(x)\lambda_{\text{old}}}{2}$$

Similarly as in fluid flow approximation, the equation takes into account both possibilities: the increase and decrease of emitted flow with probabilities determined by RED algorithm. The increase of flow is additive with $\Delta\lambda$ and the decrease of flow is multiplicative with constant $1/2$.

4   Comparison of Methods

We have already reported the ability to build and compute large network models, containing hundreds of thousands nodes in case of fluid flow and thousands of nodes in case of diffusion approximation, [12]. Here, we compare in a simple one-node model the errors of both approaches. The examples differ in the choice of buffer length, input stream, and RED parameters. The numerical comparisons were conducted in a single node in two phases. The first one (Example 1) included the analysis of queue length given by the both methods when the traffic was a predefined function of time, the same for both methods, and the second one (Examples 2 and 3) included the elements of congestion window size mechanism. The results of the approximations were compared to simulations obtained with the use of OMNET++ package [13] adapted by us to simulate transient states (automatic repetition of simulation runs 100 000 times, collection of histograms for a set of defined time moments, parameters of random number generators defined as functions of time). The number of simulation repetitions was sufficient to consider its results as almost accurate, the confidence intervals were negligible compared to the values of obtained queue lengths.

Example 1. The considered time-dependent input flow is presented in Fig. 1. The classical diffusion approximation model determines a router queue with the drop-tail (passive) algorithm. To compare this approach with the fluid flow approximation, we decided to disable the window mechanism in fluid flow node and by the choice of RED parameters – thresholds $t_{\min} = 0$ packets, $t_{\max} = 20$ packets and $p_{\max} = 0$. The buffer length in both cases was set to 20 packets and the service intensity was $\mu = C = 1.5$ packets per time-unit.

The mean queue lengths given by approximation and simulation are displayed in Fig. 2. The diffusion approach is much more accurate than fluid flow and its results are practically the same as simulations, whereas in fluid flow only the general tendency of the queue changes is preserved.

Example 2. The fluid flow approximation assumes that the input traffic at a node is defined by congestion window size and RTT time of the flow which traverses that node, whereas in the diffusion model it is given explicitly. To compare the both models, the time-dependent input stream as shown in Fig. 3 obtained

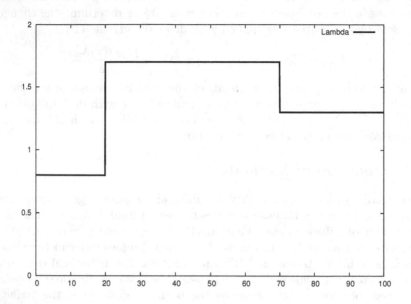

Fig. 1. *Example 1: Input stream [pac/s] as a function of time – no window mechanism*

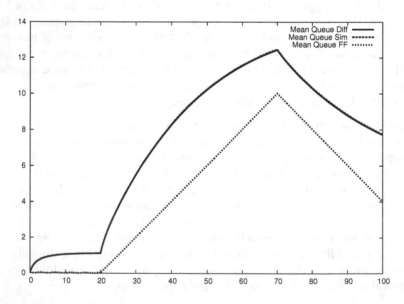

Fig. 2. *Example 1*: Comparison of the mean queue [pac] at a single node as a function of time for both approximations and simulation; results in case of disabled window mechanism

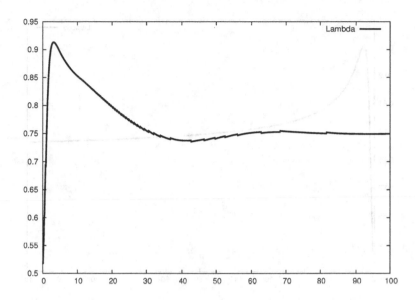

Fig. 3. *Example 2*: Input stream [pac/s] as a function of time for both models in case of window mechanism

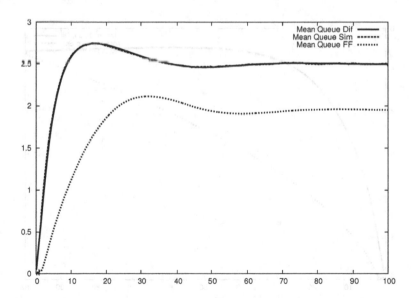

Fig. 4. *Example 2*: Comparison of the mean queues [pac] as a function of time for both approximations and simulations results in case of window mechanism

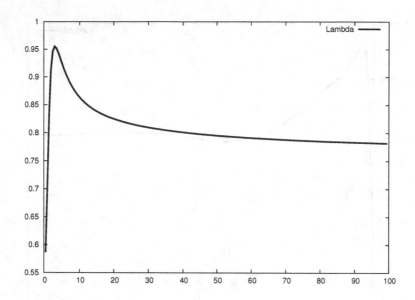

Fig. 5. *Example 3*: Input stream [pac/s] as a function of time for both approximations in case of Drop Tail algorithm

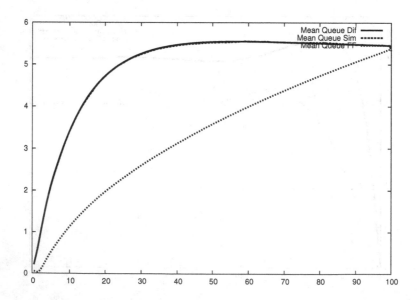

Fig. 6. *Example 3*: Comparison of the mean queues [pac] as a function of time for both approximations with simulation results in case of Drop Tail algorithm

from the fluid flow model where it was defined by the congestion window, was also given as the input flow to the diffusion model. The service intensity, C in fluid flow and μ in diffusion model, were set to the same value of 0.75 packets per time unit. The buffer size was set to 5 packets and RED linear increase range was $t_{min} = 0.75$, $t_{max} = 2.5$ packets, $p_{max} = 0.1$. The inclusion of the congestion window mechanism makes the generated fluid flow characteristics closer to the results obtained by diffusion approximation and the simulation, Fig. 4.

Example 3. in the considered diffusion model the queue management mechanism is Drop Tail algorithm – the packets are dropped only when the queue is full. For this reason we examined one more example where the RED algorithm for fluid flow was extended to the entire range of the buffer containing up to 10 packets. The maximum drop probability was chosen $p_{max} - 0$, $t_{max} = 10$, and the weight parameter w was selected as 1 (only the current queue length was considered), this way the RED algorithm in fact became the Drop Tail. As previously the service intensity was set on 0.75 packets per time unit. Like in previous case, the throughput of fluid flow node was taken also as the diffusion input stream, see Fig. 5. This approach resulted in even higher than in the Example 2 errors of the fluid flow model, Fig. 6.

The obtained results demonstrate that the fluid flow method which is frequently used in modelling because of its simplicity generates much larger errors compared to the diffusion results. However, the diffusion calculations are more time consuming and have also precision limits in its numerical computations that makes impossible to analyse this way networks with large buffers at nodes. Fluid flow model provides only a rough characteristics of network dynamics but can be used to model very large networks.

5 Conclusions

We investigated the limitations of the use of diffusion and fluid flow approximations in the transient analysis of IP router queues in presence of input flow originating from TCP congestion window algorithm. We conclude that the both approaches are useful. Fluid flow approximation generates much larger errors but is very fast and may be applied to large networks. Diffusion approximation is more accurate and may furnish not only the mean values of queues but also their distributions, therefore it is better adapted to estimate the packet losses, but the calculations are more complex. However, the most time-consuming methods – to run and to implement – is the simulation, closely followed by Markov chains. The simulation requires many repetitions to achieve the reliable results, while in Markov chains a huge equation system has to be solved. The choice of a method should depend on the goal of a particular research.

Acknowledgments. This work was supported by Polish project NCN nr 4796/B/T02/2011/40 „Models for transmissions dynamics, congestion control and quality of service in Internet" and the European Union from the European Social Fund (grant agreement number: UDA-POKL.04.01.01-00-106/09).

References

1. Reinecke, P., Krauß, T., Wolter, K.: HyperStar: Phase-Type Fitting Made Easy. In: 9th International Conference on the Quantitative Evaluation of Systems, QEST 2012, pp. 201–202. Tool Presentation (September 2012)
2. Pecka, P., Deorowicz, S., Nowak, M.: Efficient representation of transition matrix in the Markov process modeling of computer networks. In: Czachórski, T., Kozielski, S., Stańczyk, U., et al. (eds.) Man-Machine Interactions 2. AISC, vol. 103, pp. 457–464. Springer, Heidelberg (2011)
3. Misra, V., Gong, W., Towsley, D.: A Fluid-based Analysis of a Network of AQM Routers Supporting TCP Flows with an Application to RED. In: Proceedings of the Conference on Applications, Technologies, Architectures and Protocols for Computer Communication (SIGCOMM 2000), pp. 151–160 (2000)
4. Liu, Y., Lo Presti, F., Misra, V., Gu, Y.: Fluid Models and Solutions for Large-Scale IP Networks. ACM/SigMetrics (2003)
5. Hollot, K., Liu, Y., Misra, V., Towsley, D., Gong, W.B.: Fluid methods for modeling large heterogeneous networks. Tech. report AFRL-IF-RS-TR-2005-282 (2005)
6. Domański, A., Domańska, J., Czachórski, T.: Comparison of CHOKe and gCHOKe active queues management algorithms with the use of fluid flow approximation. In: Kwiecień, A., Gaj, P., Stera, P. (eds.) CN 2013. CCIS, vol. 370, pp. 363–371. Springer, Heidelberg (2013)
7. Domański, A., Domańska, J., Czachórski, T.: Comparison of AQM Control Systems with the Use of Fluid Flow Approximation. In: Kwiecień, A., Gaj, P., Stera, P. (eds.) CN 2012. CCIS, vol. 291, pp. 82–90. Springer, Heidelberg (2012)
8. Domańska, J., Domański, A., Czachórski, T.: Fluid Flow Analysis of RED Algorithm with Modified Weighted Moving Average. In: Dudin, A., Klimenok, V., Tsarenkov, G., Dudin, S. (eds.) BWWQT 2013. CCIS, vol. 356, pp. 50–58. Springer, Heidelberg (2013)
9. Gelenbe, E.: On Approximate Computer Systems Models. J. ACM 22(2) (1975)
10. Czachórski, T.: A method to solve diffusion equation with instantaneous return processes acting as boundary conditions. Bulletin of Polish Academy of Sciences, Technical Sciences 41(4) (1993)
11. Stehfest, H.: Algorithm 368: Numeric inversion of Laplace transform. Comm. of ACM 13(1), 47–49 (1970)
12. Czachórski, T., Nycz, M., Nycz, T., Pekergin, F.: Analytical and numerical means to model transient states in computer networks. In: Kwiecień, A., Gaj, P., Stera, P. (eds.) CN 2013. CCIS, vol. 370, pp. 426–435. Springer, Heidelberg (2013)
13. OMNET++ Community Site, http://www.omnetpp.org

Asymptotic Analysis of Queueing Systems with Finite Buffer Space

Evsey Morozov[1], Ruslana Nekrasova[1], Lyubov Potakhina[1],
and Oleg Tikhonenko[2]

[1] Institute of Applied Mathematical Research, Karelian Research Centre RAS,
Petrozavodsk State University
Lenin 33, 185910, Petrozavodsk, Russia
emorozov@karelia.ru, ruslana.nekrasova@mail.ru, lpotahina@gmail.com
[2] Czestochowa University of Technology, Institute of Mathematics
Al. Armii Krajowej 21, 42-200 Czestochowa, Poland
oleg.tikhonenko@gmail.com

Abstract. In the paper, we consider a single-server loss system in which each customer has both service time and a random volume. The total volume of the customers present in the system is limited by a finite constant (the system's capacity). For this system, we apply renewal theory and regenerative processes to establish a relation which connects the stationary idle probability P_0 with the limiting fraction of the lost volume, Q_{loss}, provided the service time and the volume are proportional. Moreover, we use the inspection paradox to deduce an asymptotic relation between Q_{loss} and the stationary loss probability P_{loss}. An accuracy of this approximation is verified by simulation.

Keywords: queueing system, finite capacity, customer's volume, average lost volume, loss probability, idle probability.

1 Introduction

In the paper, we study a non-conventional queueing loss systems with customers having some random volume (which can be treated as a physical parameter, for instance, space requirement), and service time (for instance, packet transmission time in the router). In general, the service time of a customer depends on his volume. Such systems have been used to model and solve various practical problems occurring in the design of computer and communication systems [1,2].

We consider a single-server $GI/G/1$ queueing system with FIFO service discipline. The system has infinite buffer for the number of the customers, while the buffer space for the total volume of customers is limited by a finite constant (system's capacity) $M > 0$. It is assumed that the n-th customer ($n \geq 1$) is characterized by the pair (S_n, V_n), where S_n is his service time and V_n is his volume. The vectors (S_n, V_n) are assumed to be independent identically distributed (i.i.d.), but, for a given n, a dependence between S_n and V_n may exist. Denote by $\{t_n, n \geq 1\}$ the arrival instants which form the i.i.d (renewal)

A. Kwiecień, P. Gaj, and P. Stera (Eds.): CN 2014, CCIS 431, pp. 223–232, 2014.

sequence of the interarrival times. Let $\mathcal{J}(n)$ be the number of the customers present in the system at time instant t_n^- (just before the n-th arrival). Then the n-th customer is lost, if and only if (see Fig. 1)

$$\sum_{i \in \mathcal{J}(n)} V_i + V_n > M \ . \tag{1}$$

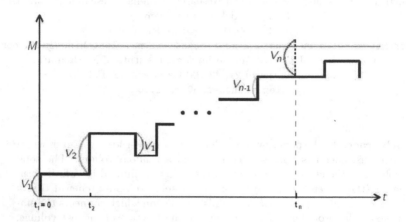

Fig. 1. Lost volume V_n as a renewal interval crossing the threshold M

It is clear that the volumes of the customers present in the system at arbitrary time instant are dependent. Indeed, assume, for instance, that the i-th and the j-th customers are in the system at some moment. Then $V_i + V_j \leq M$, and thus V_i and V_j are dependent.

The first contribution of this work is an adaptation to this system the asymptotic analysis developed in [3] for classical finite buffer system. To this end, we prove (in brief) the basic relation connecting the stationary idle probability P_0 and the stationary loss probability P_{loss} in the classical lost system. Then we adapt the analysis to original system to obtain a relation between P_0 and the limiting fraction of the lost volume, Q_{loss}, when the service time of a customer is proportional to his volume. An explicit expression for Q_{loss} is hardly available (with the exception of a few special cases [2]), and the second contribution of the paper is the obtaining of an approximate relation between Q_{loss} and P_{loss} in a general case. This relation can be used to estimate (by simulation) Q_{loss} via P_{loss}, and vise versa. As a byproduct of our analysis, the inequality $Q_{\text{loss}} \geq P_{\text{loss}}$ is proved when the volumes have *new-worse-than-used* distribution.

This work is organized as follows. In Section 2, we give some previous results on the loss probability and also recall the inspection paradox. Then, in Section 3, we describe the original loss system with the limited volume and find an important relation between Q_{loss} and the stationary idle probability P_0 provided the service time of the customer is proportional to his volume. Moreover, in Section 3, we establish approximate relation between Q_{loss} and P_{loss} using the inspection paradox. In Section 4, we present simulation results which confirm a satisfactory accuracy of the obtained approximation for some values of M, the traffic intensity $\rho := \mathsf{E}S/\mathsf{E}T$ and a few volume distributions (here S, T are the generic service time and interarrival time, respectively). Section 5 contains concluding remarks.

2   Preliminary Results

Consider a $GI/G/1/N$-type loss system with a finite buffer N for the awaiting customers. A difference with classical loss system is that customer n has both service time S_n and the volume V_n, but the space for the volumes is not limited ($M = \infty$). Note that the independence between random variables describing different customers is assumed throughout the paper.

Denote by $A(t)$, $R(t)$ the number of arrivals and the number of the lost customers in the time interval $[0, t]$, respectively, and recall that the service times $\{S_n\}$ are i.i.d. Let $W(t)$ be the remaining workload in the system at time instant t. The following balance equation holds for the described loss system:

$$V(t) = W(t) + L(t) + B(t) \ , \tag{2}$$

where

$$V(t) := \sum_{i=1}^{A(t)} S_i \ , \quad L(t) :=_{\text{st}} \sum_{i=1}^{R(t)} S_i \ , \tag{3}$$

are the arrived workload and the lost workload, respectively in the interval $[0, t]$, and $B(t)$ is the departed (served) workload in this interval. Note that, unlike the first equality, the second equality in (3) is in general *stochastic*, because it is not necessary that exactly first $R(t)$ customers will be lost. A key observation is that the departed workload can be written as $B(t) = t - I(t)$, where $I(t)$ is the empty time of the server in the interval $[0, t]$. Because the system has a finite buffer, then the regenerative workload process $\{W(t)\}$ is *positive recurrent*, that is, the mean regeneration period is finite. As a result, we obtain $W(t) = o(t)$, $t \to \infty$ [4]. (If $\rho \geq 1$ then the *regeneration* condition $\mathsf{P}(T > S) > 0$ must be added to our model to obtain positive recurrence [4].) By the positive recurrence, the following limit exists with probability 1 (w.p.1):

$$\lim_{t\to\infty} \frac{t - I(t)}{t} = 1 - P_0 \ , \tag{4}$$

and, if the interarrival time T is *non-lattice* [5], then the limit $P_0 := \lim_{t\to\infty} \mathsf{P}(W(t) = 0)$ exists and is the *stationary idle probability*. Because the renewal input rate $\lambda := \lim_{t\to\infty} A(t)/t = 1/\mathsf{E}T$, then by Strong Law of Large Numbers (SLLN),

$$\lim_{t\to\infty} \frac{V(t)}{t} = \lambda \mathsf{E}S = \rho \ . \tag{5}$$

Again, using SLLN and the regeneration theory, we obtain that (w.p.1) limit

$$\lim_{t\to\infty} \frac{L(t)}{V(t)} = \lim_{t\to\infty} \frac{R(t)}{A(t)} := P_{\text{loss}} \tag{6}$$

exists and (for non-lattice T) equals the stationary loss probability $\lim_{n\to\infty} P(I_n = 1)$, where the indicator $I_n = 1$, if the n-th customer is lost, and $I_n = 0$, otherwise. Now we deduce from (2) the following relation

$$P_{\text{loss}} = 1 - \frac{1 - P_0}{\rho} . \tag{7}$$

Now, we recall an *inspection paradox* for the renewal process

$$S_n := V_1 + \cdots + V_n , \quad n \geq 1 ,$$

generated by the i.i.d. volumes (with a generic element V) with distribution F_V and finite second moment $EV^2 < \infty$. Define the number of renewals in the interval $(0, t]$ by the relation $N(t) = \max\{n \geq 0 : S_n \leq t\}$, $S_0 := 0$. Then the renewal interval V_t covering instant t can be written out as

$$V_t := S_{N(t)+1} - S_{N(t)} . \tag{8}$$

Assume that the distribution F_V is non-lattice, then the weak limit $V_t \Rightarrow \hat{V}$ exists and has expectation (see [5,6])

$$E\hat{V} = \frac{EV^2}{EV} . \tag{9}$$

3 Finite Buffer M

Now we return to the original system with a limited volume M, in which a *dependence* between volume V_n and service time S_n is allowed. Then it is seen that the service times of the lost customers are dependent as well, and we can not apply SLLN to deduce relation (7). However, by (2), a similar expression for the *limiting fraction of the lost workload* holds:

$$\lim_{t\to\infty} \frac{L(t)}{V(t)} := P_L = 1 - \frac{1 - P_0}{\rho} , \tag{10}$$

where, in general, $P_L \neq P_{\text{loss}}$.

3.1 Service Time Proportional to Volume

Denote by $G(t) = \sum_{i=1}^{A(t)} V_i$ the total volume of the customers arrived to the system in the interval $[0, t]$. Introduce the total lost volume in $[0, t]$,

$$Q(t) := \sum_{i=1}^{R(t)} V_i . \tag{11}$$

Now we will study the limiting fraction of the lost volume

$$Q_{\text{loss}} := \lim_{t \to \infty} \frac{Q(t)}{G(t)} = \lim_{t \to \infty} \frac{\sum_{i=1}^{R(t)} V_i}{\sum_{i=1}^{A(t)} V_i} , \tag{12}$$

which exists by the positive recurrence and is an important QoS parameter of the system. If the summands in $Q(t)$ were i.i.d. we would obtain the equality $Q_{\text{loss}} = P_{\text{loss}}$ from (12), but it is not the case. (Moreover, a dependence between the lost volumes exists even if S_n and V_n are independent.) Thus, in general, we can not use SLLN to prove a relation similar to (7) for the quantity Q_{loss} but can prove it for an important special case when the service time and the volume of customer n are connected as

$$V_n = cS_n , \tag{13}$$

where $c > 0$ is a constant. Such an assumption is very natural for many settings related to telecommunication networks, expressing, in particular, a proportionality between the size (volume) of a packet and its transmission (or processing) time, see [7]. We consider the following balance equation for the volumes:

$$G(t) = W_V(t) + Q(t) + B_V(t), \ t \geq 0 , \tag{14}$$

where $W_V(t)$ is the remaining volume of the customers being in the system at instant t, and $B_V(t)$ is the departed volume in the interval $[0, t]$. Then it immediately follows from (14) that

$$1 = Q_{\text{loss}} + \lim_{t \to \infty} \frac{\sum_{i=1}^{R(t)} cS_i}{\sum_{i=1}^{A(t)} cS_i} = \lim_{t \to \infty} \frac{t - I(t)}{\sum_{i=1}^{A(t)} S_i} = Q_{\text{loss}} + \frac{1}{\rho} \frac{P_0}{} , \tag{15}$$

implying $Q_{\text{loss}} = P_L$. It is worth mentioning that relation (15) can be used to estimate the quantity Q_{loss} by means of the estimation the idle probability P_0, and vice versa. Note that relation (15) can be easily extended to an m-server loss system with the identical servers as follows:

$$Q_{\text{loss}} = 1 - \frac{m(1 - \hat{P}_0)}{\rho} , \tag{16}$$

where $\hat{P}_0$ is the stationary idle probability of an *arbitrary* server.

3.2   An Approximation of Q_{loss} for M Large

In this subsection, we find an approximate relation connecting Q_{loss} and P_{loss} provided the threshold M is large. Recall that the summands in $Q(t)$ in (11) are dependent, hence, a dependence between the lost volumes also exists. Nevertheless, based on the inspection paradox, we *assume* that the summary volume lost in the interval $[0, t]$,

$$Q(t) = \sum_{i=1}^{R(t)} \hat{V}_i , \tag{17}$$

contains the *i.i.d.* summands $\{\hat{V}_i\}$ with the generic element $\hat{V}$ and expectation EV^2/EV, see (9). Expression (17) is formally the number of the renewals in the interval $(0, t]$ in the process generated by the i.i.d. variables distributed as the stationary interval $\hat{V}$. Our purpose is, when M is large, to verify the accuracy of the following approximation:

$$Q_{\text{loss}} \approx \lim_{t \to \infty} \frac{\sum_{i=1}^{R(t)} \hat{V}_i}{\sum_{i=1}^{A(t)} V_i} = \frac{EV^2}{[EV]^2} P_{\text{loss}} \ . \tag{18}$$

Note that the coefficient

$$C := \frac{EV^2}{[EV]^2} \geq 1 \tag{19}$$

is defined by only two first moments of the typical volume V, and it simplifies verification of the accuracy of approximation (18) by simulation (see Sect. 4). Now it follows that

$$Q_{\text{loss}} \geq P_{\text{loss}} \ . \tag{20}$$

(In this regard see also [2].) Note that, if volume $V = const$, then for each t we obtain

$$\frac{Q(t)}{G(t)} = \frac{R(t)}{A(t)} \ ,$$

implying, in particular, the limit equality $Q_{\text{loss}} = P_{\text{loss}}$, which is consistent with (18). Moreover, $C = 2$ for exponential V, and, if V is Pareto, that is,

$$1 - F_V(x) := \bar{F}_V(x) = x^{-\alpha} \ x \geq 1 (= 1, \ x \in [0, 1]) \ , \quad \alpha > 2 \ , \tag{21}$$

then $C = (\alpha - 1)(\alpha - 2)^{-1}$.

Evidently, the limit inequality (20) is also an approximation. However, there exists a class of the volume distributions F_V, for which (20) can be proved strongly. Namely, assume that F_V belongs to the class of *new-worse-than-used* (NWU) distributions [8]. It means that for all $y \geq 0, x \geq 0$, the following inequality holds:

$$\bar{F}_V(y + x) \geq \bar{F}_V(y)\bar{F}_V(x) \ . \tag{22}$$

This property means that, on the event $\{V > y\}$, the excess $V - y$ is stochastically larger than or equal to the original variable V. We note that this property holds, for instance, for the heavy-tailed Weibull distribution

$$F_V(x) = 1 - e^{-x^\beta} \ , \quad x \geq 0 \ , \quad \beta \in (0, 1) \ , \tag{23}$$

and also for exponential distribution. If the n-th customer is lost, then we can write down its volume as

$$V_n = \bar{v}_n + v_n \ , \tag{24}$$

where (see Fig. 1)

$$\bar{v}_n := M - \sum_{i \in \mathcal{J}(n)} V_i \ , \quad v_n := \sum_{i \in \mathcal{J}(n)} V_i + V_n - M \ . \tag{25}$$

Returning to NWU distributions, we have the following stochastic inequality:

$$V_n = \bar{v}_n + v_n \geq_{\mathrm{st}} \bar{v}_n + V_k \geq V_k \ , \tag{26}$$

where V_k is an independent variable sampled from the distribution F_V. Denote now by $\mathcal{R}(t)$ the set of the lost customers in the interval $[0, t]$, and let $R(t) := \#\{k : k \in \mathcal{R}(t)\}$ be its capacity. It immediately follows from (26) that

$$Q_{\mathrm{loss}}(t) = \frac{\sum_{i \in \mathcal{R}(t)} V_i}{\sum_{i=1}^{A(t)} V_i} \geq_{\mathrm{st}} \frac{\sum_{k=1}^{R(t)} V_k}{\sum_{i=1}^{A(t)} V_i} \ , \tag{27}$$

where, in the last term, the summands in the numerator are i.i.d. and *independent* of the summands in the denominator. Then we obtain

$$Q_{\mathrm{loss}} = \lim_{t \to \infty} Q_{\mathrm{loss}}(t) \geq \lim_{t \to \infty} \frac{\sum_{k=1}^{R(t)} V_k}{\sum_{i=1}^{A(t)} V_i} = P_{\mathrm{loss}} \ . \tag{28}$$

Thus, for the NWU distributions F_V, the inequality (20) holds in a strong sense for any (fixed) threshold M.

4   Simulation Results

In this section, to verify an accuracy of the approximation (18), we present estimation of the coefficient C obtained by simulation the system with the Poisson input and exponential service times with rate $\mu = 1$ (implying $\rho = \lambda$). We have simulated the following volume distributions: the heavy-tailed Pareto distribution (21) with parameter $\alpha = 3.5$ (then $C = (\alpha - 1)(\alpha - 2)^{-1} = 1.19$), the light-tail Weibull distribution (23) with parameter $w = 2$ (then $C = 1.27$), and the exponential distribution with parameter $\lambda = 2$ (in this case $C = 2$). Instead of continuous time estimators (6) and (12), we apply discrete scale counting arrivals and denote $Q_n = Q(t_n^-)$, $G_n = G(t_n^-)$, $R_n = R(t_n^-)$. Note that $A(t_n^-) = n - 1$, and also denote

$$Q_{\mathrm{loss}}(n) = \frac{Q_n}{G_n} \ , \quad P_{\mathrm{loss}}(n) = \frac{R_n}{n-1} \ , \quad n > 1 \ .$$

Then, it is obvious that the new estimator $\hat{C}_n := Q_{\mathrm{loss}}(n)/P_{\mathrm{loss}}(n)$ of the coefficient C (from (19)) is strongly consistent,

$$\hat{C}_n = \frac{Q_n}{G_n} \frac{(n-1)}{R_n} \to C \ , \quad n \to \infty \ . \tag{29}$$

Figures 2–4 demonstrate the difference $|\hat{C}(n) - C|$ for a few values of the load ρ and different volume distributions F_V. In all experiments the total number of arrivals equals $1\,000\,000$. Our experiments show that the approximation based on the inspection paradox works well for the value $M \geq 30EV$ if V has

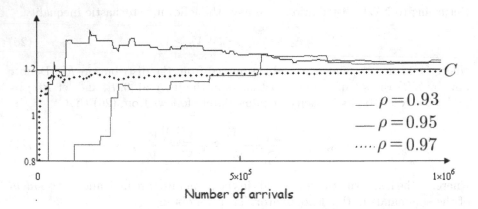

Fig. 2. $\hat{C}_n$ vs. $C = 1.19$ for Pareto V with $\alpha = 3.5$, $M = 90EV = 126$

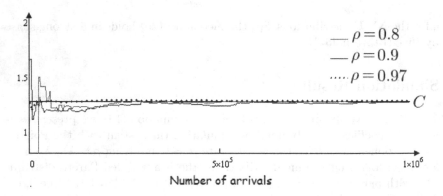

Fig. 3. $\hat{C}_n$ vs. $C = 1.27$ for Weibull V with $w = 2$, $M = 30EV = 26.59$

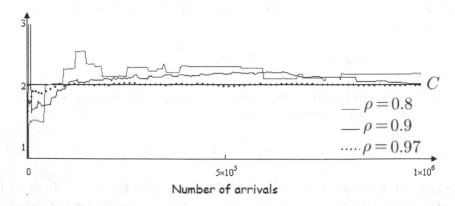

Fig. 4. $\hat{C}_n$ vs. $C = 2$ for exponential V with $EV = 1/2$, $M = 30EV = 15$

the (light-tailed) Weibull or the exponential distribution, while for the (heavy-tailed) Pareto distribution, a good agreement is achieved for $M \geq 90EV$.

Note that the inspection paradox is the asymptotic result, and it works better as M increases. At the same time, the rate of convergence crucially depends on the moment properties of the distribution F_V: the heavier the tail of F_V, the slower the rate of convergence (see [5]). This effect is well-observed on Fig. 2–4: the better result is achieved for the Weibull distribution, while the worst result is observed for the Pareto distribution (even for much larger values M). Thus our results are well-agree with the theory. (The experiments for larger M give expectedly more accurate agreement.) On the other hand, if M is large while ρ is small, then construction of a reliable estimate $\hat{C}_n$ becomes time-consuming.

5   Concluding Remarks

In this paper, for the first time, the asymptotic results of renewal theory are applied to system with limited volume. When the service time and the volume of a given customer are proportional, we establish a relation connecting the stationary idle probability P_0 with the limiting fraction of the lost volume, Q_{loss}. An approximate relation between Q_{loss} and the stationary loss probability P_{loss}, based on the inspection paradox, is considered. An accuracy of this approximation for different distributions of the volume V is verified by simulation.

In a recent work [9] diffusion approximation method (appropriately modified in [10,11]) is applied for the first time to study the system with limited volume provided that the service time and volume are independent. (In particular, the time-dependent loss probability is estimated.) We remark that the dynamics of the accumulated volume is described by a random walk rather than by a renewal process. The diffusion approximation method uses the Laplace transform and then inversion algorithm to calculate required performance measure numerically, and is time-consuming [9]. On the other hand, the direct probabilistic approach based on the inspection paradox, in general, requires considerable simulation efforts for reliable estimation. It is worth mentioning that the best results in our experiments have been achieved when the traffic intensity ρ is close to 1, in which case diffusion approximation method as well gives the most accurate results. Thus, it is important and promising to compare results predicted by renewal theory and by diffusion approximation, respectively. These two approaches are very different, and such a detailed comparison is assumed to be the purpose of a further research.

Acknowledgement. The authors thank an anonymous referee for attracting our attention to diffusion approximation of system with limited volume, and T. Czachorski who kindly supplied us with the paper [9].

The research was partially supported by the Program of Strategy Development of Petrozavodsk State University in the framework of the research activity and the Polish project NCN nr 4796/B/T02/2011/40 *Models for transmissions dynamics, congestion control and quality of service in Internet.*

References

1. Tikhonenko, O.M.: The problem of determination of the summarized messages volume in queueing systems and its applications. Journal of Information Processing and Cybernetics 32(7), 339–352 (1987)
2. Tikhonenko, O.: Computer Systems Probability Analysis. Akademicka Oficyna Wydawnicza EXIT, Warsaw (2006) (in Polish)
3. Morozov, E.V., Nekrasova, R.S.: On the estimation of the overflow probability in regenerative finite buffer queueing systems. Informtaics and their Applications 6(3), 391–399 (2012) (in Russian)
4. Morozov, E., Delgado, R.: Stability analysis of regenerative queueing systems. Automation and Remote Control 70(12), 1977–1991 (2009)
5. Asmussen, S.: Applied probability and queues. Springer, New York (2003)
6. Feller, W.: An introduction to probability theory and its applications. Wiley (1971)
7. Tanenbaum, A.S., Wetherall, D.J.: Computer Networks. Prentice Hall (2011)
8. Müller, A., Stoyan, D.: Comparisons methods for stochastic models. J. Wiley and Sons (2000)
9. Czachorski, T., Nycz, T., Pekergin, F.: Queue with limited volume, a diffusion approximation approach. In: Gelenbe, E., et al. (eds.) Computer and Information Science. LNEE, vol. 62, pp. 71–74. Springer, Heidelberg (2010)
10. Gelenbe, E.: On approximate computer system models. Journal ACM 22(2), 261–269 (1975)
11. Gelenbe, E.: Probabilistic Models of Computer Systems. Acta Informatica 12, 285–303 (1979)

Modeling of Video on Demand Systems

Slawomir Hanczewski and Maciej Stasiak

Poznan University of Technology, Chair of Communication and Computer Networks,
Polanka 3, 60-965 Poznan, Poland
{shancz,stasiak}@et.put.poznan.pl
http://nss.et.put.poznan.pl

Abstract. This paper presents a new, approximate method for modelling Video on Demand (VoD) systems. The proposed method is based on the multi-service model of Erlang's Ideal Grading (EIG). A comparison of the results obtained by the method with the results of a digital simulation has confirmed fair accuracy of all adopted theoretical assumptions. The method can be used to analyse, optimize and design VoD systems.

Keywords: VoD systems, Erlang's Ideal Grading.

1 Introduction

Currently, Video on Demand (VoD) systems are in wide use worldwide. Just a few years ago, a choice of a film or a television program was limited to the offer for scheduled program events proposed by a given television broadcaster. The only way to expand this offer was to individually make use of DVD rentals (or at earlier times, of VHS cassettes rentals). In fact, after the introduction of VoD systems that users have unlimited and adequate freedom of choice. Now, it is the user that decides by himself about a choice of viewing a specific film in the time most convenient for him. What is more, the user has a full control over the selected material, i.e. can pause viewing, fast forward, rewind or cancel his earlier order.

A contraction of a system that provides VoD services is very simple. The system consists of a node composed of the memory – most frequently in the form of disk array – where offered multimedia files (contents) are stored, and a video server that transmits demanded content to users. In the case of large VoD systems, a number of nodes is located in the network. This allows more effective usage of the network because users are serviced by the server that is located within the shortest distance from them. It is only in a case when a demanded content is missing in a local node that the call is forwarded to another node. Due to the large number of offered multimedia files it is often impossible to store all the available files on each VoD node. Another factor that influences the number of multiple copies of multimedia files in a VoD system is the popular demand for them. It turns out that in the long run only a small percentage of available files remains in high demand. Thus, a storage of multiple copies on each of the servers

A. Kwiecień, P. Gaj, and P. Stera (Eds.): CN 2014, CCIS 431, pp. 233–242, 2014.

is justified and financially viable only in the case of the most popular content. In the case of remaining files that are stored on disks, the number of copies in the system can be decidedly lower.

VoD systems have been addressed in many research studies. [1, 2] presents the general assumptions and discusses the construction of the VoD system. Issues related to modelling of systems of this type are presented, for example, in: [3–6]. In [3, 4] models constructed on the basis of the fixed-point methodology are presented. [5, 6] discusses models for VoD systems that offer files require only one transmission rate. The model shown in [5] takes advantage of the notion of Erlang's Ideal Grading, whereas the model discussed in [6] Erlang B-Formula. Other studies also focus on the problem of effective arrangement of offered content in distributed systems. For this particular purpose, in [7] genetic algorithms are used, while in [8] the Lagrange relaxation method is applied. Other publications worth mentioning here include those devoted to algorithms and mechanisms that support multicast transmission often used in VoD systems, such as for example [9–11].

This paper proposes a new model of distributed VoD systems that offer multimedia content that require different transmission rate. The model is based on the multi-service model of Erlang's Ideal Grading – EIG presented in [12, 13].

The paper is structured as follows. Section 2 presents a general characteristics of Video on Demand systems. Section 3 presents a multi-service model of Erlang's Ideal Grading with differentiated availability. Then, in Section 4 the proposed analytical model of the VoD system is discussed. Section 5 shows the results of the comparison of the analytical calculations and the results of the simulations. Section 6 sums up the paper.

2   Video on Demand Systems

Video on Demand (VoD) systems are built as either centralized or distributed systems. In centralized systems (Fig. 1a), in which one node executes and manages instances of one or more services, each user of the system is equipped with appropriate devices (set-top box or computer) that provide access to the VoD node through a public network or a private network set up by the operator. The node includes video servers and a memory system, i.e. an adequate disk space for content storage. This system can store F different multimedia files (contents). A disadvantage of a VoD system of this type is a relatively low number of concurrently serviced users resulting from the limited efficiency of the video server or the link rate between the node and the network. Distributed systems are composed of N nodes operating independently. While searching a required content, users employ these nodes for data transfer. The architecture of such a system is presented in (Fig. 1b). As in the case of the centralized system, the node is composed of a video server and the storage system. Depending on users demands, multiple copies of a given multimedia content are stored on a number of nodes. A disk system in node n can store concurrently S_n $(n = 1, 2, \ldots, N)$ different multimedia files. The total number of all offered

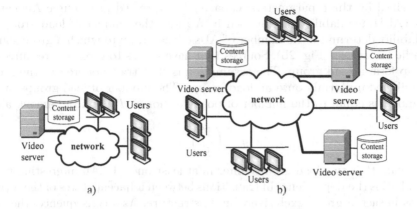

Fig. 1. Architecture of the VoD system a) centralized, b) distributed

files (without multiple copies) is equal to F. By default, users of the system are serviced by the node that is in their closest proximity. The user can be serviced by another node when the closest node does not have the demanded content or when the nearest node cannot serve a call because it is overloaded. A demand for access to a given content will be rejected (the VoD system is in the blocking state) only when none of the available nodes can service this demand. Making a given content available results in a part of the RAM being occupied, and in the usage of the processor time, which is entailed sending data with a given transmission rate. One can adopt then that the maximum transmission rate C (expressed in bps) with which the video server can transmit data will be the measure of node resources (i.e. its capacity).

Regardless of the architecture of a system, the user has full control of the presented file. This means that at any time he is in position to pause the viewing and fast forward or rewind the content. This option of full control of the viewed material can be conducive to negative influence on the operation of the network. Hence, the usual procedure is to equip users with hard disks on which the demanded file is downloaded and stored until the termination of the viewing time [1, 2].

Another important parameter that describes VoD systems is the popularity of a multimedia file [14]. This parameter determines the number of its copies in the system. In addition, the information on popularity provides an opportunity for all available files to be located in the most optimum way in line with the criterion of the uniform load N of VoD nodes.

3 Erlang's Ideal Grading

Erlang's Ideal Grading (EIG) is one of the oldest models of non-full-availability groups. The basic structure of the group (Fig. 2a) and the appropriate analytical model for single-service traffic was proposed by Erlang in 1917. The EIG is described by three parameters: capacity V expressed in resource Allocation Units (AU[1]), availability d expressed in AU and the number of load groups g. Availability determines the number of AUs of the group to which a given source of traffic has access (Fig. 2b). Sources that have access to the same resources of a group form a load group. The assumption is that traffic offered to the group is distributed uniformly onto all load groups. The number of load groups in the EIG model is equal to the number of possible choices d AUs from among all V AUs:

$$g = \binom{V}{d} \ . \tag{1}$$

This means that two load groups differ in at least one AU. An interesting property of EIG is the dependence of transitions between adjacent states of the service process in such a group exclusively on its structure. As a consequence, the conditional probability of transition can be determined in a combinatorial way (4) [13, 16, 17].

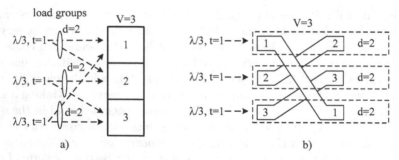

Fig. 2. Erlang's Ideal Grading: $V = 3$, $d = 2$, $g = 3$, a) offered traffic distribution, b) concept of availability [13]

Consider an EIG group with multi-rate traffic and with differentiated availability (Fig. 3). A model for this group is proposed in [12, 13]. The group is offered m independent call streams with the intensities $\lambda_1, \lambda_2, \ldots, \lambda_m$. The service time has an exponential distribution with the parameters $\mu_1, \mu_2, \ldots, \mu_m$, respectively. Calls of particular classes demand $t_1, t_2, \ldots, t_m$ AUs, whereas the values of the availability parameters are equal to $d_1, d_2, \ldots, d_m$ AUs. According to the model [12, 13], the occupancy distribution $P(n)$ is expressed by the following formula:

$$nP(n) = \sum_{i=1}^{m} a_i t_i [1 - \sigma_i(n - t_i)] P(n - t_i) \ , \tag{2}$$

where a_i is traffic offered to the group by a call of class i:

$$a_i = \lambda_i / \mu_i \ , \tag{3}$$

---

[1] The allocation unit defines a certain, and basic for a given system, bit rate expressed in bps [15]. Back in Erlang times [16] the equivalent to AU was a link with the capacity that would make a transmission of one call (voice service) possible.

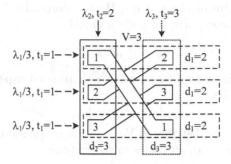

Fig. 3. Erlang's Ideal Grading with multi-rate traffic and differentiated availability [13]

$\sigma_i(n)$ is the conditional probability of transition for a traffic stream of class i in occupancy state n in the group:

$$\sigma_i(n) = 1 - \sum_{x=d_i-t_i+1}^{k} \binom{d_i}{x} \binom{V-d_i}{n-x} \Big/ \binom{V}{n} , \tag{4}$$

where:

$$k = \begin{cases} n - t_i & \text{if} & (d_i - t_i + 1) \leq (n - t_i) < d_i , \\ d_i & \text{if} & (n - t_i) \geq d_i . \end{cases} \tag{5}$$

The blocking probability for calls of class i can be determined on the basis of the following formula:

$$E_i = \sum_{i=1}^{m} [1 - \sigma_i(n)] P(n) . \tag{6}$$

Distribution (2) is an approximate distribution. Numerous studies and analyses conducted by the authors have proved high accuracy of the model [13].

4 Model of the VoD System

The adopted concept of the proposed method for modelling distributed VoD systems is to reduce the calculation of the most important characteristics of such a system in its essence to the calculation of appropriate characteristics of Erlang's Ideal Grading. To achieve that, the parameters of the VoD system should be expressed through the parameters of EIG. Consider the system shown in Fig. 1b. The system offers multimedia files that have been created with the use of m different codecs (i.e. the system delivers files with m different bit rates). By adopting the link rate (C_n) between the node and the network as the measure of the capacity of a single node, we can, in line with [15], express its capacity in AUs. In accordance with [15], AU is defined as the biggest common divisor of bit rates for all codecs.

$$c_{AU} = \text{GCD}(c_1, c_2, \ldots, c_m) , \tag{7}$$

where c_i is the bit rate of the codec i. Having determined the value c_{AU} it is possible to express the capacity of node n in AU:

$$V_n = \lfloor C_n/c_{AU} \rfloor \; , \tag{8}$$

It is possible, in a similar way, to express the demanded capacity t_i for a codec i $(i = 1, 2, \ldots, m)$ in AUs:

$$t_i = \lceil c_i/c_{AU} \rceil \; . \tag{9}$$

The total capacity of the distributed VoD system (expressed in AUs) is equal to the sum of the capacities of all nodes in the system:

$$V = \sum_{n=1}^{N} V_n \; . \tag{10}$$

Taking into consideration the number of required AUs, available files can be divided into m sub-sets:

$$F = F_1, F_2, \ldots, F_i, \ldots, F_m \; . \tag{11}$$

Let us determine now the intensities of calls for individual multimedia file. It has been proved in many previous publications (e.g. in [7, 14]) that the popularity, i.e. the intensity with which demands for access to multimedia files appear in the system has a long run nature (is of the long-tail type). It is assumed that the distribution of intensity at which demands for access to files appear can be approximated by Zipf's distribution [14]. It follows from the properties of the distribution that only a small number of offered contents is in very high demand (for example, the latest productions). The popularity of the remaining quantity of files is substantially lower. On the basis of Zipf's distribution, it is possible to determine the value of the probability p_{f_i} $(f_i = 1, 2, \ldots, F_i)$ that defines the probability of choice of a specified content from the set F_i by any user of the system. If we adopt that the intensity of calls in the whole of the VoD system for files that belong to the set F_i is equal to λ_i, then the intensity of calls related to a single content (i.e. all multiple copies of a given file) is equal to [7]:

$$\lambda_{f_i} = \lambda p_{f_i} \; , \tag{12}$$

where λ is the total intensity of calls.

In [4, 7, 8], the assumption of Poisson traffic for activities related to demands for access to content is adopted. Therefore, assuming the exponential character of the service time, demands for service related to a multimedia file (that takes into account the behaviour of the subscriber) with the mean value $t_{f_i} = 1/\mu_{f_i}$, offered traffic generated by the demands for access to the content f_i can be determined on the basis of the following formula:

$$A_{f_i} = \lambda_{f_i}/\mu_{f_i} \; . \tag{13}$$

Let us determine now the value of the availability parameter d_{f_i} for each multimedia file. The availability parameter in the proposed method is determined on the basis of the number of multiple copies of a given content in the VoD system (popularity) and the possibilities of their reproduction (viewings) by users. For the file f_i this parameter will be defined on the basis of the following reasoning. VoD system stores k_{f_i} copies of a file f_i. The capacity of the node n is equal to V_n AUs. Since the node also hosts other files and they are also offered concurrently, the real availability of the file f_i, that requires t_i AUs for the transmission, on the node n will be proportional to offered traffic:

$$d_{f_i,n} = \frac{A_{f_i,n} t_i}{\sum_{i=1}^{m} \sum_{j=1}^{F_{i,n}} A_{j,n} t_i} V_n \; , \tag{14}$$

where:
$A_{f_i,n}$ is traffic offered to the n-th VoD node by calls that demand the file f_i, $F_{i,n}$ – the number of files from the set F_i that are hosted on the node n. Since the content f_i is made available by k_{f_i} nodes, then the total availability of the multimedia file f_i in the VoD system can be determined by the formula:

$$d_{f_i} = \sum_{j=1}^{k_{f_i}} d_{f_i,j} \; . \tag{15}$$

Assuming a uniform load in the VoD nodes, traffic offered to the node n by calls that demand any randomly selected file f_i can be determined in the following way:

$$A_{f_i,n} = A_{f_i}/k_{f_i} \; . \tag{16}$$

The values of the parameters d_{f_i} and A_{f_i} are determined for each multimedia content regardless of their number offered in the system (each one of them has a different probability of its "popularity"). The computational process can be significantly made simpler by grouping these multimedia files that are characterized by similar values of the probability of "popularity" p_{f_i} and identical transmission rates (the t_i parameter). Due to a limited number of nodes in the system, one can adopt that some files are likely to have an identical number of multiple copies. Therefore, the values of the blocking probability for these files will also be similar. In such circumstances it is not necessary to consider all multimedia files separately. The files can be grouped and then make appropriate calculations for groups of files. Offered traffic generated by calls that demand access to a group of files is equal to the sum of offered traffic generated by calls that demand access to particular content that is included in the group. For a given set of files F_i we can thus write:

$$A_{i,g_z} = \sum_{f_j=1}^{F_{g_z}} A_{f_j} \; , \tag{17}$$

where F_{g_z} is the number of files that belong to the group g_z. Since we group content according to its identical availabilities, the value of this parameter (d_g) is identical as the one for a single content that belongs to the group (Equation (15)).

Now, after determining the availability parameters and traffic offered to particular multimedia content groups it is possible to determine the blocking probability in the EIG group that would correspond to the blocking in the VoD system.

5   Numerical Examples

The model of the VoD system proposed here is an approximate model. In order to verify the accuracy of its operation, the results of the analytical calculations were compared with the results of the digital simulation. The simulator of the VoD system described in Sect. 2 was implemented in the C++ language. The adopted number of series was equal to 5. The length of each of the series, expressed in the number of calls for the content that required the highest number of AUs, was 100,000 calls. This allowed the confidence interval to be determined at 95 %, at least one order of magnitude less than the values of the results of the simulation.

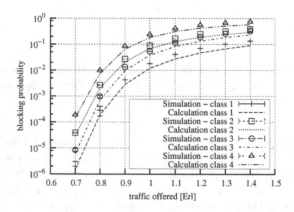

Fig. 4. Blocking probability in VoD system ($N = 5$, four classes of calls)

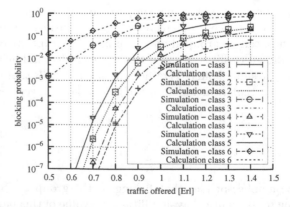

Fig. 5. Blocking probability in VoD system ($N = 6$, six classes of calls)

The blocking probability is presented in graphs in relation to the mean value of traffic a offered to a single AU in the VoD system:

$$a = \frac{\sum_{i=1}^{m} \sum_{f=1}^{F_i} A_{f_i}}{\sum_{n=1}^{N} V_n}. \tag{18}$$

Figure 4 shows the results for a VoD system composed of $N = 5$ nodes. The system offers files that requires two different transmission rates for transmission: $t_1 = 1$ and $t_2 = 2$ AUs, respectively. Each node can service concurrently $V_n = 50$ AUs. The adoption is that there are $F = 200$ contents available in the system, with the additional assumption that due to their popularity rate they have been divided into two groups. Files that belong to the most popular group have a copy on each of the nodes, while less popular files have copies on only 3 nodes. Another adoption is that, within each of the popularity groups, the number of files that demand 1 and 2 AUs is the same. The most popular files generate 60 % of all demands. From the model point of view, system services 4 classes of calls: class 1 (t_1, d_{g_1}, A_{1,g_1}), class 2 (t_1, d_{g_2}, A_{1,g_2}), class 3 (t_2, d_{g_1}, A_{2,g_1}) and class 4 (t_2, d_{g_2}, A_{2,g_2}). Figure 5 presents the results for a system built from $N = 6$ nodes. The number of offered files is equal to $F = 300$. The files have been divided into 3 groups, according to their popularity. The number of copies of files that belong to particular groups is equal to 6, 3, and 1, respectively. Exactly as in the previous case, each group includes the same number of files that require 1 or 2 AUs for transmission. In this particular case the most popular files generate 50 % of all calls, those less popular 40 % and the remaining group of files 10 %. From the model point of view, system services 6 classes of calls: class 1 (t_1, d_1, A_{1,g_1}), class 2 (t_1, d_2, A_{1,g_2}), class 3 (t_1, d_3, A_{1,g_3}), class 4 (t_2, d_1, A_{2,g_1}), class 5 (t_2, d_2, A_{2,g_2}), and class 6 (t_2, d_3, A_{2,g_3}).

6   Conclusions

This paper proposes a new method for a determination of the blocking probability in distributed Video on Demand systems that make use of codecs with different initial bit rates. In its operation, the presented method is based on the Erlang's Ideal Grading model. Its implementation is not complicated because all necessary parameters can be determined in a very simple way. The results of a comparison of the analytical calculations with the results of the simulation experiments confirm fair accuracy of the proposed method. They also confirm the universal nature of the Erlang's Ideal Grading model.

References

1. Deloddere, D., Verbiest, W., Verhille, H.: Interactive Video on Demand. IEEE Communications Magazine 32(5), 82–88 (1994)
2. Ko, M., Koo, I.: An Overview of Interactive Video on Demand System. The University of British Columbia. Technical report (1996)

3. Wong, E.M.W., Chan, S.: Performance modeling of video-on-demand systems in broadband networks. IEEE Transactions on Circuits and Systems for Video Technology 11(7), 848–859 (2001)
4. Guo, J., Wong, E.M.W., Chan, S., Taylor, P., Zukerman, M., Tang, K.-S.: Performance Analysis of Resource Selection Schemes for a Large Scale Video-on-Demand System. IEEE Transactions on Multimedia 10(1), 153–159 (2008)
5. Hanczewski, S., Stasiak, M.: Performance Modelling of Video-on-Demand Systems. In: Proc. of Asia-Pacific Conference on Communications, pp. 784–788 (2011)
6. Kanrar, S.: Analysis and implementation of the Large Scale Video-on-Demand System. IEEE Transactions on Circuits and Systems for Video Technology 2(2), 1–9 (2012)
7. Tang, K.-S., Ko, K.-T., Chan, S., Wong, E.M.W.: Optimal file placement in vod system using genetic algorithm. IEEE Transactions on Industrial Electronics 48(5), 891–897 (2001)
8. Applegate, D., Archer, A., Gopalakrishnan, V., Lee, S., Ramakrishnan, K.K.: Optimal content placement for a large-scale vod system. In: Proc. of the International Conference on Emerging Networking Experiments and Technologies, pp. 4:1–4:12 (2010)
9. Piechowiak, M., Zwierzykowski, P., Hanczewski, S.: Performance analysis of multicast heuristic algorithms. In: Proc. of the International Working Conference on Performance Modelling and Evaluation of Heterogeneous Networks, pp. 41/1–41/8 (2005)
10. Kamiyama, N., Kawahara, R., Mori, T., Hasegawa, H.: Multicast pre-distribution in vod services. In: Proc. of the IEEE International Workshop Technical Committee on Communications Quality and Reliability, pp. 1–6 (2011)
11. Stasiak, M., Piechowiak, M., Zwierzykowski, P.: On the methodology for the evaluation of unconstrained multicast routing algorithms. In: Proc. of the International Conference on Telecommunications, pp. 71–76 (2009)
12. Stasiak, M., Hanczewski, S.: A model of WCDMA radio interface with reservation. In: Proc. of the Information Theory and Its Applications (2008)
13. Glabowski, M., Hanczewski, S., Stasiak, M., Weissenberg, J.: Modeling Erlang's Ideal Grading with Multirate BPP Traffic. Mathematical Problems in Engineering 2012, 35 (2012)
14. Dan, A., Sitaram, D., Shahabuddin, P.: Scheduling Policies for an On-Demand Video Server with Batching. In: Proc. ACM International Conference on Multimedia, pp. 15–23 (1994)
15. Roberts, J. (ed.): Performance Evaluation and Design of Multiservice Networks. Final Report COST 224, Commission of the European Communities (1992)
16. Brockmeyer, E.: A survey of A.K. Erlang's mathematical works. Danish Academy of Technology Sciences 2, 101–126 (1948)
17. Lotze, A.: History and development of grading theory. In: Proc. of International Teletraffic Congress, pp. 148–161 (1967)

Performance Evaluation of Routers with the Dropping-Function Queueing

Pawel Mrozowski and Andrzej Chydzinski

Faculty of Automatic Control, Electronics and Computer Science,
Silesian University of Technology, Akademicka 16, 44-100 Gliwice, Poland

Abstract. The basic algorithm used for the active management of queues of packets in Internet routers is based on accepting an incoming packet with the probability that is a function of the queue size. An analytical model of such queueing system has been solved recently via transform techniques. However, the form of these analytical results restrict obtaining exact solutions to very small buffers (e.g. 20 packets). In this paper we study numerical techniques that allow evaluation of systems of more practical sizes – hundreds of packets. In particular, we compare accuracy and execution time of seven inversion methods needed to compute the queueing characteristics of the router.

Keywords: active queue management, dropping function, performance evaluation, transform techniques.

1 Introduction

The active queue management (AQM, see e.g. [1–6]) in Internet routers is a technique that allows obtaining a small queueing delay and a high bandwidth usage at the same time. The simplest approach to AQM, invented with the famous RED algorithm [1], is to drop incoming packets with the probability that is a function of the queue size. This function is called a dropping function. In RED, a linear dropping function is exploited, but other shapes of the dropping functions have proven to be of use as well, e.g. exponential [2] or doubly-linear [3].

The general model of a queueing system with the dropping function has been analytically solved in [7] via transform techniques. It is possible to obtain performance characteristics of a router using symbolic inversion of the transforms presented in [7], but for relatively small buffer sizes (e.g. 20 packets) only. For larger buffers, the formulas become hard in symbolic manipulation, even with the help of an appropriate software. As the buffers for tens of packets are too small in practice, we study herein numerical methods suitable for solving systems of hundreds of packets at least. A few hundreds of packets is still a rather moderate buffer. The router vendors commonly dimension the buffers using the bandwidth-delay product rule [8], with a delay of about 100 ms. For example, assuming a 1 Gb/s link, this product equals 100 Mb. If the average packet size is 1 000 bytes, the buffer for 12 500 packets is obtained. Naturally, the active queue

A. Kwiecień, P. Gaj, and P. Stera (Eds.): CN 2014, CCIS 431, pp. 243–252, 2014.

management is supposed to reduce this size by 1 or 2 orders of magnitude. Therefore, we use herein buffers for a few hundreds of packets.

In particular, we evaluate four dropping functions, each operating on the buffer of 200 packets and producing the average queue size of about 100 packets. For each of these dropping functions we first present a long-run simulation results. Then we compute the performance characteristics using seven chosen methods of numerical transform inversion. Finally, we compare the accuracy of the inversion results as well as they execution times. Contrary to our initial expectations, newest, most computationally demanding methods do not offer the best accuracy.

The remaining part of the paper is structured as follows. In Section 2, we outline the analytical results for the model of the queue with the dropping function. In Section 3, seven inversion methods chosen for numerical purposed are sketched. Then, in Section 4, we present and discuss the results obtained using simulations and the described inversion methods. Finally, remarks concluding the paper are gathered in Sect. 5.

2   Analytical Results

Consider the following model of the router's queueing algorithm. Packets arrive to the system according to the Poisson process with rate λ, the service time distribution has a distribution function $F(x)$, which can have any form (e.g. it may be determined by packet size distribution). A finite buffer of size b is assumed, which is equivalent to the maximum number of packets that can be stored in a queue (every packet incoming when b packets are present in the system is dropped). Moreover, a packet can be dropped with probability $d(n)$, even if the buffer is not full. Number n denotes the queue size observed upon packet arrival; the function $d(n)$ is called a dropping function and it can assume any form. By $\mathbf{P}$ we will denote the probability, by $X(t)$ the queue size at time t (including service position) while by X_k – the queue size left by the k-th departing packet. The load offered to the system is:

$$\rho = \lambda m ,$$

where m is the average packet transmission time.

Following [7], the steady-state distribution of the queue size, i.e. the limit $P_k = \lim_{t \to \infty} \mathbf{P}(X(t) = k)$, has the form:

$$P_b = 1 - \frac{\sum_{i=0}^{b-1} \pi_i / (1 - d(i))}{\pi_0 / (1 - d(0)) + \rho} , \quad P_k = \frac{\pi_k / (1 - d(k))}{\pi_0 / (1 - d(0)) + \rho} , \quad 0 \leq k \leq b - 1 ,$$

where π_k is the steady-state distribution of the queue size left behind by the n-th departing packet, namely $\pi_k = \lim_{n \to \infty} \mathbf{P}(X_n = k)$.

The values of π_k can be found from the following system of equations:

$$\begin{cases} \pi_k = \sum_{j=0}^{b-1} \pi_j p_{j,k} , & 0 \leq k \leq b - 1 , \\ \sum_{j=0}^{b-1} \pi_j = 1 , \end{cases} \tag{1}$$

where

$$p_{j,k} = \begin{cases} a_{1,k} & \text{if } j = 0,\ 0 \le k \le b-1, \\ a_{j,k-j+1} & \text{if } 1 \le j \le b-1,\ j-1 \le k \le b-1, \\ 0 & \text{otherwise,} \end{cases} \qquad a_{n,k} = \int_0^\infty Q_{n,k}(u)\mathrm{d}F(u),$$

and $Q_{n,k}(u)$ is the conditional probability that the first departure time is after u and k packets are accepted in time interval $(0, u]$, provided $X(0) = n$. Function $Q_{n,k}(u)$ can be computed using its Laplace transform, i.e. $q_{n,k}(s) = \int_0^\infty e^{-su} Q_{n,k}(u)\mathrm{d}u$. In [7] it is shown that

$$q_{n,k}(s) = \frac{\prod_{i=0}^{k-1} \lambda(1 - d(n + i))}{\prod_{i=0}^{k}(s + \lambda(1 - d(n + i)))}, \quad n \ge 0, \quad k \ge 0. \tag{2}$$

Formula (2) represents a rational function that can be inverted symbolically using classic results of the Laplace transform theory (see [7] for more details). However, this is practically possible for small values of b only. For large buffers, we have to invert (2) numerically.

3 Numerical Inversion Methods

According to [9], four types of inversion methods can be distinguished: methods exploiting the Fourier series expansion, exploiting the Laguerre function expansion, exploiting a combination of the Gaver functionals and exploiting a deformation the Bromwich contour.

Moreover, all methods fall into two classes: methods making use of real numbers only and methods requiring the transform in the complex plane. For the purpose of this article seven algorithms, belonging to different families, were chosen. More on numerical inversion of the Laplace transform is available in [10]. Other examples of performance evaluation problems that lead to Laplace transforms can be found in [11–13].

The following notation will be used. By $f(t)$ we will denote the original function, while by $F(s)$ its Laplace transform, namely $F(s) = \int_0^\infty f(t) e^{-st}\mathrm{d}t$.

3.1 The Gaver-Stehfest Method

The Gaver-Stehfest method [14] is one of the most popular and well known methods for inverting the Laplace transform. It does not use complex numbers. The algorithm is fast and often gives good results, especially for smooth functions. As every other method this one also suffers some limitations. The most important is that late-time exponential decline data cannot be obtained accurately. It requires even number of coefficients and is limited to functions that do not oscillate rapidly and that have no discontinuities. It is based on the idea to represent function f approximately by a finite linear combinations of the transform values according to formula:

$$f_n(t) = \frac{1}{t} \sum_{k=0}^{n} \omega_k F\left(\frac{\alpha_k}{t}\right), \tag{3}$$

where $t > 0$, nodes α_k and weights ω_k are real numbers which depend on n, but do not depend on F, nor on t. Applying the Salzer summation scheme we receive finally:

$$f(t) \approx \frac{\ln{(2)}}{t} \sum_{k=1}^{2M} \zeta_k F\left(\frac{\ln{(2)}}{t}\right) , \qquad (4)$$

where coefficients ζ_k are defined as

$$\zeta_k = (-1)^{M+k} \sum_{j=\lfloor(k+1)/2\rfloor}^{k \wedge M} \frac{j^{M+1}}{M!} \binom{M}{j} \binom{2j}{j} \binom{j}{k-j} , \qquad (5)$$

with $k \wedge M = \min\{k, M\}$, and the parameter M should be even. The default value of M is 14.

3.2 The Talbot Method

The idea of the method proposed in [15] is to use deformed standard contour (a vertical line) in the Bromwich integral $f(t) = \frac{1}{2\pi i} \int_B \exp{(ts)} F(s)\, \mathrm{d}s$. B is the contour defined as $s = r + iy$ where $-\infty < y < +\infty$ and r has a fixed value. The value of r must fulfill the requirement that all singularities of the transform are on the left side of it. The transform is analytic in the region of the complex plain on the right side of B. The contour B can be deformed into any open path wrapping all negative real axis, but the condition is that no singularity of f is crossed in the deformation. The Talbot inversion formula has a form:

$$f(t) \approx \frac{2}{5t} \sum_{k=0}^{M-1} \mathrm{Re}\left[\gamma_k F\left(\frac{\delta_k}{t}\right)\right] , \qquad (6)$$

where

$$\delta_0 = \frac{2M}{5} , \quad \delta_k = \frac{2k\pi}{5}\left(\cot\left(\frac{k\pi}{M}\right) + i\right) ,$$

$$\gamma_0 = \frac{1}{2}e^{\delta_0} , \quad \gamma_k = \left[1 + i\left(\frac{k\pi}{M}\right)\left(1 + \left[\cot\left(\frac{k\pi}{M}\right)\right]^2\right) - i\cot\left(\frac{k\pi}{M}\right)\right]e^{\delta_k} .$$

The default value of M is 24.

3.3 The Zakian Method

In the Zakian method [16], a formula similar to (3) is used, namely:

$$f(t) \approx \frac{2}{t} \sum_{k=0}^{n} \mathrm{Re}\left[\omega_k F\left(\frac{\alpha_k}{t}\right)\right] . \qquad (7)$$

Contrary to the Gaver-Stehfest method, the nodes α_k and the weights ω_k are complex now. The most difficult part is a proper choice of the values of n, α_k and ω_k. With $n = 4$, the following values of α_k and ω_k are commonly used:

k	α_k	ω_k
0	$12.83767675 + i\ 1.666063445$	$-36902.08210 + i\ 196990.4257$
1	$12.22613209 + i\ 5.012718792$	$61277.02524 - i\ 95408.62551$
2	$10.93430308 + i\ 8.409673116$	$-28916.56288 + i\ 18169.18531$
3	$8.776434715 + i\ 11.92185389$	$4655.361138 - i\ 1.901528642$
4	$5.225453361 + i\ 15.72952905$	$-118.7414011 - i\ 141.3036911$

3.4   The Weeks Method

Another method was proposed in its basic form in [17]. Several modifications and improvements were proposed to date; we use the one of [18]. The method origins from the Laguerre method based on Tricomi-Widder theorem. The original function is represented as a weighted sum of the Laguerre functions, where the weights are coefficients of a generating function constructed from the Laplace transform using bilinear transformation. The method dictates the approximation of $f(t)$ in following form:

$$f(t) \approx e^{\sigma t} \sum_{n=0}^{M} a_n e^{-bt/2} L_n(bt) , \qquad (8)$$

where $L_n(t)$ are the Laguerre polynomial of degree n, b and σ are positive real parameters while expansion coefficients a_n are given by:

$$a_n = \frac{1}{mr^n} \sum_{j=1}^{m} \phi(re^{2\pi ij/m})e^{-2\pi ijn/m} , \quad \psi(z) = \frac{b}{1-z} \Gamma\left(\sigma + \frac{b}{1-z} - b/2\right) .$$

We use the following values of parameters: $M = 60$, $\sigma = 1$ and $b = 1$. The values of m and r are changed dynamically, depending on the required accuracy (see [19] for more details).

3.5   The Piessens Method

In the method of [20], Chebyshev polynomials are exploited instead of Laguerre. The following inversion formula is used:

$$f(t) \approx e^{\beta t} \frac{t^{a+1}}{\Gamma(a)} \sum_{k=0}^{N} c_k^* \varphi_k\left(\frac{bt}{2}\right) , \qquad (9)$$

where c_k^* are approximations of the Chebyshev coefficients, $\varphi_k(x)$ is a generalized hypergeometric function (defined below) whilst β, a and b are parameters. To obtain c_k^*, the following system must be solved: $\left(\frac{b}{1-u}\right)^a F\left(\frac{b}{1-u} + \beta\right) = \sum_{k=0}^{N} c_k^* T_k(u)$, where T_k are Chebyshev polynomials of the first kind and k degree. The generalized hypergeometric function can be determined as

$$\varphi_0(x) = 1 , \quad \varphi_1(x) = 1 - 2xa^{-1} , \quad \varphi_2(x) = 1 - 8xa^{-1} + 8x^2 a^{-1}(a+1)^{-1} ,$$

and for $n > 2$:

$$\varphi_n(x) = (A_n + B_n x)\varphi_{n-1}(x) + (C_n + D_n x)\varphi_{n-2}(x) + E_n \varphi_{n-3}(x) ,$$

where

$$A_n = \frac{3n^2 - 9n + an - 3a + 6}{(n + a - 1)(n - 2)}, \quad B_n = \frac{-4}{n + a - 1}, \quad C_n = -\frac{3n^2 - 9n - an + 6}{(n - 2)(n + a - 1)},$$

$$D_n = -\frac{4(n - 1)}{(n + a - 1)(n - 2)}, \quad E_n = -\frac{(n - 1)(n - a - 2)}{(n + a - 1)(n - 2)}.$$

We use the following values of parameters: $b = 0.3$, $N = 110$, $a = 1$, $\beta = 1$.

3.6   The Dubner Abate Method

Dubner and Abate proposed in [21] a method that belongs to the Fourier series category. The method works when $f(t)$ is a real function, which is not in conflict with our applications. The authors applied the Fourier cosine transform with the attenuation factor and exploited the trapezoidal rule for computation of the integrals. Finally, they obtained:

$$f(t) \approx \frac{2e^{ct}}{T} \left[\frac{1}{2} \mathrm{Re} \left[F(c) \right] + \sum_{k=1}^{N} \mathrm{Re} \left[F \left(c + \frac{i\pi k}{T} \right) \right] \cos \frac{k\pi t}{T} \right]. \qquad (10)$$

The default values of parameters are the following: $T = 20$, $c = 0.5$ and $N = 200$.

3.7   The Abate-Choudhury-Whitt (ACW) Method

The newest method used herein is the one proposed in [22]. The idea is to use the Euler summation formula for approximation of infinite series with alternating signs. Namely, we have $f(t) \approx \sum_{k=0}^{m} \sum_{j=0}^{n+k} \binom{m}{k} 2^{-m} (-1)^j a_j(t)$, where functions $a_k(t)$ are given by:

$$a_k(t) = \frac{e^{A/2t}}{2lt} b_k(t), \quad b_0(t) = F \left(\frac{A}{2lt} \right) + 2 \sum_{j=1}^{l} \mathrm{Re} \left[F \left(\frac{A}{2lt} + \frac{ij\pi}{lt} \right) e^{ij\pi/t} \right],$$

$$b_k(t) = 2 \sum_{j=1}^{l} \mathrm{Re} \left[F \left(\frac{A}{2lt} + \frac{ij\pi}{lt} + \frac{ik\pi}{t} \right) e^{ij\pi/t} \right], \quad k \geq 1.$$

The default vales of parameters are: $n = 38$, $m = 11$, $A = 19$ and $l = 1$. In [23] an example of the usage of this method can be found.

4   Results

In our experiments we assumed that the router line card of $1\,\mathrm{Gb/s}$ capacity serves a Poisson traffic of $1.1\,\mathrm{Gb/s}$ bitrate. Therefore we had $\rho = 1.1$. Moreover, $b = 200$ and constant service time were assumed. The following four dropping functions were considered (each of them maintains the average queue size of about 100 packets):

(a) linear dropping function:

$$d(n) = \begin{cases} 0 & \text{for } n \leq 50 , \\ 0.00167n - 0.08389 & \text{for } 50 < n < 200 , \\ 1 & \text{for } n \geq 200 ; \end{cases}$$

(b) convex dropping function:

$$d(n) = \begin{cases} 0 & \text{for } n \leq 65 , \\ \frac{\sqrt{n-65}}{65} & \text{for } 65 < n < 200 , \\ 1 & \text{for } n \geq 200 ; \end{cases}$$

(c) concave dropping function:

$$d(n) = \begin{cases} 0 & \text{for } n \leq 70 , \\ 0.00002n^2 - 0.1 & \text{for } 70 < n < 200 , \\ 1 & \text{for } n \geq 200 ; \end{cases}$$

(d) step dropping function:

$$d(n) = \begin{cases} 0.2 & \text{for } n \leq 10 , \\ 0.01 & \text{for } 10 < n \leq 60 , \\ 0.1 & \text{for } 60 < n \leq 75 , \\ 0.05 & \text{for } 75 < n \leq 111 , \\ 0.4 & \text{for } 111 < n \leq 130 , \\ 0.45 & \text{for } 130 < n \leq 165 , \\ 0.01 & \text{for } 165 < n < 200 , \\ 1 & \text{for } n \geq 200 . \end{cases}$$

The last function, which is of minor practical importance, was added in order to check if the monotonicity (which is present in functions (a)–(c)) can influence the accuracy and execution time of inversion methods.

For all of the presented dropping functions a long-run simulation (72 h) of the queue was performed using Omnet++ simulator, ver. 4.2. In each simulation about 10^9 sample queue sizes were collected. Then the average queue size, the standard deviation of the queue size, the loss ratio, the empty buffer probability, the full buffer probability and the probability that the queue size exceeds 150 packets were computed. A large number of samples provided small confidence intervals for the average queue size. On the 0.999 confidence level, the confidence intervals for the average queue size were ±0.05341, ±0.05924, ±0.03528, ±0.04402 for dropping functions (a)–(d), respectively.

Secondly, the same performance characteristics were computed using the seven described inversion methods. All the results are presented in Tables 1–4. Moreover, in Table 5 the computation times for all methods are summarized.

All calculations were done in double precision, on one core Xeon E5410 2.33 GHz, using Java VM 1.7 32bit, with no extraordinary optimizations. In

Table 1. Performance of the router with the linear dropping function

Method	Average queue size	Standard deviation	Loss ratio	Empty queue probab., P_0	Full buffer probab., P_{200}	Tail prob., $P(X > 150)$
Simulation	1.0419E+02	1.6465E+01	9.0902E-02	4.0756E-09	1.0450E-09	2.2615E-03
Gaver-Stehfest	1.0415E+02	1.6881E+01	9.0910E-02	2.1952E-06	1.6115E-06	3.3996E-03
Talbot	1.0418E+02	1.6468E+01	9.0909E-02	3.7867E-09	4.8989E-10	2.2677E-03
Zakian	1.0418E+02	1.6470E+01	9.0909E-02	1.2709E-08	7.1067E-09	2.2724E-03
Weeks	1.0418E+02	1.6468E+01	9.0909E-02	3.8655E-09	5.4835E-10	2.2677E-03
Piessens	1.0418E+02	1.6470E+01	9.0910E-02	1.4959E-08	8.7772E-09	2.2737E-03
Dubner-Abate	1.0418E+02	1.6468E+01	9.1136E-02	3.4986E-07	2.6856E-07	2.7291E-03
ACW	1,0418E+02	1.6468E+01	9.0909E-02	3.0486E-09	-5.7469E-10	2.2673E-03

Table 2. Performance of the router with the convex dropping function

Method	Average queue size	Standard deviation	Loss ratio	Empty queue probab., P_0	Full buffer probab., P_{200}	Tail prob., $P(X > 150)$
Simulation	1.0263E+02	1.8306E+01	9.0908E-02	2.5436E-09	2.1021E-07	7.7061E-03
Gaver-Stehfest	1.0261E+02	1.8318E+01	9.0910E-02	5.0121E-08	4.9225E-07	7.7494E-03
Talbot	1.0261E+02	1.8308E+01	9.0909E-02	3.8199E-09	3.8455E-07	7.7088E-03
Zakian	1.0261E+02	1.8310E+01	9.0909E-02	1.2662E-08	4.0512E-07	7.7165E-03
Weeks	1.0261E+02	1.8308E+01	9.0909E-02	3.8199E-09	3.8455E-07	7.7088E-03
Piessens	1.0261E+02	1.8308E+01	9.0909E-02	3.7984E-09	3.8450E-07	7.7088E-03
Dubner-Abate	1.0261E+02	1.8308E+01	9.0909E-02	3.8432E-09	3.8458E-07	7.7085E-03
ACW	1.0261E+02	1.8308E+01	9.0909E-02	3.0031E-09	3.8265E-07	7.7081E-03

Table 3. Performance of the router with the concave dropping function

Method	Average queue size	Standard deviation	Loss ratio	Empty queue probab., P_0	Full buffer probab., P_{200}	Tail prob., $P(X > 150)$
Simulation	9.7052E+01	1.0950E+01	9.0908E-02	1.3573E-09	0.0000E+00	3.0493E-08
Gaver-Stehfest	9.7045E+01	1.0959E+01	9.0914E-02	5.2576E-08	8.6210E-09	9.9012E-06
Talbot	9.7045E+01	1.0951E+01	9.0909E-02	1.3244E-09	3.3149E-16	1.6080E-08
Zakian	9.7045E+01	1.0959E+01	9.0910E-02	5.2576E-08	8.6210E-09	9.9012E-06
Weeks	9.7045E+01	1.0951E+01	9.0909E-02	1.3244E-09	4.1073E-15	1.6084E-08
Piessens	9.7045E+01	1.0952E+01	9.0911E-02	1.2418E-08	1.8665E-09	2.1563E-06
Dubner-Abate	9.7116E+01	1.0937E+01	9.1168E-02	3.4734E-07	5.8363E-08	6,9295E-05
ACW	9.7045E+01	1.0951E+01	9.0909E-02	5.0757E-10	-1.3741E-10	-1.4148E-07

Table 4. Performance of the router with the step dropping function

Method	Average queue size	Standard deviation	Loss ratio	Empty queue probab., P_0	Full buffer probab., P_{200}	Tail prob., $P(X > 150)$
Simulation	1.0002E+02	1.3605E+01	9.0903E-02	4.2275E-06	0.0000E+00	0.0000E+00
Gaver-Stehfest	1.0001E+02	1.3622E+01	9.0919E-02	6.7963E-06	7.0451E-09	7.5282E-06
Talbot	1.0001E+02	1.3612E+01	9.0914E-02	4.8955E-06	3.9534E-16	3.0894E-12
Zakian	1.0001E+02	1.3615E+01	9.0915E-02	5.3949E-06	1.8478E-09	1.9745E-06
Weeks	1.0001E+02	1.3612E+01	9.0914E-02	4.8955E-06	1.6158E-16	2.8396E-12
Piessens	1.0001E+02	1.3615E+01	9.0916E-02	5.5207E-06	2.3184E-09	2.4774E-06
Dubner-Abate	1.0005E+02	1.3619E+01	9.1182E-02	2.4326E-05	2.8539E-08	7.5300E-05
ACW	1.0001E+02	1.3611E+01	9.0913E-02	4.8494E-06	-1.7069E-10	-1.8239E-07

Table 5. Computation time for the linear dropping function

Method	Execution time [s]
Simulation	259200
Gaver-Stehfest	4
Talbot	7
Zakian	2
Weeks	10
Piessens	30
Dubner-Abate	51
ACW	1468

the methods which require complex numbers, the real and imaginary parts were also in double precision. As the ACW method requires extended precision[1], the BigInteger, BigDecimal and BigComplex implementations were used.

As we can see, all the exploited methods produce quite accurate numerical values of the average queue size and its standard deviation. Only in the case of the Gaver-Stehfest method, a slight inaccuracy can be observed when the linear dropping function is used. As regards the packet loss ratio, all the methods again produce accurate numerical values, with a slightly larger error in the case of the Dubner-Abate method used for the concave and step dropping function.

Much more demanding are calculations of the empty and full buffer probabilities, i.e. P_0 and P_{200}, which are very small in the considered examples. The simulation results are not fully reliable in this case, as the investigated events are very rare. However, we may distinguish the methods that produce obviously wrong results, such as negative probabilities, or likely wrong results, which are inconsistent with the results obtained by other methods. As for the former, the ACW method produces negative probabilities for three dropping functions, namely the linear, concave and step function. As for the latter, the Dubner-Abate method gives the empty buffer probability which is inconsistent with all other methods, so it is likely to be erroneous.

As regards the execution time, the Zakian, Gaver-Stehfest and Talbot are fast methods; the Weeks, Piessens and Dubner-Abate are moderate, while the ACW method is very slow. The worst result of the ACW method is partially connected with the usage of extended precision numbers. However, the usage of such numbers is forced by the method itself, as it causes range overflow when used with standard double precision numbers. (This is not the case of other methods).

Now, combining the offered accuracies with the execution times we see that the most computationally demanding methods, i.e. the ACW and Dubner-Abate do not offer the best accuracy. On the contrary, their accuracy is poor in some cases. On the other hand, the fastest methods, like Zakian or Talbot, produce accurate values, so they are worth recommending.

5   Conclusions

We compared the accuracy and the execution time of seven inversion methods needed to compute the queueing characteristics of a router equipped with the active queue management based on the dropping function. Surprisingly, the newest, computationally demanding methods do not offer the best accuracy, which in fact is offered by classic, fast algorithms (e.g. the Zakian method).

It must be stressed that this conclusion is specific to the inversion problem we deal with (i.e. to the form of function (2)), therefore it may not apply when other types of inversion problems are considered.

Acknowledgement. This work was partially supported by the Polish National Science Centre under Grant No. N N516 479240.

---

[1] Using standard double precision numbers in this method results in an overflow error.

References

1. Floyd, S., Jacobson, V.: Random early detection gateways for congestion avoidance. IEEE/ACM Tran. Netw. 1, 397–413 (1993)
2. Athuraliya, S., et al.: REM: active queue management. IEEE Network 15(3), 48–53 (2001)
3. Rosolen, V., et al.: A RED discard strategy for ATM networks and its performance evaluation with TCP/IP traffic. ACM SIGCOMM Computer Communication Review 29(3), 23–43 (1999)
4. Farzaneh, N., et al.: A novel congestion control protocol with AQM support for IP-based networks. Telecommunication Systems 52(1), 229–244 (2013)
5. Suzer, M.H., Kang, K.-D., Basaran, C.: Active queue management via event-driven feedback control. Computer Communications 35(4), 517–529 (2012)
6. Na, Z., et al.: A novel adaptive traffic prediction AQM algorithm Journal. Telecommunication Systems 49(1), 149–160 (2012)
7. Chydzinski, A., Chrost, L.: Analysis of AQM queues with queue-size based packet dropping. Int. J. Applied Mathematics and Comp. Science 21(3), 567–577 (2011)
8. Dhamdhere, A., et al.: Buffer sizing for congested internet links. In: INFOCOM 2005, vol. 2, pp. 1072–1083 (2005)
9. Abate, J., Valko, P.P.: Multi-precision Laplace transform inversion. Int. J. Numer. Meth. Engng. 60, 979–993 (2004)
10. Cohen, A.M.: Numerical Methods for Laplace Transform Inversion. Springer, New York (2007)
11. Chydzinski, A.: The M/G-G/1 oscillating queueing system. Queueing Systems 42(3), 255–268 (2002)
12. Chydzinski, A.: The oscillating queue with finite buffer. Performance Evaluation 57(3), 341–355 (2004)
13. Chydzinski, A.: Duration of the buffer overflow period in a batch arrival queue. Performance Evaluation 63(4-5), 493–508 (2006)
14. Stehfest, H.: Numerical inversion of Laplace transforms. Communications of the ACM 13, 47–49 (1970)
15. Talbot, A.: The Accurate Numerical Inversion of Laplace Transforms. J. Inst. Maths. Applics. 23, 97–120 (1979)
16. Zakian, V.: Numerical inversion of Laplace transform. Electronic Letters 5, 120–121 (1969)
17. Weeks, W.T.: Numerical Inversion of Laplace Transforms Using Laguerre Functions. JACM 13, 419–426 (1966)
18. Lyness, J.N., Giunta, G.: A modification of the Weeks method for numerical inversion of the Laplace transform. Math. Comput. 47(175), 313–322 (1986)
19. Garbow, B.S., et al.: Software for an implementation of Weeks' method for the inverse Laplace transform problem. ACM Trans. Math. Software 14(2), 163–170 (1988)
20. Piessens, R.: A New Numerical Method for the Inversion of the Laplace Transform. J. Inst. Maths. Applics. 10, 185–192 (1972)
21. Dubner, H., Abate, J.: Numerical Inversion of Laplace Transforms by Relating them to the Finite Fourier Cosinus Transform. Journ. ACM 15, 115–123 (1968)
22. Abate, J., Choudhury, G.L., Whitt, W.: An introduction to numerical transform inversion and its application to probability models. In: Computational Probability, pp. 257–323. Kluwer (2000)
23. Chydzinski, A.: Transient Analysis of the MMPP/G/1/K Queue. Telecommunication Systems 32(4), 247–262 (2006)

Deterministic Control of the Scalable High Performance Distributed Computations

Zdzislaw Onderka

Dept. of Geoinformatics and Applied Computer Science,
AGH University of Science and Technology, Krakow, Poland
zonderka@agh.edu.pl

Abstract. The paper presents the description of deterministic policies
for distribution tasks of scalable high performance applications computed
in the distributed environment. The model of dynamic behavior of state
of background load for the computers in the network was used (described
in [1]). The analysis of execution time of the set of identical tasks of the
master-slave application was described. Two problems of static policy of
task distribution was discused, first for selecting the optimal subset of
machines and the second one for the choice of optimal partition of the set
of tasks. For the second problem of static policy the algorithm of task
distribution was presented based on the proven property.[1]

Keywords: deterministic control, distributed computation, optimal
task distribution, static policy.

1 Introduction

The policies of the distributed computation control in a heterogeneous environment can be devided into two groups: *deterministic policies* and *stochastic policies*. Under the stochastic policies in the [1] was defined and theoretical analysed
the open-loop control algorithm based on the Markov model for the additional
workload of the computers and was defined and theoretical analysed closed-loop
control algorithm based on the Markov Decision Process theory. Similarly, in
[2] was defined and analysed the Markov model for the additional workload of
the network communication and was defined open-loop and closed-loop control
algorithms of task distribution.

In this work was presented and analysed a group of the deterministic policies
leads to minimum execution time of the distrbuted application where the subset
of task will be allocated to the machines of the heterogeneosus network with
fluctuating load.

The problem of optimal N tasks allocation to the heterogeneous network consisting of m machines is NP-complete. Therefore, many authors propose algorithms with additional restrictions. For example in [3] was presented the solution

---

[1] In the statutory framework of the Dept. of Geoinformatics and Applied Computer
Science, AGH.

A. Kwiecień, P. Gaj, and P. Stera (Eds.): CN 2014, CCIS 431, pp. 253–264, 2014.
© Springer International Publishing Switzerland 2014

with the time complexity $O(N^3m)$ for any m and N and for linear task graph and for linear network architecture . This algorithm was improved in [4] ($O(N^2m)$) and in in [5] with the time complexity $O(m^2N)$. Moreover, the authors of [4] gives algorithm of time complexity $O(mN\log(N))$ for homogeneous processors with certain restrictions imposed on the tasks execution times and communication times and solved the problem of linear task graph allocation to the architecture with the shared memory with the time complexity $O(N\log(N))$ and memory complexity $O(N)$. All these algorithms have polynomial time complexity.

Also, there are a lot of algorithms involved with static and dynamic scheduling but from the point of view of development operating systems for multiprocessor systems. Investigated distributed computation control problems do not use any information on forecast changes in the parameters of a distributed system. For example, in [6] were presented several policies of task distribution, most of which is a type of work-greedy (time complexity $O(N^2m)$ or worst). Another stated policy is to DFBN (Depth First Breadth Next) based on two assumptions: independent tasks are assigned to different processors and dependent task on the same processor. The achieved complexity is $O((n+m)\log(m)+e)$, where e is the amount of edges of the task graf. In [7] the authors proved optimality of the policy which selects the subset of machines that will be assigned for the parallel applications composed of the same tasks. The authors consider the case of uniform load at different rates of the duty cycle (the ratio of the number of cycles allocated to local tasks to the theoretically available total number of cycles) and the case of unequal machines load.

2   Deterministic Policies of the Task Distribution

Let's assume, that the control policies of the task distribution will be realized under the finite horizon, it means consecutive decisions of the control policy will be realized in the time epochs $\Delta_n = [\tau_n, \tau_{n+1})$, $n = 0, 1, \ldots, Z-1$, $Z < \infty$

According to the assumtions made in [1]:

- heterogeneous network is a set of computers H such that $H = \{M_0, \ldots, M_m\}$, where M_i, $i = 1, \ldots, m$ denotes compuers in that network,
- application A is composed of sequential part t_0 and distributed part consisting of N identical tasks $V_D = \{t_1, \ldots, t_N\}$. Each task t_i is executed to the end by exactly one machine without any communication and synchronization with other tasks (possible communication only at the initialiation phase and after computation completion). The computation model of the application A is so called *scheduled task graph* $G_A = (V, E)$, which has one input node and one output node identified with the taks t_0 and for each task $t_i, i = 1, \ldots, N$ exists the path from t_0 to t_i and path from t_i to t_0.
  $E \subseteq V \times V$, $V = V_D \cup \{t_0\}$ and $(t_i, t_k) \in E$ if and only if the task t_i is executed before the task t_k according to the relationships between data and synchronizacyjnymi requirements of the application A.

– The mapping $F : V \to H$ allocates tasks to machines in the network in such a way that:

$$V_{\mathrm{D}} = \bigcup_{i=1}^{m} F^{-1}(M_i) \ , \quad F^{-1}(M_0) = \{t_0\} \ . \tag{1}$$

In case of the dynamic task allocation F is defined as a sequence of the partial mappings $\{F_n\}_{n=0,1,\dots}$ i.e.

$$F = \bigcup_{n} F_n \quad \text{and} \quad F_n^{-1}(M_j) \cap F_k^{-1}(M_j) = \emptyset \ , \quad \forall n \neq k \tag{2}$$

$$\bigcup_{n} F_n^{-1}(M_i) = F^{-1}(M_i) \ \ \forall M_i \in H \ , \quad \mathrm{card}\left(\bigcup_{i=1}^{m}\bigcup_{n} F_n^{-1}(M_i)\right) = N \ .$$

2.1 The *Single-Task* Distribution Policy

Very simple and low effective policy, consists in the fact that in each step task t_0 (*master task*) allocates m tasks, one task to one machine

$$\forall M \in H \ \ \forall n = \mu, \mu+1, \dots, \mu+Z-2 \ , \quad \mathrm{card}(F_n^{-1}(M)) = 1 \ , \tag{3}$$

$$Z = \lfloor N/m \rfloor \ , \quad \forall M \in H \ \ \mathrm{card}(F_{\mu+Z-1}^{-1}(M)) \leq 1$$

At the Fig. 1(I) is presented the *scheduled task graph* for this policy. Let $W(i) \subseteq V_{\mathrm{D}} : W(i) = \{F_i^{-1}(M_1), \dots, F_i^{-1}(M_m)\}$, $i = \mu, \mu+1, \dots$ is a set of tasks distributed in step i and $\mathrm{card}(W(i)) = m$, $i = \mu, \dots, \mu+Z-2$, $\mathrm{card}(W(\mu+Z-1)) < m$ then the execution time $\overline{T}_{W(i)}^{H}$ of the task set $W(i)$ with the additional background load of the network H can be evaluated by the formula

$$\overline{T}_{W(i)}^{H} = \max_{t_{ij} \in W(i)} \left[\overline{T}_{ij}^{j} + \overline{T}_{(ij,0)}^{\mathrm{net}}\right] \tag{4}$$

where $\overline{T}_{(ij,0)}^{\mathrm{net}}$ is the communication time between the task t_0 and the task t_{ij} with additional background load of the communication network and $\overline{T}_{ij}^{j}$ is the execution time of the task t_{ij} with the additional background load of the machine M_j.

Property 1. The execution time $\overline{T}_{V_{\mathrm{D}}}^{H}$ of the distributed part V_{D} of application A on the network H applying the single-task distribution policy is as follows:

$$\overline{T}_{V_{\mathrm{D}}}^{H} = \sum_{i=1}^{\lfloor N/m \rfloor} \overline{T}_{W(i)}^{H} + \overline{T}_{V'}^{H} \tag{5}$$

where

$$V' = V_{\mathrm{D}} - \bigcup_{i} W(i) \ , \quad \mathrm{card}(V') = N - \lfloor N/m \rfloor \ .$$

The value of time $\overline{T}_{W(i)}^{H}$ additionally includes the synchronization time, because the master task waits for the completion of all tasks on all machines of the network in each time epoch of the realized policy.

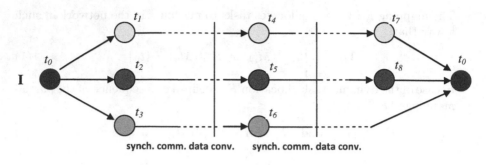

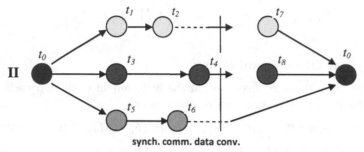

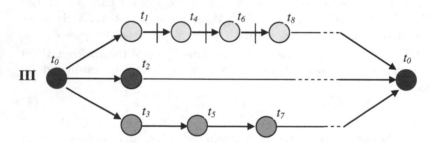

Fig. 1. *Scheduled task graphs* for the deterministic policies (*single task, multiple task, dynamic single*). Identically shaded vertices belong to the same machine.

2.2   The *Multiple Task* Distribution Policy

This policy divides the entire collection of tasks V_D into m equinumerable subsets each of which is mapped to another machine, that is:

$$F^{-1}(M_j) = \left\{ t^j_{i_1}, \ldots, t^j_{i_{k(j)}} \right\} \quad \forall j = 1, \ldots, m \quad \text{and} \quad k = k(j) = \lfloor N/m \rfloor , \quad (6)$$

$$W = \bigcup_{j=1}^{m} F^{-1}(M_j) , \quad V' = V_D - W , \quad \text{card}(V') = N - m \cdot k$$

and if the set V' is not empty, it can be allocated under the previous policy. At the Fig. 1(II) is presented the *scheduled task graph* for this policy. The execution time for the set of tasks W can be calculated according to the formula

$$\overline{T}_W^H = \max_j \left\{ \overline{T}_{F^{-1}(M_j)}^j + \overline{T}_{(i_{k(j)},0)}^{\text{net}} \right\} \tag{7}$$

where

$$\overline{T}_{F^{-1}(M_j)}^j = \left(1 + \varphi_1^j \left(X_\mu^j \right) \right) \cdot \left(\sum_{r=1}^{k(j)} T_{i_r}^j \right) \tag{8}$$

The mapping φ_1^j is the bijection between the set of states of background load for machine M_j and the set of slowing down parameters and the stochastic proces X_μ^j describes the dynamic behavior of state of background load for M_j. The mappings φ_1^j and stochastic proceses X_μ^j for all M_j were defined precisely in [1].

Property 2. The execution time of the application A on the network H using multiple task distribution policy is as follows: $\overline{T}_{V_D}^H = \overline{T}_W^H + \overline{T}_{V'}^H$.

This is two-step policy (as far as $\mathbf{mod}(n, m) \neq 0$) and after the first step occurs the synchronization of the calculations. Of course, one could modify the policy resigning from the second step and adding one other task to each subset of tasks introducing additional unbalance. This can be done in an arbitrary manner but the gain in execution time is little or this can be done by sorting of the values $\overline{T}_t^j$ and by increasing these subsets of tasks which are allocated to the faster machines (according to the sorted list of task execution times). In this case the gain in execution time is little too (loses on sorting).

Compared with the previous policy, the communication and synchronization time $\overline{T}_W^H$ was highly reduced.

2.3   The *Dynamic Single Task* Policy

This is the modification of the first policy. This policy is that the master task t_0 does not synchronize the calculations at each step on all computers in the network. So, the master task do not wait for the completion of the tasks on all machines in each step but dynamically distributes tasks, which means that on machine $M_j \in H$ is initiated the new task execution just immediately after the previous task completion on that machine. This is the *work-greedy* policy because neglecting the communication time does not allow any machine to remain free (does not perform any tasks of application A) if there is still some task to calculate [6]. At the Fig. 1(III) is presented the *scheduled task graph* fot this policy.

This policy applied to homogeneous network is a special case of the presented policy by Kruscal [8] for the allocation of independent tasks to machines of type MIMD.

Let $W_j = \bigcup_{n=\mu} F_n^{-1}(M_j)$, $j = 1, \dots, m$, and $\forall n = \mu, \mu+1, \dots$ and $\text{card}(F_n^{-1}(M_j)) = 1$ is a set of tasks, which are computed on machine M_j during execution of application A and

$$\overline{T}_{W_j}^{j} = \sum_{n=\mu}^{\text{card}(W_j)} \left(\overline{T}_{F_n^{-1}(M_j)}^{j} \right) + \overline{T}_{(i_n,0)}^{\text{net}} , \quad F_n^{-1}(M_j) = t_{i_n} \in V_{\text{D}} . \tag{9}$$

That is

$$\overline{T}_{W_j}^{j} = \sum_{n} \left(1 + \varphi_1^j(X_n^j) \right) \cdot T_t^j + \sum_{n} \overline{T}_{(i_n,0)}^{\text{net}} . \tag{10}$$

The time $\overline{T}_{(i_n,0)}^{\text{net}}$ is the communication time (and if needed the data conversion time) with the additional workload of the communication network deriving from the other processes.

Property 3. With the above determinations the execution time of distributed part V_{D} of the application A in the network H is as follows:

$$\overline{T}_{V_{\text{D}}}^{H} = \max_{M_j \in H} \overline{T}_{W_j}^{j} . \tag{11}$$

Unlike the previous two policies significant gains in the time associated with synchronization of tasks especially in the case of a large diversity of computing power or background load of the machines in the network.

It is easy to make some simple modifications for the described policy, for example, dynamically not allocating tasks one by one but two or more tasks or time-varying number of tasks.

2.4   The *Static* Policy

This is the modification of the multiple task policy. In this case it was assumed that the background workload of each machine will be constant throughout considered time epoch, that is the forecast of the state of the load of each machine is based on the current state of the background workload and do not take into account any changes in background load during the calculations in the time epoch.

Let's consider two problems of task distribution

- how to select the optimal subset of the machines on which the calculated application will reach its maximum value of *speedup*,
- how to divide the whole set of tasks V_{D} into subsets of tasks i.e. $V_{\text{D}} = \bigcup_{j=1}^{m} F^{-1}(M_j)$ and how to map these subsets to machines to obtain the maximum value of *speedup*.

Selecting the Optimal Subset of Machines. First, let's assume that the network H is homogeneous and on each machine $M_j \in H$ can be executed the process that looks after the state of load of the machine and state of load of communication network and based on those informations sends the couple $(\eta_j, \overline{T}_j^{\text{net}})$

to the other one process which prepares the decision which performs the master task of the application A concerning the number tasks that shoud be allocated to machine M_j. Such tracking processes can be a simple agent system composed of multiple *state agents* cooperating with one *decision agent* [1].

The goal is to determine the subset B of machines of the network ($B \in 2^H$, where 2^H denotes the family of subsets of the set H), which will be assigned to the distributed application to obtain the best speedup value.

For this case the *speedup* is the mapping $S_A = 2^H \to \mathcal{R}_+$ and is defined in [7] as:

$$S_A(B) = \frac{\min\{\overline{T}^j_{V_D}, j = 1, \ldots, m\}}{g(B)} \tag{12}$$

$$g(B) = \max_{M_j \in B} \overline{T}^{net}_j + \left(1 + \max_{M_j \in B}\left(\varphi^j_1\left(X^j_n\right)\right)\right) \cdot T^B_A .$$

The algorithm leading to the optimal distribution of tasks have to minimize the function $g(B)$. In the [7] was defined algorithm under the above assumtions with the time complexity $O(m^2)$ and the memory complexity $O(m)$ calculating $\min_{B \in 2^H} g(B) = g(\overline{B}_d)$, $\mathrm{card}(\overline{B}_d = d)$. It is based on the following theorem:

Theorem 1. *Let's L_k be the sorted list consisting of elements of the set* $\left\{\eta_j : \overline{T}^{net}_j < \overline{T}^{net}_k, 1 \le j \le m\right\}$ *for a fixed k, $1 \le k \le m$ and let's $\mathrm{card}(L_k) \ge d - 1$. Let's $\eta_{i_1}, \ldots, \eta_{i_{d-1}}$ will be $d - 1$ first elements of the L_k set inscribed in the asceding order. Then the set $\overline{B}_d = \{M_{i_1}, \ldots, M_{i_{d-1}}, M_k\}$.*

This algorithm $\forall k \colon 1 \le k \le m$ performs the following steps:

1. creates a list L_k,
2. constructs sets $B_{k,d}$,
3. searchs $\min_d g(B_{k,d}) = g(\overline{B}_{k,d})$,

and then selects the best set $\overline{B}_d$ in terms of $\min_k g(\overline{B}_{k,d})$.

Corollary 1. *In the case of the heterogeneous computer network can not be directly applied this algorithm because in the definition (12) the mapping $g(B)$ should be defined in a different way, namely:*

$$g(B) = \max_{M_j \in B} \left(\overline{T}^{net}_j + \left(1 + \varphi^j_1\left(X^j_n\right)\right) \cdot T^j_t\right) .$$

Therefore, the algorithm should create the list L_k consisting of elements:

$$\left\{\alpha_j : \alpha_j = \overline{T}^{net}_j + (1 + \varphi^j_1(X^j_n)) \cdot T^j_t, \alpha_j < \alpha_k, 1 \le j \le m\right\} \tag{13}$$

and the set $\overline{B}_d$ should contain first d machines from L_k.

The Choice of Optimal Partition of the Set of Tasks V_D. In this case the entire set V_D of identical tasks must be devided into m disjoint subsets $F^{-1}(M_j)$, $j = 1, \ldots, m$, for which the greatest speedup is archived. Let's denote

$$W_F(V_D) = \left\{ F^{-1}(M_j) \subset V_D : \sum_{j=1}^{m} \text{card}(F^{-1}(M_j)) = \text{card}(V_D) = N \right\} . \quad (14)$$

To simplify the notation let's assume that in case of very intesive and long calculations the communication time $\overline{T}^{\text{net}}$ is insignificant or can be approximated by the sum of tasks distribution and computation initialization times for the sets $F^{-1}(M_j)$ and times of results receiving by the master task.

If we denote by $\mathcal{A} = \{W_F(V_D)\}_F$ then the speedup is the mapping $S_A : \mathcal{A} \to \mathcal{R}_+$ and is defined as:

$$S_A(W_F(V_D)) = \frac{\min_j \left\{ \overline{T}_{V_D}^j \right\}}{\overline{T}^{\text{net}} + g(W_F(V_D))} \quad (15)$$

$$g(W_F(V_D)) = \max_{F^{-1}(M_j) \in W_F(V_D)} \left(\text{card}(F^{-1}(M_j)) \cdot \overline{T}_t^j \right) .$$

The construction of partition of the tasks set V_D has to minimize $g(W_F(V_D))$ (of course, whereas the previous assumtions related to communication time). The algorithm described below in each iteration step (Algorithm 1 – the loop do... while()) adds to each set $F^{-1}(M_j)$ at most one task, but so that the number of tasks already separated to the sets $F^{-1}(M_j)$ the value

$$\max_{F^{-1}(M_j) \in W_F(V_D)} \left(\text{card}(F^{-1}(M_j)) \cdot \overline{T}_t^j \right) \quad (16)$$

was minimal. So, in each iteration step the algorithm increases $\text{card}(F^{-1}(M_{\sigma(1)}))$ (σ is the permutation, such that $\overline{T}_t^{\sigma(1)} \leq \ldots \leq \overline{T}_t^{\sigma(m)}$) and then increases the value $\text{card}(F^{-1}(M_{\sigma(j)}))$, $j = 2, \ldots, m$ for such indexes j for which the value of calculation time on the machine $M_{\sigma(j)}$ for the current set of tasks $F^{-1}(M_{\sigma(j)})$ is less then or equal to the value of calculation time on the fastest machine $M_{\sigma(1)}$ for the current set of tasks $F^{-1}(M_{\sigma(1)})$:

$$\text{card}\left(F^{-1}\left(M_{\sigma(j)} \right) \right) \cdot \overline{T}_t^{\sigma(j)} \leq \text{card}\left(F^{-1}\left(M_{\sigma(1)} \right) \right) \cdot \overline{T}_t^{\sigma(1)} . \quad (17)$$

This algorithm has the computation time complexity $O(m \log(m) + m \frac{2(N-1)}{m+1})$ (in one step to the following algorithm on average can be distributed $\frac{1}{m} \sum_{j=1}^{m} j = \frac{m+1}{2}$ tasks, and so the average number of steps for the loop do... while() (Algorithm 1) is equal to $\frac{2(N-1)}{m+1}$).

In the Table 1 is presented the example of creation of subsets of tasks for each of four machines ($m = 4$) in the network for $N = 15$ tasks. In the first row of the table for each machine $M_j, j = 1, \ldots, 4$ were given the times needed to perform

Algorithm 1.

Let σ be a permutation of $(1,\ldots,m)$, so that $\overline{T}_t^{\sigma(1)} \leq \ldots \leq \overline{T}_t^{\sigma(m)}$,
/* σ can be evaluated in time $O(m\log(m))$ */
$k_{\sigma(j)} \leftarrow 0$, $j = 2,\ldots,m$
$k_{\sigma(1)} \leftarrow 1$,
/* $W_{\sigma(j)} = \emptyset$, $j = 1,\ldots,m$ and $W_{\sigma(1)} = \{t\}$ */
$Rest \leftarrow N - 1$
do
    for($\sigma(2) \leq j \leq \sigma(m)$)
        if(for M_j : $\overline{T}_{W_j}^j + \overline{T}_t^j \leq \overline{T}_{W_{\sigma(1)}}^{\sigma(1)} + \overline{t}_t^{\sigma(1)}$) $k_j \leftarrow k_j + 1$
    if($\sum_j k_j < N$)
        $k_{\sigma(1)} \leftarrow k_{\sigma(1)} + 1$
        $Rest \leftarrow (Rest - \sum_j k_j)$
while($Rest$ div $m = 0$)

Let L be the sorted list (π is a permutation of $1\ldots,m$) consisting of elements given in the asceding ordder:
$\overline{T}_{W_{\pi(1)}}^{\pi(1)} + \overline{T}_t^{\pi(1)}, \ldots, +\overline{T}_{W_{\pi(m)}}^{\pi(m)} + \overline{T}_t^{\pi(m)}$
for($l = 1,\ldots,(N - Rest)$)
    $k_{\pi(l)} \leftarrow k_{\pi(l)} + 1$

---

a single task and the initial number of tasks (k_j) assigned to the set $F^{-1}(M_j)$. The next rows of the table corresponds to the following iteration steps af the algorithm. In each row, for each machine are given the number of tasks assigned to the created sets and in brackets are given times required to execute these sets of tasks on each machine.

Table 1. The example of the set of 15 tasks partition for the static policy

M_1: $\overline{T}_t^1 = 10$ s, $k_1 = 0$ $\left(\overline{T}_{F^{-1}(M_1)}^1 = 0\right)$	M_2: $\overline{T}_t^2 = 8$ s, $k_2 = 0$ $\left(\overline{T}_{F^{-1}(M_2)}^2 = 0\right)$	M_3: $\overline{T}_t^3 = 7$ s, $k_3 = 0$ $\left(\overline{T}_{F^{-1}(M_3)}^3 = 0\right)$	M_4: $\overline{T}_t^4 = 6$ s, $k_4 = 1$ $\left(\overline{T}_{F^{-1}(M_4)}^4 = 6\right)$
1 (10)	1 (8)	1(7)	2(12)
1 (10)	2 (16)	2(14)	3(18)
2 (20)	3 (24)	3(21)	4(24)
3 (30)	3 (24)	4(28)	5(30)

Another algorithm divides the set V_D in such way that the cardinal numbers of subsets $F^{-1}(M_j)$ were proportional to current state of workload of machines, that is to $\forall M_j \in H$ the value

$$\left| \frac{\text{card}(F^{-1}(M_j))}{\text{card}(V_D)} - \varrho_j \right| \tag{18}$$

was minimal and

$$\varrho_j = \frac{\lambda_j}{\Lambda} \ , \quad \sum_{j=1}^{m} \varrho_j = 1 \ . \tag{19}$$

Definition 1. *The magnitude λ_j is called the* power coefficient for the machine *$M_j \in H, j = 1, \ldots, m$ and the magnitude Λ is called the* power coefficient for the network H. *They are defined, respectively, as*

$$\lambda_j = \frac{1}{\overline{T}_t^j} \quad \Lambda = \sum_{j=1}^{m} \lambda_j \ . \tag{20}$$

where $\overline{T}_t^j$ is the current value of time required for single task computation on machine M_j with the additional load (the background load).

The algorithm that meets the above assumptions is based on the following property:

Property 4. Let's L be the sorted list of elements $\{\beta_j \in \mathcal{R}_+ : \beta_1 \le \beta_2 \le \ldots \le \beta_m, \ \sum_j \beta_j = N, \ N \in \mathcal{N}\}$ then $\exists \gamma_j \in (0,1] \cap \mathcal{R} : \gamma_1 = 1 \ge \gamma_2 \ge \ldots \ge \gamma_m$, such that $\beta_j = \beta_1 \cdot \gamma_j$ and $\forall j = 1, \ldots, m$ the value $\left| \frac{k_j}{N} - \frac{1}{\sum_{j=1}^{m} \gamma_j} \right|$ is minimal for $k_j = \lfloor \beta j \rfloor$.

Let σ is a permutation of numbers $(1, \ldots, m)$, such that $\overline{T}_t^{\sigma(1)} \le \ldots \le \overline{T}_t^{\sigma(m)}$ and let's all machines compute independently consecutively identical tasks.

Denote by $\beta_{\sigma(j)} \in \mathcal{R}_+, \ j = 1, \ldots, m$ the part of tasks computed on all machines at the same time. The computation time is that

$$\sum_j \beta_{\sigma(j)} = N = \operatorname{card}(V_D) \ . \tag{21}$$

The time at which the fastest machine M_j compute one task then the other machines compute only the part of their executed tasks which are equal to

$$\gamma_{\sigma(j)} = \frac{\overline{T}_t^{\sigma(1)}}{\overline{T}_t^{\sigma(j)}} \ , \quad j = 2, \ldots, m \ . \tag{22}$$

Because the tasks are identical then

$$\forall j = 2, \ldots, m \quad \beta_{\sigma(j)} = \beta_{\sigma(1)} \cdot \gamma_{\sigma(j)}$$

$$\gamma_{\sigma(1)} = 1 \ge \gamma_{\sigma(2)} \ge \ldots \ge \gamma_{\sigma(m)} \ . \tag{23}$$

Substituting (23) to (21) and then using the (22) we obtain

$$\beta_{\sigma(1)} \left(1 + \sum_{j=2} \gamma_{\sigma(j)} \right) = N \tag{24}$$

$$\left| \frac{\beta_{\sigma(1)}}{N} - \frac{1}{1 + \sum_{j=2} \gamma_{\sigma(j)}} \right| = 0 \tag{25}$$

$$\left| \frac{\beta_{\sigma(1)}}{N} - \frac{\frac{1}{T_t^{\sigma(1)}}}{\sum_{j=1} T_t^{\sigma(j)}} \right| = 0 . \tag{26}$$

Now, substituting $k_{\sigma(j)} = \lfloor \beta_{\sigma(j)} \rfloor$ we obtain either equation such as (26) if $k_{\sigma(j)} = \beta_{\sigma(j)}$ either if $k_{\sigma(j)} < \beta_{\sigma(j)}$ then the left side of Equality (26) is minimal.

Taking into account in (26) definitions of the power coefficients λ_j and Λ respectively for the machine M_j and for the network H (20) we obtain (18). The computation complexity of this algorithm is the same as the complexity of sorting algorithm (for example $O(m \log(m))$).

Below was presented the example of the action of the above algorithm for the data like in the Table 1:

$$\beta_4 \left(1 + \frac{6}{7} + \frac{6}{8} + \frac{6}{10} \right) = 15$$

$$\beta_j = \beta_4 \cdot \gamma_j \quad j = 1, 2, 3$$

$$\beta_4 = \frac{\lambda_4}{\Lambda} \cdot 15$$

$$\lambda_4 = \frac{1}{6} \quad \Lambda = \frac{449}{840}$$

$$\beta_4 = 4.6771 \quad k_4 = 4$$

$$\beta_3 = 4.0089 \quad k_3 = 4$$

$$\beta_2 = 3.5078 \quad k_2 = 3$$

$$\beta_1 = 2.8063 \quad k_1 = 2 .$$

The last phase of the algorithm will increase the values k_1, k_4 and therefore we get the same result as in the last row of Table 1.

3 Summary

The article presents the description and the analysis of the deterministic policies for the realization of the calculations of the set of tasks in the distributed heterogeneous environment. First, very simple algorithms of deterministic task distribution were presented and the formulas for the execution times were defined. Then two possibilities of the static policy of tasks distribution were discused. First one was the modification of known algorithm for the selecting the optimal subset of machines and the second one was the new algorithm for the problem of choice of the optimal partition of the set of tasks. For this new algorithm the property which defines how to choose the optimal subsets of task was proven.

All described algorithms could be used also for the case of the set of non-identical tasks. In this case the *single task, multiple task* and *dynamic single* policies could be utilized directly. In case of static policy the problem may be in case of preparing the permutation such that $\overline{T}_{t_1}^{\sigma(1)} \leq \ldots \leq \overline{T}_{t_m}^{\sigma(m)}$. But the execution times of different tasks can be seen as the execution times of the identical task on different loaded machines (with the different background load).

Presented policies of task distribution are much easier to apply to control the distributed application then the *closed loop control* and *open loop control* presented in [1,2] and in the case of a small number of tasks (card(V_D)) and may be more efficient to use. Moreover, the stochastic policies requires the existence of the special agent system tracking the load of the computers and the load of the communication network and preparing the forecast of task time execution for the following time step of computations.

Moreover, the presented task distribution policies are easier to apply than ather published algorithms for the case of fluctuating additional load and fluctuating network load. The proposed solutions can be simply applied to the most time complex parts of SBS computation [9] – matrix formulation and non linear algebraic solver.

References

1. Onderka, Z.: Stochastic Control of the Scalable High Performance Distributed Computations. In: Wyrzykowski, R., Dongarra, J., Karczewski, K., Waśniewski, J. (eds.) PPAM 2011, Part II. LNCS, vol. 7204, pp. 181–190. Springer, Heidelberg (2012)
2. Onderka, Z.: Stochastic Model of the Network Communication for the Control of the Distributed Computations. Studia Informatica 33(3A), 79–90 (2012) (in Polish)
3. Bokhari, S.H.: Partitioning Problems in Parallel, Pipelined and Distributed Computing. IEEE Trans. Software Engrg. 37(1), 48–57 (1988)
4. Nicol, D.M., O'Hallaron, D.R.: Improved Algorithms for Mapping Pipelined and Parallel Computations. IEEE Trans. Comput. 40(3), 295–306 (1991)
5. Hansen, P., Lih, K.-W.: Improved Algorithms for Partitioning Problems in Parallel, Pipelined and Distributed Computing. IEEE Trans. Comp. 41(6), 769–771 (1992)
6. Manoharan, S., Topham, N.P.: An Assessment of Assignment Schemes for Dependency Graphs. Parallel Computing 21, 85–107 (1995)
7. Atallah, M.J., Black, C.L., Marinescu, D.C., Siegel, H.J., Casavant, T.L.: Models and Algorithms for Coscheduling Compute-Intensive Tasks on a Network of Workstations. Journal of Parallel and Distributed Computing 16, 319–327 (1992)
8. Kruskal, C.P., Weiss, A.: Allocating Independent Subtasks on Parallel Processors. IEEE Trans. on Software Engrg. SE-11(10), 1001–1016 (1985)
9. Onderka, Z., Schaefer, R.: Markov Chain Based Management of Large Scale Distributed Computations of Earthen Dam Leakages. In: Palma, J.M.L.M., Dongarra, J. (eds.) VECPAR 1996. LNCS, vol. 1215, pp. 49–64. Springer, Heidelberg (1997)

A Technique for Detection of Bots Which Are Using Polymorphic Code

Oksana Pomorova, Oleg Savenko, Sergii Lysenko,
Andrii Kryshchuk, and Andrii Nicheporuk

Department of System Programming, Khmelnitsky National University,
Instytutska, 11, Khmelnitsky, Ukraine
o.pomorova@gmail.com, {savenko_oleg_st,sirogyk}@ukr.net,
rtandrey@rambler.ru, paunni@mail.ru
http://spr.khnu.km.ua

Abstract. The new technique of botnet detection which bots use polymorphic code was proposed. Performed detection is based on the multi-agent system by means of antiviral agents that contain sensors. For detection of botnet, which bots use polymorphic code, the levels of polymorphism were investigated and its models were built. A new sensor for polymorphic code detection within antivirus agent of multi-agent system was developed. Developed sensor performs provocative actions against probably infected file, restarts of the suspicious file for probably modified code detection, behavior analysis for modified code detection, based on the principles of known levels of polymorphism.

Keywords: bot, botnet, polymorphic code, levels of the polymorphism, multi-agent system, agent, sensor.

1 Introduction

Today the problem of cyber security is very important because the data protection problem is extremely relevant.

Virus detection is a very important task because the information pilfering, anonymous access to network, spy actions, spamming are observed.

The most dangerous occurrence in the virus elaboration is botnet – network of private computers infected with malicious software and controlled as a group without the owners' knowledge, e.g. to send spam [1].

Some examples of the most dangerous botnet are Virut, which have infected over 3 million unique users computer systems according to the report "Kaspersky Lab" [2]; botnet Zeus, is designed to attack servers and intercept personal data (damage for European customers is about € 36 million); botnet Kelihos, which performs abduction of passwords stored in the browser, sends spam, steals users' credentials. All mentioned botnets include the self-defense modules, which have contributed to their rapid spread and inefficient detection [3,4].

These facts indicate a lack of effectiveness of the known detection methods. That is why an important task is to build new techniques and approaches to

A. Kwiecień, P. Gaj, and P. Stera (Eds.): CN 2014, CCIS 431, pp. 265–276, 2014.

identify botnet, which will take into account its properties and availability of hiding module.

2   Related Works

To conceal the presence of a botnet, the polymorphism technology is used.

For polymorphic malwares, the decryptor part of the virus is mutated at each infection, thanks to common obfuscation techniques: "garbage-commands" insertion, register reassignment, and instruction replacement. In order to detect such high-mutating viruses, several solutions have been developed.

Byte-level detection solutions. Current antiviral solutions use different techniques in order to detect malicious files. The techniques are: pattern-matching, emulation, dynamic behavioral detection, and various heuristics [5]. Authors focused on pattern-matching techniques (heuristics are aimed at new malware detection and are subject to a high false positive rates, while emulation may not always succeed). Because they have time and complexity constraints, the models and detection algorithms used in today's antiviral products are relatively simple. The detection algorithm consists of determining whether a given binary program is recognized by one of the viral signature. Since regular expressions are used as signature descriptions, antivirus products may use finite state automatons to perform linear-time detection.

Another emerging approach consists of using machinelearning techniques in order to detect malicious files [6]. Several models have been tested: data-mining [7], Markov chains on n-grams [8,9], Naive Bayes as well as decision trees [10]. These methods provide an automatic way to extract signature from malicious executables. But while the experiments have shown good results, the false positive and negative rates are still not negligible.

Structural and semantic models. In [11,12], graphs are used as a model for malwares. The control flow graph (CFG) of a malware is computed (when possible) and reduced. Then, subsets of this graph are used as a signature. Detection of a malware is done by comparing a suspicious file against these sub-CFGs, and seeing if any part of the CFG of the file is equivalent (with a semantics-aware equivalence relation for [11]) to a sub-CFG in the viral database. The idea is that most of mutation engines' obfuscations do not alter the control flow graph of the malware.

CTPL (Computation Tree PredicateLogic) is a variant of CTL (Computation Tree Logic) [13], able to handle register-renaming obfuscation. Detection is done via model checking of API call sequence, while signatures extraction is done manually.

A promising approach was initiated by Preda et. al [14] and followed by [15]. It consists of using the semantics of a metamorphic malware as a viral signature. None of them provides an automated process to extract this grammar for a given malware.

3   Previous Work

For new botnet detection in [16] the technique for determining the degree of presence of botnet based on multi-agent systems was proposed. Offered method is based on analyzing the bots actions demonstration in corporate area network. The technique proposes the construction of a schematic map of connections which is formed by corresponding records in each antiviral agent of multi-agent systems for some corporate area network. All agents based on this information can perform communicative exchange data to each other. Proposed method is based on analyzing the bots actions demonstration in situations of intentional change of connection type is probably infected computer system.

During computer system (CS) is functioning the antivirus detection via sensors available in an each agent is performed. The antivirus diagnosis results are analyzed in order to define which of sensors have triggered and what suspicion degree it has produced. If triggering sensors are signature S_1 or checksum S_2 analyzers, the results R_{S_1} and R_{S_2} are interpreted as a 100 % malware detection. In this situation, the blocking of software implementation and its subsequent removal are performed.

For situations when the sensors of heuristic S_3 and behavioral S_4 analyzers have triggered, the suspicion degrees R_{S_3} and R_{S_4} are analyzed, and in the case of overcoming of the defined certain threshold n, $n \leq \max(R_{S_3}, R_{S_4}) \leq 100$, the blocking of software implementation and its subsequent removal are performed. If the specified threshold hasn't overcome the results R_{S_3}, R_{S_4} are analyzed whether they belong to range $m \leq \max(R_{S_3}, R_{S_4}) < n$ in order to make the final decision about malware presence in CS. If the value is $\max(R_{S_3}, R_{S_4}) > m$ than the new antivirus results from sensors are expected. In all cases the antiviral agents information of infection or suspicion software behavior in CS must be sent out to other agents.

The important point of this approach is to research the situation where the results of antivirus diagnosis belong to range $m \leq \max(R_{S_3}, R_{S_4}) < n$. In this case, the antiviral agent of CS asks other agents in the corporate area network about the similarity of suspicion behavior of some software that is similar to the botnet. After that the analysis of botnet demonstrations on computer systems of the corporate area network and the definition of the degree of a new botnet presence in the network was determined. The presence of botnet in the corporate area network was concluded by the fuzzy expert system that confirmed or disproved this fact.

The developed system has been demonstrating the efficiency of botnet detection at about 88–96 %.

As some botnets use the technology of hiding malicious code (polymorphic code) today, we have tested our multi-agent systems for botnet detection, where bots used such technology. Test results were unexpected. It turned out that the developed system was not fully adapted to detect polymorphic code, and efficiency decreased by 7–12 %. Also after retest some bots were detected, which had been previously identified and removed (all bots contained the polymorphic code).

That is why the actual problem is a development of a new botnet detection technique that will find out the polymorphic code in bots.

4 Technique for Bots Detection Which Use Polymorphic Code

In order to develop the technique for bot detection we have to investigate the properties of polymorphic viruses. They create varied (though fully functional) copies of themselves as a way to avoid detection by anti-virus software. Some polymorphic virus use different encryption schemes and require different decryption routines. Thus, the same virus (bot) may look completely different on different systems or even within different files. Other polymorphic viruses vary instruction sequences and use false commands in the attempt to thwart anti-virus software. One of the most advanced polymorphic viruses uses a mutation engine and random-number generators to change the virus code and its decryption routine [17].

4.1 Levels of Polymorphism

Today 6 polymorphism levels are known [18]. Let us build models for all the levels of polymorphism.

Viruses of the first polymorphism level use the constant set of actions for different decryption modules. They can be detected by some areas of permanent code decryption.

Let us present the virus model for the first level of polymorphism as a tuple

$$M_1 = (A, V, X, G, U, \xi, Q, P, R) \tag{1}$$

where A – a set of commands of some program which can be infected with virus, $A = \{a_1, \ldots, a_n\}$; V – a set of virus commands for selection of one of the present decryption modules in virus, $V = \{v_1, \ldots, v_m\}$; X – a set of decryption modules which are present in virus, $X = \{x_1, \ldots, x_y\}$; G – set of virus commands of the x_i decryption module, $G = \{g_{1_{x_i}}, \ldots, g_{\theta_{x_i}}\}$; U – a set of malicious commands (virus body), $U = \{u_1, \ldots, u_w\}$; ξ – a function for selection of decryption module x_i, $\xi : V \rightarrow X$, $x_i \in X$; Q – a function of creation the malicious commands (virus body) by the means of commands $g_{x_i} \in G$ of the decryption's module x_i, $Q : G_{x_i} \rightarrow U$; P – a function of creation the polymorphic virus behavior R by the means of inserting the malicious commands U into program's commands A, $P : A \times U \rightarrow R$; function of creation the polymorphic virus behavior R without inserting malicious commands U into program's commands A just by the means of the decryption virus body U appears as $Q : U \rightarrow R$.

Thus, polymorphic virus has its behavior that is formed by some sequences of commands. Based on this we can build the virus behavior as sequences.

Virus behavior R_1^A of the first polymorphism level which is created by the means of inserting the malicious commands U into program's commands A

and virus behavior R_1 which is created without inserting the malicious commands U into program's commands A can be presented as sequences: $R_1^A = g_{\phi_{x_\varepsilon}} \ldots g_{\eta_{x_\varepsilon}} a_1 \ldots a_n u_1 \ldots u_w$, $R_1 = g_{\phi_{x_\varepsilon}} \ldots g_{\eta_{x_\varepsilon}} u_1 \ldots u_w$, where values ϕ, η indicate that possible virus commands of decryption module $g_{\phi_{x_\varepsilon}} \ldots g_{\eta_{x_\varepsilon}}$ can vary for different decryption modules x_ε, ε – number of the selected decryption module.

The second level of polymorphism includes viruses, which decryption module has constant one or more instructions. For example, they may use different registers or instructions in some alternative decryption module. These viruses can also be identified by a specific signature in the decryption module [18].

Let us present the virus model for the second level of polymorphism as a tuple

$$M_2 = (A, E, U, P, Z, R) \tag{2}$$

where A – a set of commands of some program which can be infected with virus, $A = \{a_1, \ldots, a_n\}$; E – a set of virus commands of the decryption module, $E = \{e_1, \ldots, e_\theta\}$; U – a set of malicious commands (virus body), $U = \{u_1, \ldots, u_w\}$; Z – a function of creation the malicious commands (virus body) by the means of selection of the present decryption module's commands, $Z : E \rightarrow U$; P – a function of creation the polymorphic virus behavior R by the means of inserting malicious commands U into program's commands A, $P : A \times U \rightarrow R$; function of creation the polymorphic virus behavior R without inserting malicious commands U into program's commands A appears as: $Z : E \times U \rightarrow R$.

Virus behaviors R_2^A and R_2 of the second polymorphism level can be presented as sequences: $R_2^A = e_\kappa \ldots e_\lambda a_1 \ldots a_n u_1 \ldots u_w$; $R_2 = e_\kappa \ldots e_\lambda u_1 \ldots u_w$, where values κ, λ indicate that possible virus commands $e_\kappa \ldots e_\lambda$ of decryption module can vary for each new start of virus.

Viruses that use decryption commands and do not decrypt the virus code and have a "garbage-commands" refer to the third level of polymorphism. These viruses can be determined using a signature if all the "garbage-commands" are discarded. Viruses of the fourth level use the interchangeable or "mixed" instructions for the decryption without changing the decryption algorithm [18].

Let us present the virus model for the third and fourth levels of the polymorphism as a tuple

$$M_{3,4} = (A, E, U, B, Y, D, R) \tag{3}$$

where A – a set of commands of some program which can be infected with virus, $A = \{a_1, \ldots, a_n\}$; E – a set of virus commands of the decryption module, $E = \{e_1, \ldots, e_\theta\}$; U – a set of malicious commands (virus body), $U = \{u_1, \ldots, u_w\}$; B – a set of the "garbage-commands", $B = \{b_1, \ldots, b_t\}$; Y – a function of creation the malicious commands (virus body) by the means of decryption module, that integrates the "garbage-commands" into malicious commands, $Y : E \times B \rightarrow U$; D – a function of creation the polymorphic virus behavior R by the means of virus body inserting malicious commands U into program's commands A, $D : A \times U \rightarrow R$; function of creation the polymorphic virus behavior R without inserting malicious commands U into program's commands A appears as: $Y : E \times B \rightarrow R$.

Virus behaviors R_3^A, R_4^A of the third and fourth polymorphism levels which are created by the means of decryption module, which integrates the "garbage-commands" into malicious commands and by the means of inserting malicious commands U into program's commands A and virus behaviors R_3, R_4 without inserting malicious commands U into program's commands A can be presented as sequences: $R_3^A = e_1 \ldots e_\theta a_1 \ldots a_n u_1 b_\rho \ldots u_w b_\zeta$; $R_3 = e_1 \ldots e_\theta u_1 b_\rho \ldots u_w b_\zeta$; $R_4^A = e_1 \ldots e_\theta a_1 \ldots a_n u_\vartheta b_\rho \ldots u_\sigma b_\zeta$; $R_4 = e_1 \ldots e_\theta u_\vartheta b_\rho \ldots u_\sigma b_\zeta$, where values ρ, ζ, ϑ, σ indicate that possible "garbage-commands" and virus commands $u_\vartheta b_\rho \ldots u_\sigma b_\zeta$ can vary for each new start of virus.

The fifth level of polymorphism includes all properties of the above levels, and the decryption module may use different algorithms for decrypting the virus code [18].

Let us present the virus model for the fifth level of polymorphism as a tuple

$$M_5 = (A, B, X, G, U, \xi, H, D, R) \tag{4}$$

where A – a set of commands of some program which can be infected with virus, $A = \{a_1, \ldots, a_n\}$; V – a set of commands for selection of one of the present decryption modules in virus, $V = \{v_1, \ldots, v_m\}$; X – a set of decryption modules which are present in virus, $X = \{x_1, \ldots, x_y\}$; G – a set of virus commands of the decryption module x_i, $G = \{g_{1_{x_i}}, \ldots, g_{\theta_{x_i}}\}$; U – a set of malicious commands (virus body), $U = \{u_1, \ldots, u_w\}$; ξ – a function for x_i decryption module selection, $\xi : V \to X$, $x_i \in X$; B – a set of the "garbage-commands", $B = \{b_1, \ldots, b_t\}$; H – a function of creation the malicious commands (virus body) by the means of selection of the present decryption module's x_i commands $g_{x_i} \in G$ and generation the order of its execution, $H : B \times G_{x_i} \to U$; D – a function of creation the polymorphic virus behavior R by the means of virus body inserting malicious commands U program's commands A, $D : A \times U \to R$; function of creation the polymorphic virus behavior R without inserting malicious commands U into program's commands A by the means of selection of the present decryption module's x_i commands $g_{x_i} \in G$ and generation order of its execution appears as $D : U \to R$.

Virus behaviors R_5^A and R_5 of the fifth polymorphism level can be presented as sequences: $R_5^A = g_{\phi_{x_\varepsilon}} \ldots g_{\eta_{x_\varepsilon}} a_1 \ldots a_n u_\vartheta b_\rho \ldots u_\sigma b_\zeta$; $R_5 = g_{\phi_{x_\varepsilon}} \ldots g_{\eta_{x_\varepsilon}} u_\vartheta b_\rho \ldots u_\sigma b_\zeta$, where values ρ, ζ indicate that possible virus commands of decryption module $g_{\phi_{x_\varepsilon}} \ldots g_{\eta_{x_\varepsilon}}$ can vary for different decryption modules x_ε, ϑ – number of the selected decryption module, values ρ, ζ, ϑ, σ indicate that possible "garbage-commands" and virus commands $u_\vartheta b_\rho \ldots u_\sigma b_\zeta$ can vary for each new start of virus.

Viruses of the sixth level of polymorphism consist of software units and parts that "move" within the body of the virus. These viruses are also called permutating [18].

Let us present the virus model for the sixth level of polymorphism as a tuple

$$M_6 = (A, E, U, C, R) \tag{5}$$

where A – a set of commands of some program which can be infected with virus, $A = \{a_1, \ldots, a_n\}$; E – a set of decryption module's commands, $E = \{e_1, \ldots, e_\theta\}$;

U – a set of malicious commands (virus body), $U = \{u_1, \ldots, u_w\}$; C – a function of creation the polymorphic virus behavior R formed by program's commands, decryption commands and malicious commands as blocks in some order, $C : A \times E \times U \rightarrow R$; a function of creation the polymorphic virus behavior R formed only by the decryption commands and malicious command as blocks in some sequence appears as: $C : E \times U \rightarrow R$.

Virus behaviors R_6^A and R_6 of the sixth polymorphism level can be presented as sequences: $R_6^A = a_1 e_\phi \ldots a_i e_\eta a_{i+1} u_\vartheta \ldots a_n u_\sigma$; $R_6^A = a_1 e_\phi u_\vartheta a_2 \ldots a_n e_\eta u_\sigma$; $R_6 = e_\phi \ldots e_\eta u_\vartheta \ldots u_\sigma$; $R_6 = e_\phi u_\vartheta \ldots e_\eta u_\sigma$, where values ϕ, η, ϑ, σ indicate that possible virus commands of decryption module and malicious commands $e_\phi \ldots e_\eta u_\vartheta \ldots u_\sigma$ can vary for each new start of virus.

4.2 Polymorhic Code Detection Sensor

To detect botnet that use polymorphic code, the inclusion of a new sensor S_7 for agent of the multi-agent system is proposed. This sensor must be a virtual environment that allows the emulation of execution some specific action towards the potentially malicious software. Responses to the actions allow to conclude that polymorphic code is present in it. Taking into account the properties of polymorphic viruses, sensor S_7 have to perform:

- provocative actions against probably infected file;
- restarts of the suspicious file for probably modified code detection;
- behavior analysis for modified code detection, based on the principles of known levels of polymorphism.

Provocative actions mean the identification of the polymorphic viruses' properties to create their own copies and to change their body when they are removed. This property often leads to the fact that the original virus can be found and removed, and its new copy will be invisible to antivirus.

Restarts of the suspicious software can show the possible change of the program body as a result of decryption. Detection such change is possible due to the construction of "fingerprints" of reference K and modified K' files and their subsequent comparison. "Fingerprints" K and K' are formed by a defined binary sequence $K = \alpha, \beta, \chi, \delta, \varepsilon$, where α – file name; β – file size; χ – last modification time; δ – system attribute; ε – 128 byte code of MD5.

Restarts of the suspicious software are performed by the algorithm:

```
Form the "fingerprint" K for file_M for i=1 to p times do
    execute file_M
    Form the "fingerprint" K'
    if (K ≠ K') then
        then sensor S7 notifies processor of the agent_i to block file_M;
    end
end
```

Algorithm 1. Detection the polymorphic mutation in a suspicious file by its restarts

Sensor S_7 also provides the behavioral analyzer, which evaluates the program's actions with taking into account the models of polymorphic viruses of different levels. Based on knowledge of the polymorphic viruses' behaviors and botnet behaviors it is possible the botnet detection by comparing the known behaviors with the new ones. Identification of polymorphic code is performed with taking into account the rejection of possible "garbage-commands", the permutations of commands, commands for decryptor selection, decryptor's commands etc. Behaviors are represented by sequences that are compared.

In order to perform the comparison the reference behaviors with the potentially malicious behavior, the approximate string matching algorithm, developed by Tarhio and Ukkonen [19], is used. It solves the k differences problem. Given two strings, text $\Psi = \phi_1\phi_2\ldots\phi_n$ and pattern $\Sigma = \varepsilon_1\varepsilon_2\ldots\varepsilon_m$ and integer k, the task is to find the end points of all approximate occurrences of Σ in Ψ. An approximate occurrence means a substring Σ' of such that almost k editing operations (insertions, deletions, changes) are needed to convert Σ' to Σ. The algorithm has scanning and checking phases. The scanning phase based on a Boyer Moore idea repeatedly applies two operations: mark and shift. Checking is done by enhanced dynamic programming algorithm. The algorithm needs time $O(kn)$ [20].

Based on knowledge of the possible botnet behaviors of bots there were generated 200 bots' behaviors. Taking into account the knowledge of the polymorphism levels there were generated 10 000 polymorphic behaviors. Each of them is represented by a sequence. The alphabet of sequences is defined by a set of API-functions $\Omega = \omega_1\ldots\omega_f$, which are the base for malware construction. For the experiment the behavior of three well-known bots [21] were constructed and investigated, which were "unknown" with respect to our base behaviors. These bots use three levels of polymorphism.

The experimental results of the approximate string matching are presented in Table 1.

The results showed that an exact match ($k = 0$) had not found a solution, however, when $k = 2$, $k = 3$ the number of solutions was sufficiently small. With increasing k the number of solutions was growing rapidly, but the search time for matches also increased. Thus, the experiments proved that for the detection of the similar suspicious behavior it was enough to lay down parameter $k = 4$. In practice, the sensor S_7 stops the search approximate matches when it detects the first match.

Based on the concept of antivirus multi-agent system functioning, each agent is waiting for triggering of heuristic S_3 or behavioral S_4 sensors. If one of them have triggered or the fact of file unpacking has been detected then the suspicious file is placed in the sensor emulator S_7.

Algorithm of sensor S_7 functioning is shown below.

Sensor S_7 functioning in agent of the multi-agent system is presented in Fig. 1.

Table 1. The experimental results of the approximate string matching for different values of length R and parameter k

	Alphabet Ω	Length of sequence R	k – difference parameter	Number of found strings
	300	38	2	0
P1	300	38	3	0
	300	38	4	1
	300	38	5	2
	300	93	2	0
P2	300	93	3	1
	300	93	4	2
	300	93	5	7
	300	71	2	1
P3	300	71	3	4
	300	71	4	14
	300	71	5	22

for $i=1$ to k agents **do**

    while $(agent_i$ is_ on$)$ **do**

        if $((R_{S_3} = true)or(R_{S_4} = true))and(m \leq \max(R_{S_3}, R_{S_4}) \leq n$ **then**

        | **state** probably infected $file\_M$ is placed in sensor S_7;

        end

        if $(file\_M$ makes unpacking $)$ **then**

        | $file\_M$ is blocked and is placed in sensor S_7;

        end

    end

    while $(file\_M$ is in sensor S_7 $)$ **do**

        if $(provocative$ actions regarding to $file\_M$ have detected the new file creation $)$ or $(new$ file creation with mutation$)$ **then**

        | sensor S_7 notifies processor of the $agent_i$ to block $file\_M$;

        | collected information about $file\_M$ is sent to other agents;

        end

        if $(restarts$ have detected the $file\_M$ body mutation$)$ **then**

        | sensor S_7 notifies processor of the $agent_i$ to block $file\_M$;

        | collected information about $file\_M$ is sent to other agents;

        else

        | behavior analysis is being performed;

        end

        if $(result$ of behavior analysis $R_{S_7} = true$ $)$ **then**

        | sensor S_7 notifies processor of the $agent_i$ to block $file\_M$;

        | collected information about $file\_M$ is sent to other agents;

        else

        | $file\_M$ leaves the sensor S_7

        end

    end

end

Algorithm 2. Sensor S_7 functioning algorithm

5   Experiments

In order to determine the efficiency of the proposed technique for botnet detection several experiments were held. Bots used polymorphic code. Experiments were carried out on the base of developed multi-agent system that is functioning in the corporate area network. The main aim of the experiment was to determine the effectiveness of the botnet detection with the use of sensor S_7 and without it.

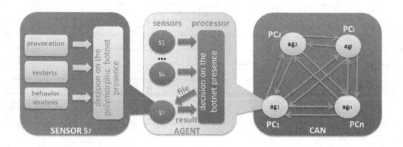

Fig. 1. Sensor S_7 functioning in agent of the multi-agent system

For the implementation of an experiment 50 programs with the botnet properties (Agobot, SDBot and GT-Bot) without polymorphic code were generated. Also 50 programs (its analogs) with polymorphic code were generated (programs contained only first four levels of polymorphism). During the experiment computer systems in the corporate area network were infected only by one botnet and experiment was lasting during 24 hours. As a virtual environment for sensor S_7 functioning the virtual machine Oracle VirtualBox [22] was used; as a host operating system MS Windows 7 was used.

The results of the experiment are shown in Table 2.

Table 2. The results of the experiment for 50 programs

	Detection		Fault positives
	%	number	number
Results of detection without sensor S_7; Programs do not use polymorphic code	90	45	5
Results of detection without sensor S_7; Programs use polymorphic code	76	38	5
Results of detection with sensor S_7; Programs use polymorphic code	92	46	6

Experimental results showed the growth of the botnet detection efficiency which bots used polymorphic code by means of the multi-agent system including sensor S_7.

In order to compare developed antiviral multi-agent system (AMAS) with other antiviruses some experiments were held. We have tested 5 antiviruses with 50 generated bots, which contained polymorphic code. Results are presented in Fig. 2.

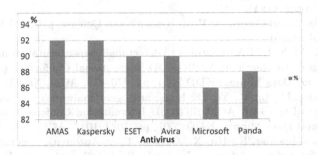

Fig. 2. Test results (14–24.12.2013)

6 Conclusions

The new technique of botnet detection which bots use polymorphic code was proposed. Performed detection is based on the multi-agent system by means of antiviral agents that contain sensors.

For detection of botnet, which bots use polymorphic code, the levels of polymorphism were investigated and its models were built.

The new sensor for polymorphic code detection within antivirus agent of multi-agent system was developed.

Developed sensor performs provocative actions against probably infected file, restarts of the suspicious file for probably modified code detection, behavior analysis for modified code detection, based on the principles of known levels of polymorphism.

Results of the experiments have demonstrated the increase of the botnet detection efficiency by 16 % with involving the sensor S_7 compared to its absence. Thus the growth of false positives is not significant.

The disadvantage of the proposed technique is sufficiently large computational complexity of the behavior analysis that is based on the principles of polymorphism levels.

References

1. Oxford Dictionaries,
   http://www.oxforddictionaries.com/definition/english/botnet?q=botnet
2. Nikitina, T.: In Virut have taken away key domains (2013),
   http://www.securelist.com/ru/blog/207764413/U_Virut_otobrali_
   klyuchevye_domeny#page_top

3. Yaneza, J.: ZeuS/ZBOT Malware Shapes Up in 2013 (2013),
   http://blog.trendmicro.com/trendlabs-security-intelligence/
   zeuszbot-malware-shapes-up-in-2013/
4. Scott, M.E.: Boston Marathon/West, Texas Spam Campaigns (2013),
   http://mrpdchief.blogspot.com/2013/04/boston-marathonwest-texas-
   spam-campaigns.html
5. Szor, P.: The Art of Computer Virus Research and Defense, p. 744. Addison-Wesley
   Professional (2005)
6. Kolter, J.Z., Maloof, M.A.: Learning to detect malicious executables in the wild. In:
   Proceedings of the tenth ACM SIGKDD International Conference on Knowledge
   Discovery and Data Mining, KDD 2004, pp. 470–478. ACM, New York (2004)
7. Ye, Y., Wang, D., Li, T., Ye, D.: Imds: intelligent malware detection system. In:
   Proceedings of the 13th ACM SIGKDD International Conference on Knowledge
   Discovery and Data Mining, KDD 2007, pp. 1043–1047. ACM, New York (2007)
8. Griffin, K., Schneider, S., Hu, X., Chiueh, T.-c.: Automatic generation of string
   signatures for malware detection. In: Kirda, E., Jha, S., Balzarotti, D. (eds.) RAID
   2009. LNCS, vol. 5758, pp. 101–120. Springer, Heidelberg (2009)
9. Yan, W., Wu, E.: Toward automatic discovery of malware signature for anti-virus
   cloud computing. In: Zhou, J. (ed.) Complex 2009. LNICST, vol. 4, pp. 724–728.
   Springer, Heidelberg (2009)
10. Wang, J.-H., Deng, P., Fan, Y.-S., Jaw, L.-J., Liu, Y.-C.: Virus detection using data
    mining techinques. In: Proceedings of the IEEE 37th Annual 2003 International
    Carnahan Conference on Security Technology (2003)
11. Christodorescu, M., Jha, S.: Static analysis of executables to detect malicious pat-
    terns. In: Proc. of the 12th USENIX Security Symposium, pp. 169–186 (2003)
12. Bonfante, G., Kaczmarek, M., Marion, J.-Y.: Control flow graphs as malware sig-
    natures. In: Filiol, E., Marion, J.-Y., Bonfante, G. (eds.) International Workshop
    on the Theory of Computer Viruses TCV 2007, Nancy, France (2007)
13. Clarke, E., Emerson, E.: Design and synthesis of synchronization skeletons using
    branching time temporal logic. In: Kozen, D. (ed.) Logic of Programs 1981. LNCS,
    vol. 131, pp. 52–71. Springer, Heidelberg (1982)
14. Leder, F., Steinbock, B., Martini, P.: Classification and detection of metamorphic
    malware using value set analysis. In: 2009 4th International Malicious and Un-
    wanted Software (MALWARE), pp. 39–46 (2009)
15. Preda, M.D., Christodorescu, M., Jha, S., Debray, S.: A semanticsbased approach
    to malware detection. ACM Trans. Program. Lang. Syst. 30(5), 1–54 (2008)
16. Pomorova, O., Savenko, O., Lysenko, S., Kryshchuk, A.: Multi-Agent Based Ap-
    proach for Botnet Detection in a Corporate Area Network Using Fuzzy Logic.
    In: Kwiecień, A., Gaj, P., Stera, P. (eds.) CN 2013. CCIS, vol. 370, pp. 146–156.
    Springer, Heidelberg (2013)
17. Glossary (2014), http://home.mcafee.com/virusinfo/glossary?ctst=1#P
18. Kaspersky, E.: Computer viruses. SK-Press, Moscow (1998)
19. Jokinen, P., Tarhio, J., Ukkonen, E.: A Comparison of Approximate String Match-
    ing Algorithms. Software: Practice and Experience 26(12), 1439–1458 (1996),
    http://onlinelibrary.wiley.com/doi/10.1002/SICI1097-024X19961226:
    121439:AID-SPE713.0.CO;2-1/abstract
20. Smyth, B.: Computing Patterns in Strings, p. 496. Williams, Moscow (2006)
21. http://security.ludost.net/exploits/index.php?dir=bots/
22. https://www.virtualbox.org

Applications of QR Codes
in Secure Mobile Data Exchange

Artur Hłobaż, Krzysztof Podlaski, and Piotr Milczarski

University of Lodz, Faculty of Physics and Applied Informatics, Lodz, Poland

Abstract. In the paper new method of secure data transmission between mobile devices is proposed. Already existing applications demand on the user to care about secure issues himself or herself. Derived method focuses on secure channel creation using well known QR codes technology, which allows prevent the eavesdropper from interception transmitted data during connection establishing process.

Keywords: QR codes, data security, secure transmission.

1 Introduction

In present times, when science has become a driving force in the development of civilization, it is difficult to imagine life without cryptography and data security. Mobile phones, tablets, laptops, digital TV, Internet and many other achievements of scientific and technical issues relating to the flow of information should be based on cryptography and enforce use of it. From the other side hacking methods have changed a lot, too like phishing and re often related to human curiosity and weakness of the security solutions for example in mobile technologies [1, 2]. Hackers use more and more sophisticated methods and very quickly incorporate new technologies. One of the example is QRcode-initiated phishing attacks, or QRishing [3, 4].

One of the points where we are very vulnerable to hacking attacks is the moment when we start to exchange the secure data session. There are some measures that can be taken to protect the communications like [5, 6] but the security problem has its origin in starting the session. In Diffie-Hellman key exchange two users can communicate privately over a public medium with an encryption method of their choice using the decryption key. But the most serious limitation of Diffie-Hellman is the lack of authentication that is why data transmission using Diffie-Hellman are vulnerable to man in the middle attacks. Of course the increase in computing power led to the need to develop more and more secure data transmission encryption algorithms, which breaking would be impossible, at least in the near future. When constructing methods and algorithms for secure transmission of information, it is important to find a reasonable compromise between facility of implementation of the algorithm and the difficulty of breaking it.

As we mentioned nowadays methods of secure data transmission consists of two steps: establishing of symmetric key and transmission with data encryption

A. Kwiecień, P. Gaj, and P. Stera (Eds.): CN 2014, CCIS 431, pp. 277–286, 2014.

using a symmetric cryptographic algorithms with agreed key in the first step. This article is focused on the novel method of reconciliation of transmission key in which no information related to the key will not be possible to eavesdrop during the session establishing. The method uses QR codes to exchange the session keys. Because QR code exchange data is based on "visual transmission" it cannot be hijacked. The method can be adapted for Private Area Networks (PAN).

In the paper we shortly describe data transmission technologies in Sect. 2. Then in Sect. 3 QR codes are described and decision is made which version of QR code will be used in our solution of key exchange or session key establishment presented in Sect. 4. In Section 5 the chosen encryption method is introduced. At the end Conclusions are presented.

2   Mobile Data Exchange

In everyday personal or business life it is important be able to exchange data, mobile devices like smartphones are good candidates for this task. How often we want to share pictures, files on site just by direct connection without using any sophisticated methods. When we share personal data like family pictures we don't care about security opposite to business data we have to take into account data encryption.

There are a few methods of data exchange between modern mobile devices and they have some pros and corns [7, 8]. Each of them usually have effective signal range longer than 100 meters, this means that anyone in this radius can easily record our transmissions. If somebody would be able to record all encrypted transmission and on the other hand be able to intercept encryption key all send data can be decrypted. Wireless technologies used in modern devices (Bluetooth, WiFi) have this weakness.

It is worth to mention one outdated technology – IrDA. It was really secure – there was no way to eavesdrop sent data. Unfortunately this technology was not easy to use everywhere, devices should be stabilized for example laid on table, user could not move the devices during transmission etc.

2.1   Bluetooth

Bluetooth is the most popular standard to mobile data exchange. It is wild spread and used in almost every mobile devices. One of its advantages is low demands on user, this easiness of use leads unfortunately to lowering security demands. In this paper security is most important issue.

At the beginning Bluetooth standards, prior to 2.1 specifications, does not required any data encryption. Encryption of Bluetooth connection is based on short key (up to 128 bits). All connections from a given device use the same key generated usually once in factory and its mostly vendor dependent.

One of most important step in order to start communication is device pairing, this usually demand from user to input some code and agree to start communication. At this stage devices exchange encryption keys. Having this keys any paired device (with ours) can easily eavesdrops our future Bluetooth conversations [9–11].

2.2 WiFi

Second way of communication in mobile devices is usage of WiFi Networks. When somebody wants to establish WiFi one-to-one communication for data exchange adhoc networks or WiFI-Direct can be used. In order to use WiFi for data/files exchange in mobile devices one have to use additional applications to create server-client environment.

Nowadays WiFi communication uses WPA2 security protocol. All encryption keys are established in the negotiation phase and eavesdropping on this phase is most dangerous [12, 13]. Usually during negotiation phase Diffie-Hellman key exchange algorithm is used. We cannot forget vulnerability of that algorithm to man in the middle type of attacks [5, 6].

2.3 Cloud Services

One can always use cloud services for data exchange between devices. If the device can use Internet communication then usage of cloud services is simple. There are two problems with this type of data exchange:

- we not always have the Internet connection on site or it can be costly (i.e. roaming networks),
- cloud services usually aren't developed for incident data exchange, usually are for long term data sharing (i.e. Dropbox, SkyDrive, ...).

Cloud services need also additional applications installed. On the other hand cloud services usually have high level of security implemented.

3 QR Code [14, 15]

Quick Response Codes, QR codes, are present in our life on almost everywhere. We use them to label places, so as to mobile devices could find very quickly the data about the labelled place as:

- a text display to the user,
- a vCard contact to the user's device,
- a Uniform Resource Identifier (URI),
- an e-mail or text message.

They can be also used in: starting chats with blackberry users, connecting to WI-FI networks, getting coupons, viewing videos, purchasing items, processing orders, advertising products, a parking reservation system [16], online banking [17], support handicapped people [18] etc.

The QR codes are known since 1994 and the standards are shown in Table 1. This phenomenon is supported by mobile devices manufacturers like Apple, Samsung, HTC, Blackberry etc. There are different types of QR codes, see Table 2 and that is the reason why the devices' recognition scope can vary mainly due to applications that are accessible for the devices. The structure of the QR code shows Fig. 1. The QR code is divided into modules – the more modules the more data encoded.

Table 1. QR Code Standardization

Year	Standard
1997	Automatic Identification Manufacturers International standard(ISS-QR Code)
1998	Japanese Electronic Industry Development Association standard (JEIDA-55)
1999	Japanese Industrial Standards standard (JIS X 0510)
2000	ISO international standard (ISO/IEC18004)
2004	Micro QR Code is Approved as JIS (Japanese Industrial Standards) standard (JIS X 0510)
2011	Approved by GS1, an international standardization organization,as a standard for mobile phones

Table 2. QR code types and their capacity

Parameters	QR code type			
	QR code Model 1	QR code Model 2	Micro QR code	iQR code
Max Size [modules]	73×73	177×177	17×17	422×422
Max Capacity in numerals	1101	7089	35	40637
Max Capacity in alphanumeric	667	4296	21	$\sim$24626
Max Capacity in binary	458	2953	15	$\sim$16928

There are several types of QR codes:

- QR code Model 1, available in 14 versions,
- QR code Model 2, available in 40 versions,
- Micro QR code, available in 4 variations form M1 to M4,
- iQR code,
- SQRC,
- LogoQ.

Because of the great possibility of damaging or smearing of the QR code the error correction based on Reed-Solomon algorithm is used. There are usually four levels of error correction algorithm, the higher the level the less storage capacity. The approximate error correction level and number in percent of the damaged codewords that can be restored is: L(7 %), M(15 %), Q(25 %), H(30 %). The 5th one (S) is used in iQR codes and there is possible to restore up to 50 % of the entire code.

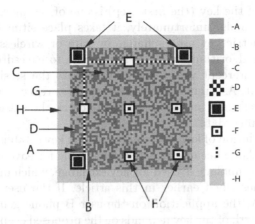

Fig. 1. QR code structure: A – version , B – format, C – data and error correction keys, D – pattern of QR code, E – position (required), F – alignment (required), G – timing, vertical and horizontal synchronization (required), H – quiet zone

In our method we propose using Model 1 or Model 2 version 14. It allows us to encode up to 458 bytes.

4   Method Description

The proposed method is dedicated for point to point connections in Private Area Networks (PAN), in which users create a direct connection between their mobile devices. Therefore, it can be applied in devices such as smartphones, tablets or even a laptops (devices which have screen and camera). The method consists of two steps, which are also conventionally used in the present methods for secure data exchange between devices (Fig. 2):

1. Reconciliation of symmetric key (transmission key).
2. Transmission with data encryption using a symmetric cryptographic algorithms and agreed key in the previous step.

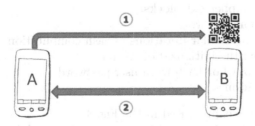

Fig. 2. Steps of secure data exchange between two mobile devices

Reconciliation of the key (the first step) is one of the most important steps. In the present methods, unfortunately, it takes place situation that the data to establish symmetric key are exchange by wire or wireless, for example using Diffie-Hellman algorithm, which is susceptible to eavesdropping. This may of course lead to intercept values used to generate the session key, and later attempts to break it. Assuming that the attacker has registered the entire transmission of encrypted data and he managed to crack the session key, he will be able to read encrypted data.

The proposed solution of key exchange or session key establishment is different from standard methods because it is not susceptible for eavesdropping. For this purpose QR code technology is used for key exchange, which until now had other uses which were described earlier in this article. If the user B wants to send data to the user A, the application on the user B phone generates a one-time session key. The length of the key depends on the proposed cryptographic method (details in Sect. 5). After that the key is encoded into a QR code. The user A gets the transmission key from QR code displayed on the user B device using to do this built-in camera and public application to read QR codes. This means that no information related to the key will not be possible to eavesdrop. The attacker will be able to capture only the data already encrypted (step 2) with which he will not be able to do anything.

5   The Choice of Data Encryption Method

At the present time we have available a lot of encryption methods, both private (where the cryptographic algorithm is kept secret) and public (where the algorithm is public). Today's cryptography prefers this second group. It is important that the algorithm must be public to show its strength. One of the algorithms that could be used in the presented method (Fig. 2, Step 2) is AES, which is the current cryptography standard. However, the authors of the article recommend the use of more secure and faster algorithm based on SHA-256 and SHA-512 hash functions [19, 20] presented in the Fig. 3. The method is easily scalable and allows for the future use of new hash functions. An additional advantage of the method is the possibility of concurrent transmitted data authentication.

The common symbols used in article:

- $M_1, M_2, \ldots M_i$ – plaintext blocks,
- $C_1, C_2, \ldots C_i$ – ciphertext blocks,
- IV – initialization vector,
- $h_1, h_2, \ldots h_k$ – particular iterations of hash computation,
- H_{IV} – hash from the initialization vector,
- H_U – user password or hash from user password,
- $\oplus$ – XOR operation.

Decryption process is presented in the Fig. 4.

The next figures (Fig. 5 and 6) show the encryption and decryption for an odd number of blocks of the explicit message M_i.

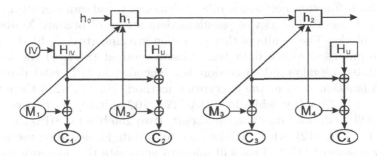

Fig. 3. Encryption for even number of blocks

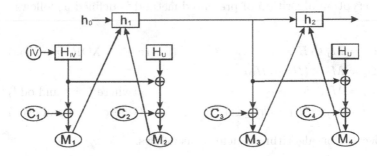

Fig. 4. Decryption for even number of blocks

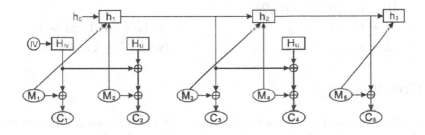

Fig. 5. Encryption for odd number of blocks

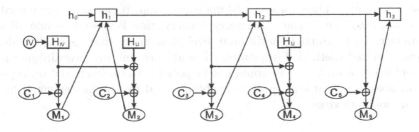

Fig. 6. Decryption for odd number of blocks

Used hash function perform a role of dynamic pseudorandom string of bits generator. Generated blocks of pseudorandom string of bits are XORed with plaintext blocks. The results of that operation are ciphertext C_i blocks.

In the proposed secure data transmission method (Sect. 4) the length of the generated pseudo-random session key (encoded in QR code) depends on the hash function used in the encryption method. For SHA-256 there will be generated a 512-bit key, while for SHA-512 a 1024 bit key, respectively. Then this key will be divided into two equal parts (two 256 bits for SHA-256 and two 512 bits for SHA-512), which will be used to initiate initialization vector (H_{IV}) and user password (H_U). This will allow to automate the authentication process and does not affect the security of the transmission – for each new session between devices there will be different key generated.

The encryption algorithm of presented method is defined as follows:

$$C_1 = M_1 \oplus H_{IV} \qquad \text{(where } \oplus - \text{XOR operation)} \qquad (1)$$
$$C_2 = M_2 \oplus (H_U \oplus H_{IV}) \qquad\qquad\qquad (2)$$
$$C_i = M_i \oplus h_{(i-1)/2} \qquad \text{(where } i \geq 3 \text{ and odd)} \qquad (3)$$
$$C_j = M_j \oplus (H_U \oplus h_{j/2-1}) \qquad \text{(where } j \geq 4 \text{ and even)} \qquad (4)$$

The decryption algorithm is defined as follows:

$$M_1 = C_1 \oplus H_{IV} \qquad \text{(where } \oplus - \text{XOR operation)} \qquad (5)$$
$$M_2 = C_2 \oplus (H_U \oplus H_{IV}) \qquad\qquad\qquad (6)$$
$$M_i = C_i \oplus h_{(i-1)/2} \qquad \text{(where } i \geq 3 \text{ and odd)} \qquad (7)$$
$$M_j = C_j \oplus (H_U \oplus h_{j/2-1}) \qquad \text{(where } j \geq 4 \text{ and even)} \qquad (8)$$

6   Conclusions

In nowadays easy and secure method of data exchange on spot is very important and often needed. Mobile devices are best candidates for this purpose, unfortunately existing devices allows us for easy or secure data exchange. Even though there are secure standards for wireless communications like WPA2, they all have one security leak – phase of establishing connection. If eavesdropper record all data transmitted he or she can intercept encryption keys and decode all sent information. In this article additional level of security that uses QR codes is propose. Derived method if implemented would prevent any eavesdropper from intercept encryption keys – it would not be possible to read sent data. Proposed method does not need any special capabilities and can be implemented for all modern mobile devices.

References

1. Zhang, Y., Hong, J., Cranor, L.: Cantina: a content-based approach to detecting phishing web sites. In: Proceedings of the 16th International Conference on World Wide Web, pp. 639–648. ACM (2007)
2. Vidas, T., Votipka, D., Christin, N.: All your droid are belong to us: A survey of current android attacks. In: Proceedings of the 5th USENIX WOOT, p. 10. USENIX Association (2011)
3. Vidas, T., Owusu, E., Wang, S., Zeng, C., Cranor, L.F., Christin, N.: QRishing: The Susceptibility of Smartphone Users to QR Code Phishing Attacks. In: Adams, A.A., Brenner, M., Smith, M. (eds.) FC 2013. LNCS, vol. 7862, pp. 52–69. Springer, Heidelberg (2013)
4. Tamir, C.: AVG (AU/NZ) Cautions: Beware of Malicious QR Codes. PCWorld (June 2011)
5. Nikiforakis, N., Meert, W., Younan, Y., Johns, M., Joosen, W.: Sessionshield: lightweight protection against session hijacking. In: Erlingsson, Ú., Wieringa, R., Zannone, N. (eds.) ESSoS 2011. LNCS, vol. 6542, pp. 87–100. Springer, Heidelberg (2011)
6. Adid, B.: Sessionlock: securing web sessions against eavesdropping. In: Proceedings of the 17th International Conference on World Wide Web, pp. 517–524 (2008)
7. Lin, H.: The study and implementation of the wireless network data security model. In: Fifth International Conference on Machine Vision (ICMV 12). International Society for Optics and Photonics (2013)
8. Al-Haiqi, A., Jumari, K., Ismail, M.: Insider Threats | Disruptive Smart Phone Technology = New Challenges to Corporate Security. Research Journal of Applied Sciences 8(3), 161–166 (2013)
9. Ryan, M.: Bluetooth: with low energy comes low security. Presented as part of the 7th USENIX Workshop on Offensive Technologies, USENIX (2013)
10. Ahuja, M.S.: A Review of Security Weaknesses in Bluetooth. International Journal of Computers & Distributed Systems 1(3), 50–53 (2012)
11. Haataja, K., Hyppönen, K., Pasanen, S., Toivanen, P.: Reasons for Bluetooth Network Vulnerabilities. In: Bluetooth Security Attacks, Springer, Heidelberg (2013)
12. Sriram, V., Sahoo, G.: A Mobile Agent Based Architecture for Securing WLANs. International Journal of Recent Trends in Engineering (2009)
13. Huang, J.: A study of the security technology and a new security model for WiFi network. In: Fifth International Conference on Digital Image Processing. International Society for Optics and Photonics (2013)
14. BS ISO/IEC 18004:2006. Information technology. Automatic identification and data capture techniques. QR Code 2005 bar code symbology specification
15. http://www.qrcode.com/en/codes/
16. Khang, S., Hong, T., Chin, T., Wang, S.: Wireless Mobile-Based Shopping Mall Car Parking System (WMCPS). In: IEEE Asia-Pacific Services Computing Conferences (APSCC), pp. 573–577 (2010)
17. Lee, Y.S., Kim, N.H., Lim, H., Jo, H.K., Lee, H.J.: Online banking authentication system using mobile-OTP with QR-code. In: 2010 5th International Conference on Computer Sciences and Convergence Information Technology (ICCIT), pp. 644–648 (2011)

18. Al-Kalifa, H.: Utilizing QR Code and Mobile Phones for Blinds and Visually Impaired People. In: Proceedings of the 11th International Conference on Computers Helping People with Special Needs (2008)
19. Hłobaż, A.: Security of measurement data transmission – message encryption method with concurrent hash counting. Przegląd Telekomunikacyjny i Wiadomości Telekomunikacyjne 80(1), 13–15 (2007) (in Polish)
20. Hłobaż, A.: Security of measurement data transmission – modifications of the message encryption method along with concurrent hash counting. Przegląd Włókienniczy – Włókno, Odzież, 39–42 (2008) (in Polish)

Industrial Implementation of Failure Detection Algorithm in Communication System

Błażej Kwiecień[1], Marcin Sidzina[2], and Edward Hrynkiewicz[3]

[1] Silesian University of Technology, Institute of Informatics
[2] University of Bielsko-Biala, Department of Mechanical Engineering Fundamentals
[3] Silesian University of Technology, Institute of Electronics
blazej.kwiecien@gmail.com, msidzina@ath.bielsko.pl,
edward.hrynkiewicz@polsl.pl

Abstract. The paper presents the results of empiric research into testing software algorithm for failure detection of transmission line and network node in industrial communication system. After implemented this algorithm in PLC the results referring to measurements of duration of basic transaction in a system and duration of failure detection on bus A (B) and Slave station were presented.

Keywords: PLC, distributed real time system, industrial computer network, time cycle of exchange data, PLC programming, avalanche of events, dual bus, network redundancy.

1 Introduction

In current automation systems essential components are information systems working in real time. The purpose of these systems is data organization process and processing that is necessary for the proper functioning of the devices. Basic features of RT systems are to response correctly when events occur, analyzing and generating a response within a specified time. Commonly used RT systems are created together with cooporate with other RT systems, and they are created together with distributed real-time systems (DRT). In DRT system it is necessary to use communication model that guarantees the correctness of the exchange of messages and time determinism [1]. DRT systems are built based on the well-known exchange models e.g. Master-Slave, token-ring, Producer-Distributor-Consumer [2–6] and their modifications or hybrids [7–9]. There is a large group of protocols, whose main goal is to obtain time determinism in networks based on Ethernet. These protocols are compatible with Ethernet in the layer of software and hardware e.g. Modbus/TCP, EtherCAT, Profinet, EthernetIP, Ethernet PowerLink, Foundation Fieldbus HSE, CC-Link IE [2, 9–17]. One of the method of increasing the reliability of RT is to use the redundancy [18–23]. A specific case is redundant system based on Master-Slave model [24–26]. Authors tried to use bus redundancy as a way of improving the performance of transmission time [26], through the division of exchanges between the two buses, in contrast to the classical redundancy [27] in which

A. Kwiecień, P. Gaj, and P. Stera (Eds.): CN 2014, CCIS 431, pp. 287–297, 2014.

the same information is sent on the main bus (primary) and on the backup bus (secondary).

It is proposed to split the scenario of exchanges so that at each bus are transmitted different transactions. The conditions under which the division of transaction between two buses can be done are durations of each transaction in transition scenario. It should be noted that the correct construction of the scenario in Master-Slave networks is crucial in aspect of minimizing the number of transactions and to shorten the data exchange cycle time. Thus constructed system can reduce data exchange cycle time when all devices are working properly. If a device or bus failure occurs system must perform some additional activities to make RT system stable. In this article authors implemented algorithm of transmission and device failure detection using industrial devices (PLC). In previous tests [24] the authors implemented self failure detection algorithm using microcontrollers ATMega2560. For this purpose authors build its own redundant system based on Master-Slave network using microcontrollers equipped with two interfaces UART $\rightarrow$ RS-485. Moreover, authors modeled an algorithm which detect the bus, device or whole system failure. Positive results of previous works resulted in implementing the algorithm into commercial system based on PLC. For the purpose of this study, authors built and configured test bench based on industrial machinery PACSystems RX3i GE IP and F&F. Authors implemented algorithms for failure detection and adapted them to the specified devices. The main problem was how to change dynamically the scenario of exchanges in case of failure. Therefore, the coprocessor can be analyzed, in most cases, only as a black box with specific access methods . It was necessary to use commercial coprocessor functions that could be used to implement own algorithm to bus and device failure detection.

2   Scenario of Transmission

The term scenario of transmission refers to a list that includes names of variables whose values are supposed to be transmitted, transmission direction and the time of their refreshment. The very essential issue is the appropriate design of a scenario of transmission and its time dependencies. If they are assigned in such a way that in case of the smallest delay in a system the system fails then applying any of the algorithms does not make sense. For these reasons, the phase of planning and designing the real time system, and next implementing in it the algorithm for failure detection should be precisely considered. Moreover, it is also necessary to determine the time of system response to a failure. Another important step it to determine whether each element of a system requires the constant checking activities (subscriber, data bus). Perhaps, the systems is able to maintain the steady state of operation without all components. Thus, at the early stage of designing, the designer is required to determine the absolute 'minimum' that provides the correct system work. In real-time systems the main parameter is duration of data exchange. In Master-Slave networks, often occur that duration of data exchange is equal with network cycle. The main aspect in Master-Slave networks is to transmit all exchanges in scenario during data exchange cycle so as to not exceed max transmission time T_G. T_G time is individually set depending on communication system. Devices failure in communication system (network's gap) cause many delays in data exchange times. In Master-Slave networks any perturbation in data transmission is very unfavorable, so it's important to prevent such situation, detect problem and provide solution e.g. by using redundant buses connected with mechanism that can switch transmission between buses.

3   The Failure Detection Algorithm

During the proper work of a communication system the doubled Master stations execute, in a given order transactions request/instruction – reply. Data is exchanged among subscribers via two communication buses. Algorithm detects the failure of communication medium (a single interface), subscriber failures or it detects errors informing about the entire system collapse. Some distinctive feature of computer systems is the fact that in some cases the symptoms of a failure (especially internal errors) increase gradually, and appear as a momentary delays, e.g. in communication. In some cases, not to say in critical ones, the system break down is rapid. In most cases, such failures are caused by damages that are generated by external factors. The presented algorithm [24] is able to detect temporary communication loss and inform the rest of the system about the existing situation. As mentioned in papers [24] and [25], in case of incorrect system work as the first symptom appears timeout that is exceeding the time for response T_{OD}. The results in paper [24] and current measurements indicate that it is necessary to pay attention on determining timeout parameters, because such delay has a significant impact on executing time of the rest of exchanges [2]. Below a block diagram of failure detection algorith is presented in (Fig. 1).

The task of presented algorithm is to detect failure of buses A or B and failure of Slave (subscriber) (Fig. 2). Efficient failure detection of the system allows the continuous updating the system status and to take further decisions related to transactions in distributed system.

Time dependencies for individual variants are as follows:

– In case of failure detection of one data bus or interface, the delay is as follow:

$$T_{ALG1} = T_{OD} + T_W \tag{1}$$

where:
T_{ALG1} – time of executing the algorithm used for bus failure detection,
T_{OD} – time of correct exchange on the second data bus,
T_W – the maximum response time for request frame.

– In case of subscriber failure, the delay is as follows:

$$T_{ALG2} = T_{OD1} + T_{OD2} + T_W \tag{2}$$

where:

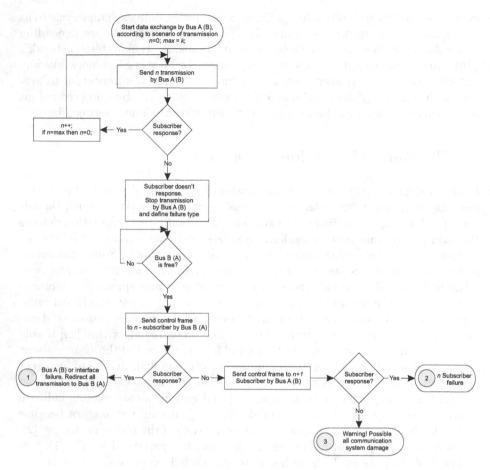

Fig. 1. The failure detection algorithm

T_{ALG2} – time of executing the algorithm used for node (subscriber) failure detection,

T_{OD} – the maximum response time on bus A (B),

T_W – the maximum response time on bus B (A).

— In case of communication system failure:

$$T_{ALG3} = T_{OD1} + T_{OD2} + T_{OD3} \tag{3}$$

where:

T_{ALG3} – time of executing the algorithm used for system failure detection,

T_{OD3} – the maximum response time for the next subscriber on bus B (A).

4 Test Bench

In this paper, being the continuation of previous work [24] and [25], the measurements of algorithm for software failure detection of bus (interface), node

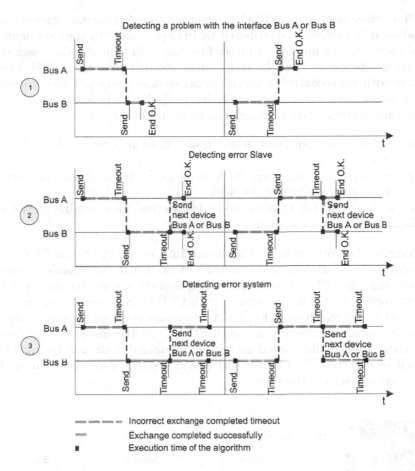

Fig. 2. Time dependencies in algorithm for detecting link failure and subscriber failure

or the entire communication system were conducted. The research was carried out for industrial control-measuring circuit based on Modbus/RTU protocol. The discussed conception required to applying solutions used in industrial systems. The measurement of failure detection was conducted on a test bench, which was prepared straight for this purpose, and was provided with special software. The measurements of software were conducted on test bench consisting of several devices. The most essential is to mention that basic assumption of the system was its work in a redundant system. As Master station, the authors used PAC RX3i Intelligent Platforms controller. The central unit of controller was equipped with RS-485 interface. In order to double interface, it could be possible to complement the device with RS-232/RS-485 converter for port 1 or additional network coprocessor. However, in the discussed case the PAC controller was equipped with additional network coprocessor CMM004 with RS-485 interface. This two coprocessors have different functionality. Coprocessor with

RS-485 interface is in-built in PLC and has implemented communication driver to Modbus RTU. Driver is distributed by PLC producer. To start communication it's necessary to implement driver function in PLC application, each data exchange has to be declared by programmer. Second coprocessor CMM004 is a module with microcontroller which realizes data exchanges. This coprocessor is independent device and programmer has to declare data exchanges scenario during configuration. Data exchange can be realize in three ways:

– data exchange in Continuous mode, transactions are executed in cycle without break,
– data exchange in Continuous Bit Control mode, transactions are executed in cycle only when Control Bit is set,
– data exchange in Bit Control mode, transaction are executing on demand when Control Bit are changing.

In mentioned algorithm mode Continuous Bit Control and Single Bit Control was used. Mode Continuous was rejected because it make impossible to control data exchanges in CPU. In such designed configuration, a solution based on two different coprocessors (one integrated with CPU-BUS A and second external CMM004-BUS B) was tested (Fig. 3). Another important aspect was to provide two communication interfaces for each SLAVE. In the discussed test bench, the authors applied distributed system with doubled addresses for each bus. External signals connected to the distributed systems were equalized so that the reading values were the same.

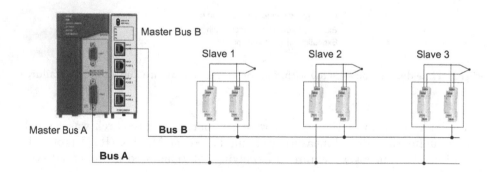

Fig. 3. Test bench equipped with PLC GE Intelligent Platforms

5   Research Results

The principle tests of algorithm were conducted on test bench presented in Sect. 5. The aim of the research was to verify the effectiveness of algorithm and duration of detecting subscriber and data bus failure. There were 100 measurement tests conducted consisting of three phases. Research scenario:

- Basic measurement tests of duration of request and reply transactions for control frame,
- Measurement tests of duration of algorithm for detecting failure in data bus A (B) – for various values of timeout 1 s, 0.5 s, 0.1 s,
- Measurement tests of duration of algorithm for detecting subscriber failure – for various values of timeout 1 s, 0.5 s, 0.1 s.

The results of measurement are presented in figure below (Fig. 4).

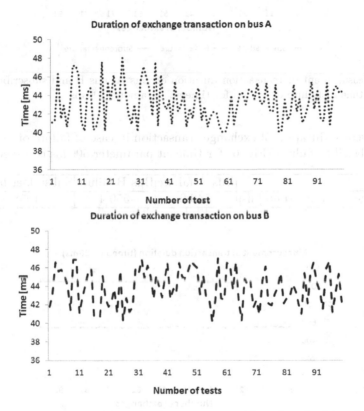

Fig. 4. Measurement of duration for transaction (a) on bus A, (b) on bus B

The average duration of exchange transaction of bus A and B is respectively:

	BUS A	BUS B
Average exchange time [ms]	43.1	43.8

The average duration of exchange transaction in case of failure of bus A and B and subscriber failure (Fig. 5) for timeout parameter 100 ms is respectively:

	Bus A failure	Bus B failure	Subscriber failure
Average exchange time [ms]	147.9	151	257.9

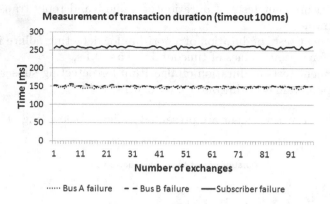

Fig. 5. Measurement of transaction duration on bus A, bus B and subscriber failure including time of failure detection for timeout parameter 100 ms

The average duration of exchange transaction in case of failure of bus A and B and subscriber failure (Fig. 6) for timeout parameter 500 ms is respectively:

	Bus A failure	Bus B failure	Subscriber failure
Average exchange time [ms]	548.9	550.4	1058.5

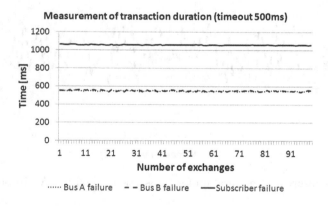

Fig. 6. Measurement of transaction duration on bus A, bus B and subscriber failure including time of failure detection for timeout parameter 500 ms

The average duration of exchange transaction in case of failure of bus A and B and subscriber failure (Fig. 7) for timeout parameter 1000 ms is respectively:

	Bus A failure	Bus B failure	Subscriber failure
Average exchange time [ms]	1050.3	1050.5	2059.4

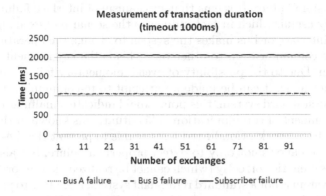

Fig. 7. Measurement of transaction duration on bus A, bus B and subscriber failure including time of failure detection for timeout parameter 1000 ms

Presented research results show the time dependencies that occur in proposed algorithm designed for detecting the failure of interface link or the entire subscriber in the redundant network based on Master-Slave model of exchange. Having analyzed the results it can be clearly seen that the greatest delay is generated by Timeout parameter. Regardless of its value, time of algorithm realization and time for a single exchange was similar in all cases. It is worth to notice that in case of a failure detection, the time of algorithm realization is very small (single milliseconds), and it does not make any significant delay into the real time system. Delay resulting from the selection of value for Timeout parameter, influences essentially the time of switching the communication between buses of the redundant system. When time of the maximum response is selected incorrectly, then it can practically collapse all procedures related to security improvement of the system because it no longer fulfills the assumptions established at the stage of design.

Discussed in this paper, empirical results are significantly different from the results obtained for the test bench based on the ATmega2560 microcontroller. The main reason for this is the fact that the duration of the application microcontroller was much shorter than in the PLC. Another important fact is that the duration of the exchange in PLC was burden with Modbus RTU protocol overhead and diagnostic functions used in PLC devices. Inordinate influence on the implementation of the algorithm has time limit timeout. For both positions of maximum time delay introduced respectively 0.1, 0.5, 1 s.

6 Conclusion

The aim of the algorithm presented in this paper is to detect various failures in the systems. The presented idea of applying redundant bus for increasing the frequency of exchanges in distributed real time systems must be strongly related to the methods of failure detection. If these algorithm are omitted, then one should expect the system failure. It seems that mechanism of interface failure detection is particularly useful, which in fact influences the scenario of exchanges. Failure detection of interface or bus makes the system to change the operation mode by switching the communication to other bus. This is the key moment for the discussed system. Due to the possibility of even a momentary failure on the bus or in the interface node, it can be made an attempt to re-communicate on a different bus. In the standard system, this point would indicate a malfunction and lead to the abandonment of communication. This situation is known in the literature as "a hole in the network". Conducted tests enable to present the thesis, that discussed algorithm allows not only to detect properly a failure, but also to realize the exchanges scenario correctly (which cannot be realized via the corrupted bus). Despite the system is not a standard redundant system, it enables to proper operation in case of failure of communication bus, even if in some failures the periodicity of micro cycle realization may be exceeded. However, it does not mean completion of the task, but only makes the system to automatic reconfiguration so that the task could be executed on the efficient bus.

References

1. Gaj, P., Jasperneite, J., Felser, M.: Computer Communication Within Industrial Distributed Environment-a Survey. IEEE Transactions on Industrial Informatics 9(1), 182–189 (2013), doi:10.1109/TII.2012.2209668
2. Kwiecień, A.: The improvement of working parameters of the industrial computer networks with cyclic transactions of data exchange by simulation in the physical model. In: The 29th Annual Conference of the IEEE Industrial Electronics Society, Roanoke, Virginia, USA, vol. 2, pp. 1282–1289 (November 2003) (Curr. ver. April 2004)
3. Miorandi, D., Vitturi, S.: Analysis of master-slave protocols for real-time-industrial communications over IEEE802.11 WLAN. In: 2nd IEEE International Conference on Industrial Informatics 2004, pp. 143–148 (June 2004)
4. Conti, M., Donatiello, L., Furini, M.: Design and Analysis of RT-Ring: a protocol for supporting real-time communications. IEEE Transactions on Industrial Electronics, 1214–1226 (December 2002)
5. Raja, P., Ruiz, L., Decotignie, J.D.: On the necessary real-time conditions for the producer-distributor-consumer model. In: IEEE International Workshop on Factory Communication Systems, WFCS 1995, pp. 125–133 (October 1995)
6. Modbus-IDA. Modbus Application Protocol Specification V1.1b3, http://modbus.org/docs/ (December 2006)
7. PROFIBUS Nutzerorganisation e.V. (PNO), PROFIBUS System Description – Technology and Application, Order number 4.332 (ver. November 2010)
8. PACSystemsTM Hot Standby CPU Redundancy, GE Fanuc Intelligent Platforms. doc. no: GFK-2308C (March 2009)
9. Genius I/O System and Communications. GE Fanuc Automation, doc. no: GEK-90486f1 (November 1994)
10. Decotignie, J.-D.: Ethernet-Based Real-Time and Industrial Communications. Proceedings of the IEEE, 1102–1117 (June 2005)
11. Siemens. Simatic PROFINET description of the system, Siemens, doc. no: A5E00298288-04, Warsaw (2009)

12. Modbus-IDA. Modbus Messaging on TCP/IP Implementation Guide V1.0b (October 2006), http://modbus.org/docs/
13. Parker Automation, EtherNet/IP Specification: ACR Series Products (March 2005)
14. Ethernet POWERLINK Standardisation Group, EPSG Draft Standard 301, Ethernet POWERLINK, Communication Profile Specification, Ver. 1.20. EPSG (2013)
15. CC-Link Partner Association, CC-Link IE Field, Ethernet-based Open Network, CLPA (June 2010)
16. Sean, J.: Vincent Fieldbus Inc., Foundation Fieldbus High Speed Ethernet Control System. Fieldbus Inc. (2001)
17. Felser, M.: Real-Time Ethernet-Industry Prospective. Proceedings of the IEEE 93(6), 1118–1129 (2005)
18. Kirrmann, H., Weber, K., Kleinberg, O., Weibel, H.: Seamless and low-cost redundancy for substation automation systems (high availability seamless redundancy, HSR). In: 2011 IEEE Power and Energy Society General Meeting, pp. 1–7 (July 2011)
19. Neves, F.G.R., Saotome, O.: Comparison between Redundancy Techniques for Real Time Applications. In: Fifth International Conference on Information Technology: New Generations, ITNG 2008, pp. 1299–1300 (April 2008)
20. Kirrmann, H., Hansson, M., Muri, P.: IEC 62439 PRP: Bumpless recovery for highly available, hard real-time industrial networks. In: IEEE Conference on Emerging Technologies and Factory Automation, ETFA 2007, pp. 1396–1399 (September 2007)
21. Wisniewski, L., Hameed, M., Schriegel, S., Jasperneite, J.: A survey of ethernet redundancy methods forreal-time ethernet networks and its possible improvements. In: Proc. Fieldbuses and Networks in Industrial and Embedded Systems (FET 2009), vol. 8, pp. 163–170 (May 2009)
22. IEC 62439, Committee Draft for Vote (CDV): Industrial communication networks: high availability automation networks. Entitled Parallel Redundancy Protocol, ch. 6 (April 2007)
23. IEC 62439, Committee Draft for Vote (CDV): Industrial communication networks: high availability automation networks. entitled Media Redundancy Protocol based on a ring topology, ch. 5 (April 2007)
24. Sidzina, M., Kwiecień, B.: Basic research of failure detection algorithms of transmission line and equipment in a communication system with a dual bus. In: Kwiecień, A., Gaj, P., Stera, P. (eds.) CN 2013. CCIS, vol. 370, pp. 166–176. Springer, Heidelberg (2013)
25. Sidzina, M., Kwiecień, B.z.: The Algorithms of Transmission Failure Detection in Master-Slave Networks. In: Kwiecień, A., Gaj, P., Stera, P. (eds.) CN 2012. CCIS, vol. 291, pp. 289–298. Springer, Heidelberg (2012)
26. Kwiecień, A., Sidzina, M.: Dual Bus as a Method for Data Interchange Transaction Acceleration in distributed Real Time Systems. In: Kwiecień, A., Gaj, P., Stera, P. (eds.) CN 2009. CCIS, vol. 39, pp. 252–263. Springer, Heidelberg (2009)
27. Kwiecień, A., Stój, J.: The Cost of Redundancy in Distributed Real-Time Systems in Steady State. In: Kwiecień, A., Gaj, P., Stera, P. (eds.) CN 2010. CCIS, vol. 79, pp. 106–120. Springer, Heidelberg (2010)

Influence of Electromagnetic Disturbances on Multi-network Interface Node

Andrzej Kwiecień[1], Michał Maćkowski[1], Jacek Stój[1], and Marcin Sidzina[2]

[1] Silesian University of Technology, Institute of Computer Science,
[2] University of Bielsko-Biała, Department of Mechanical Engineering Fundamentals
{akwiecien,michal.mackowski,jacek.stoj}@polsl.pl,
msidzina@ath.bielsko.pl

Abstract. The main idea presented in the paper focuses on testing behavior of multi-network interface node while electromagnetic disturbances in transmission line occur. Such node is able to unload an increased traffic in network or provide the security of transmission, especially in cases of temporary overloads, system failure – damage of transmission system, or events occurring in industry area. The authors marked the potential possibilities of using equipment for EMC (Electromagnetic Compatibility) measurements in the process of testing and simulating the real threats (disturbances) that may appear in the object. The final results point out the correct work of developed query distributor system and the opportunities of automatic adjustment of the network topology in case of electromagnetic disturbances.

Keywords: distributed control system, redundancy, network node, industrial networks, distributed real-time systems, medium access, electromagnetic compatibility, electromagnetic disturbances.

1   Introduction

Modern electronic devices are more and more prone to the electromagnetic disturbances that come mostly from radio emitters or power supply converters. What is more, the growing number of electronic equipment used in the industry, results in increasing importance of the mutual influence of such devices.

One of the features of industrial real time systems is a set of strict requirements as to the reliability. In the previous work [1], the authors proposed the idea of multiprotocol node in the distributed real time systems in order to increase the communication reliability. Moreover, the paper considered also the method applying such nodes for increasing the speed of data transmission in distributed control system. The proposed method implied hardware and software solution, and is based on a new node communication protocol. Such protocol may serve various protocols allowing, at the same time data transmission via various buses. Thanks to the developed method not only the bandwidth increased, but also the security of data transmission.

Taking into consideration the reliability of the system (which can be increased thanks to the redundancy) or network overload (permanent or temporary) there

A. Kwiecień, P. Gaj, and P. Stera (Eds.): CN 2014, CCIS 431, pp. 298–307, 2014.

is a problem how to increase the flexibility of network connections in order to improve the time of data exchange. The mentioned redundancy [2–5] can also be used as a way of improving the transmission time parameters [3] and it also may increase the bandwidth [2, 6]. However, it should be noted that the greatest advantages of using redundant communication link to increase the performance of transmission time, occur when the redundancy already exists in the system (for example because of the reliability) [7, 8]. This is due to the high cost of system with redundancy.

A very interesting solution may be the use of communication via multiprotocol node. In many cases, especially during temporary network overloads, system failure, such as damage of transmission system (wire, optical fiber, network co-processor etc.), or events in an industry area, such node would be able to unload the increased network traffic [9, 10], or ensure the safety of communication.

Figure 1 shows the idea of multiprotocol network with one distinguished node A0, which additionally contains three communication interfaces with networks A, B and C. Subscribers S1, S2 and S4 are connected to network A, subscribers S2 and S3 to network B, and subscribers S1, S2, and S3 to network C. It is concluded that node A0 is the Master station for three various networks A, B and C. Moreover, it is obvious that networks A, B and C can use various protocols and various transmission media.

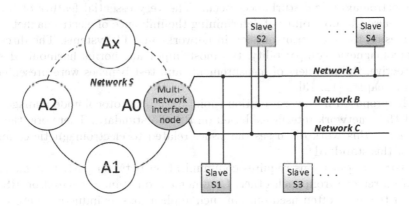

Fig. 1. The idea of network with multiprotocol nodes

The further part of the paper presents a model of an actual solution of network node equipped with more than one communication interface. The solution was developed on the example of 8-bits microcontroller AVR ATMega2560, where each microcontroller is equipped with 4 UART (*Universal Asynchronous Receiver and Transmitter*) circuits, which allow the microcontroller to communicate with other devices.

The current paper discusses the problems related to electromagnetic safety of the proposed solution. Apart from this, it also indicates the potential possibilities of applying the lab equipment for EMC (*ElectroMagnetic Compatibility*) measurements for testing a device (multi-network interface node), and simulating the real threats (disturbances) that may appear around the object. Such tests will allow to verify the correctness of proposed device platform, and also check the effectiveness of developed algorithm of query scheduler.

2 Electromagnetic Compatibility

Definition of electromagnetic compatibility included in EMC directive 2004/108/EC, determines in a very general way the requirements for devices. "Electromagnetic Compatibility means the ability of equipment to function satisfactorily in its electromagnetic environment without introducing intolerable electromagnetic disturbances to other equipment in that environment" [11].

It means, that each device is required to work properly, especially when it is exposed to electromagnetic disturbances and simultaneously, the device cannot emit too high disturbances, which could affect the other devices work. Thus, a system on the one hand must be immune to disturbances, and on the other it cannot emit too high disturbances. There is always a distinction between required electromagnetic immunity (EMI) and acceptable electromagnetic interference – these are two basic aspects of electromagnetic compatibility.

In addition, EMC tests are also supposed to increase the reliability of electronic devices and IT systems, which operate in an environment where more and more electromagnetic disturbances occur. The very essential feature of the research is also the possibility of determining the influence of electromagnetic disturbances on the data transmission in networks and IT systems. The directive of electromagnetic compatibility, the most important norms harmonized with the directive, the manners of measurements and test benches were presented in details in papers [12, 13].

In the proposed hardware solution (point 3), multiprotocol node manages the flow of three network interfaces, based on RS-485 standard. Therefore, the further part of this work discusses the issues related to electromagnetic compatibility of this standard.

Industrial applications requires communication between particular modules, very often remote from each other. Communication standard based on RS-485 is one of the most often used ones in such applications as industrial automatic, control of technological processes. Standard TIA/EIA-485-A describes the physical layer of RS-485 interface and it is usually used together with some protocol of higher layers, such as Profibus, Interbus or Modbus [14, 15].

In the real industrial applications, networks based on RS-485 operate usually in difficult conditions of electromagnetic disturbances. Significant surges in the lines, caused by electrostatic discharges or other phenomena, are able to damage ports of devices in the network. In order to operate properly in a system, the devices have to fulfill the requirements set by EMC standards. These requirements are divided into three main categories – they refer to electrostatic discharges, electrical fast transients and surges. Figure 2 presents the division of disturbances in electromagnetic environment (all electromagnetic phenomena observable in a given location) into conducted and radiated disturbances.

An electromagnetic disturbance may be electromagnetic noise, an unwanted signal or a change in the propagation medium itself.

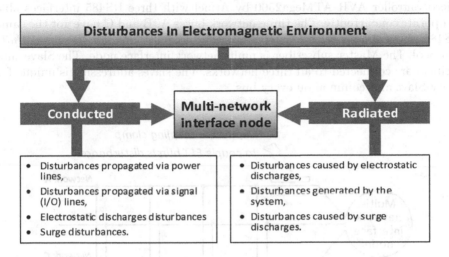

Fig. 2. Disturbances in electromagnetic environment – radiated (propagation via electromagnetic field) and conducted (propagation via power and signal lines)

In the real applications, the disturbances caused by atmospheric discharges, power supply fluctuations, and yet electrostatic and high-frequency discharges (e.g. power-inverters) generate temporary over voltage of considerable value in RS-485 network. Such voltages can damage transceivers of the network or disturb data transmission. A project of circuits operating in this network needs to include the above phenomena, because even the best interface may not be able to operate in real world.

As far, there are many papers [16–18], where authors focus on testing the immunity of RS-485 standard to electromagnetic disturbances. Moreover the authors propose some solutions to this problem. Regardless of the used transmission medium and protocol of data transmission, there is always some element of the system which while being exposed to disturbances of appropriate frequency and amplitude, causes disorders of the entire system or its part [19, 20]. The extortion of error that appear on data buses as a result of the influence of disturbing signals on system, will allow examining the behaviors of the proposed system/algorithms of exchanges and choosing the best communication interface (bus/buses).

3 Test Bench and Research Procedure

To determine the influence of electromagnetic disturbances on Multi-Network Interface Node, a new test bench was developed (Fig. 3). Because of the complex aspects of proposed solution, the authors build a test bench based on open platform Arduino for programming microcontroller. For the needs of the experimental research a system consisting of one Multi-Network Interface Node and four slave stations was constructed (Fig. 3). All system nodes were based on 8-bit microcontroller AVR ATMega2560 by Atmel with three RS485 interfaces able to operate concurrently. The three network buses A, B and C have got the same RS485 physical layer with Master-Slave Modbus RTU (*Remote Terminal Unit*) protocol. The Master subscriber is multi-network interface node. The Slave subscribers are connected to all three networks. The slaves addressing is unique for every Slave and common on every bus.

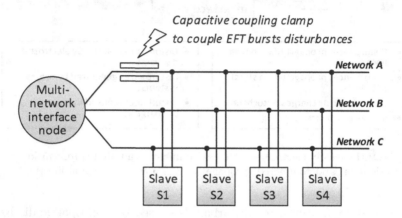

Fig. 3. The schema of test bench – transmission in network A is interfered via capacitive coupling clamp

From the point of view of empirical research, this solution will allow in the future to examine some methods using various access forms to the link as Master-Slave, Token-Ring, etc. Another advantage of the solution is the possibility to extend the circuit to communicate in Ethernet network, CAN (*Controller Area Network*) and wireless communication according to ZigBee (according to protocols specification of data transmission in mesh wireless network). The developed test bench is a strong point of the discussed issue. In terms of organization, it is very flexible, and the applied solution are innovative and, contrary to expensive PLC (*Programmable Logic Controller*) systems, accessible almost for everyone. Using Arduino open platform allows controlling the entire communication in a system.

In order to force the errors on one of the buses, a special test bench for electrical fast transient (EFT) was prepared. Electrical fast transient testing involves coupling a number of extremely fast transient impulses onto the signal lines to represent transient disturbances associated with external switching circuits that are capacitively coupled onto the communication ports, which may include relay and switch contact bounce or transients originating from the switching of inductive or capacitive loads (all of which are very common in industrial environments). This type of disturbance can propagate either by a direct link from the source of the disturbance to the equipment via power cables, or by induction onto signal or power cables running close to the source or its cables. The EFT test defined in EN 61000-4-4 attempts to simulate the interference resulting from these types of events [21].

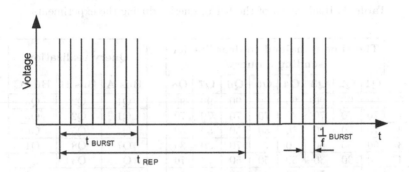

Fig. 4. Electrical fast transient (Bursts) disturbances according to the EN 61000-4-4 standard used during the test

During the test BUS A (Fig. 3) was inserted into capacitive coupling clamp that is used to couple EFT bursts onto I/O lines. Buses B and C were not disturbed during the test because each of them constitutes a separate link. Moreover each of these buses can be arranged in various forms. The output waveform (Fig. 4) consists of a 15 ms burst (t_{BURST}) of 5 kHz high voltage transients repeated at 300 ms intervals (t_{REP}). Each individual pulse has a rise time of 5 ns and pulse duration of 50 ns. The total energy in a single EFT pulse is similar to electrostatic discharge pulse. Voltages applied to the data ports can be as high as ±2 kV.

The master station was configured to realize 8 queries (two for each subscriber) with periodicity of 90 ms. The time needed for realization of every query t_{EX} was about 20 ms.

In contrast to typical implementations of master-slave communication, in the presented solution the schedule of the queries realization is not predefined at the system startup. Quite the opposite, it is dynamic and it adjusts to the current state of the communication buses. The Query Scheduler (QS) selects the consecutive queries for realization basing on the current communication buses status and the timestamps of the last successfully realized query of every kind. In case of communication fault on the network A cause by electromagnetic disturbances, it is expected to realize all the queries on the networks B and C. The next section presents the results of the influence of disturbances on the behavior of multiprotocol node.

4   The Research Results

In Table 1 the realization of the defined list of queries is presented. The numbers in rows corresponds to the periodicity of the queries and define the maximum time in which a given query should be send. Initially it is equal to the periodicity of the queries (row 0).

Table 1. Realization of the list of queries during the experiments

	The time remained to deadline for sending a query								Query realization		
	Q1	Q2	Q3	Q4	Q5	Q6	Q7	Q8	Bus A	Bus B	Bus C
0	90	90	90	90	90	90	90	90	Q1	Q2	Q3
1	70	70	70	70	70	70	70	70	Q1	Q2	Q3
2	70	70	70	50	50	50	50	50	Q4	Q5	Q6
3	50	50	50	70	70	70	30	30	Q7	Q8	Q1
4	70	30	30	50	50	50	70	70	Q2	Q3	Q4
5	50	70	70	70	30	30	50	50	Q5	Q6	Q1
6	70	50	50	50	70	70	30	30	Q7	Q8	Q2
7	50	70	30	30	50	50	70	70	Q3	Q4	Q1
8	70	50	70	70	30	30	50	50	Q5	Q6	Q2
	...				...				...		
15	50	70	50	50	70	70	30	30	Diag.	Q7	Q8
16	30	50	30	30	50	50	70	70	Diag.	Q1	Q3
17	70	30	70	10	30	30	50	50	Diag.	Q4	Q2
18	50	70	50	70	10	10	30	30	Diag.	Q5	Q6
19	30	50	30	50	70	70	10	10	Diag.	Q7	Q8
20	10	30	10	30	50	50	70	70	Diag.	Q1	Q3
21	70	10	70	10	30	30	50	50	Diag.	Q2	Q4
22	50	70	50	70	10	10	30	30	Diag.	Q5	Q6
	...				...				...		
41	70	10	70	10	30	30	50	50	Q2	Q4	Q5
42	50	70	50	70	70	10	30	30	Q6	Q7	Q8
43	30	50	30	50	50	70	70	70	Q1	Q3	Q2
44	70	70	70	30	30	50	50	50	Q4	Q5	Q6

At the communication system start up, the periodicity values of all the queries are equal. Therefore, three first queries in the list are chosen for realization ($Q1$–$Q3$), one for every communication bus. The deadlines of all the queries are set to the periodicity less the query duration $t_{EX} = 20$ ms. In the next step (row 1), all the deadline are equal again (70 ms) and one more the queries $Q1$–$Q3$ are chosen. They get new deadline values being a result of the periodicity subtracted by the query duration, i.e. 90 ms – 20 ms in the row 2. As the rest of the queries were not realized, their deadline is decreased again by 20 ms and set to 50 ms. Therefore in the next step the queries $Q4$ to $Q6$ are chosen for realization (row 2).

This procedure continues until electromagnetic disturbances (electrical fast transients) are coupled to the communication bus A (Fig. 3). It happens after 0.28 s from communication startup row 15). The Master detects the communication fault and starts realizing the communication basing only on two busses – bus B and C.

Meanwhile, on the bus A, some diagnostic queries are being sent – one query from the defined list of queries for every network subscriber sent in turns. It allows a detection of the end of the disturbances at 0.8 s from the communication startup (row 41). From that point, the communication is realized on all three busses again.

During the experimental research there were some deviation in the query duration t_{EX}. They had however no influence on the final results so in Table 1 values rounded to milliseconds were presented for better legibility.

Despite of disturbances introduced to bus A, the multi-network node was able to execute all instructions (execute all exchanges) on two effective communication buses. However, time for network exchanges increased in this case from 3 micro cycles (one micro cycle stands for one line in the table, e.g., rows 6 to 8) up to 4 micro cycles (rows 15 to 18).

5   Conclusion

The main idea presented in the paper focuses on testing behavior of multi network interface node while electromagnetic disturbances occur in the transmission line.

The proposed solutions includes both hardware and software implementation together with algorithms for multiplied communication interfaces management. The goal of this solution is to take advantage of redundant communication interfaces with real-time communication protocols. It allows for both, increased reliability and greater communication subsystem efficiency.

The research results presented in the previous section indicate the proper work of developed query distributor algorithm. In case of disturbances on bus A, communication between network nodes is via buses B and C. In this situation, the change of the network topology is unplanned and was imposed because of failures affecting the accessibility of the system nodes. Any changes in the topology are automatically serviced by the query distributor (in the master node) so that the communication link is maintained using the available resources in most efficient way. The biggest advantage of proposed solution, besides increasing the network bandwidth, is the possibility of automatic adjustment of network topology when disturbances occur so that, to provide the communication with all network nodes.

In the future, the authors intend to extend the test bench. It would be achieved by adding the other ways of communication between nodes, e.g. Ethernet, CAN or wireless. Each transmission medium (e.g. wires – twisted-pair cables, optical fibers) or communication standards are prone to different types of electromagnetic disturbances (frequency, amplitude). Therefore, developing a multi-network

node based on various communication interfaces should allow, in case of disturbances, to choose a type of bus (network) with an optimal and undisturbed communication. The use of measurement equipment in laboratory of Electromagnetic
Compatibility enables to induct such electromagnetic disturbances that can appear in industry area. At the same time, it is possible to examine and adjust the
behavior of discussed system to real conditions that can appear in an industrial
environment, and not only by software simulation.

Acknowledgments. This work was supported by the European Union from
the European Social Fund (grant agreement number: UDA-POKL.04.01.01-00-
106/09).

References

1. Kwiecień, A., Sidzina, M., Maćkowski, M.: The concept of using multi-protocol
   nodes in real-time distributed systems for increasing communication reliability.
   In: Kwiecień, A., Gaj, P., Stera, P. (eds.) CN 2013. CCIS, vol. 370, pp. 177–188.
   Springer, Heidelberg (2013)
2. Sidzina, M., Kwiecień, B.z.: The algorithms of transmission failure detection in
   Master-Slave networks. In: Kwiecień, A., Gaj, P., Stera, P. (eds.) CN 2012. CCIS,
   vol. 291, pp. 289–298. Springer, Heidelberg (2012)
3. Kwiecień, A., Sidzina, M.: Dual bus as a method for data interchange transaction
   acceleration in distributed Real Time systems. In: Kwiecień, A., Gaj, P., Stera, P.
   (eds.) CN 2009. CCIS, vol. 39, pp. 252–263. Springer, Heidelberg (2009)
4. Wei, L., Xiao, Q., Xian-Chun, T., et al.: Exploiting redundancies to enhance
   schedulability in fault-tolerant and real-time distributed systems. IEEE Transactions on Systems, Man and Cybernetics, Part A: Systems and Humans 39(3),
   626–639 (2009)
5. Neves, F.G.R., Saotome, O.: Comparison between redundancy techniques for real
   time applications. In: Information Technology: New Generations, pp. 1299–1310.
   IEEE, Las Vegas (2008)
6. Kirrmann, H., Weber, K., Kleineberg, O., et al.: Seamless and low-cost redundancy
   for substation automation systems (high availability seamless redundancy, HSR).
   In: Power and Energy Society General Meeting, pp. 1–7. IEEE (2011)
7. IEC 62439, Committee Draft for Vote (CDV): Industrial communication networks:
   high availability automation networks, entitled Parallel Redundancy Protocol, ch.
   6 (2007)
8. IEC 62439, Committee Draft for Vote (CDV): Industrial communication networks:
   high availability automation networks, entitled Media Redundancy Protocol based
   on a ring topology, ch. 5 (2007)
9. Gaj, P.: The concept of a Multi-Network approach for a dynamic distribution of
   application relationships. In: Kwiecień, A., Gaj, P., Stera, P. (eds.) CN 2011. CCIS,
   vol. 160, pp. 328–337. Springer, Heidelberg (2011)
10. Gaj, P., Jasperneite, J., Felser, M.: Computer communication within industrial dis-
    tributed environment – a survey. IEEE Transactions on Industrial Informatics 9(1),
    182–189 (2013)
11. Directive 2004/108/EC of the European Parliament and of the Council,
    http://europa.eu.int

12. Montrose, M.I., Nakauchi, E.M.: Testing for EMC compliance: approaches and techniques. John Wiley and Sons, Institute of Electrical and Electronics Engineers, Canada (2004)
13. Williams, T.: EMC for product designers. Elsevier Ltd., Oxford (2001)
14. Pereira, C.E., Neumann, P.: Industrial communication protocols. In: Springer Handbook of Automation, pp. 981–999. Springer, Heidelberg (2009)
15. Krist, P.: Advanced industrial communications. In: Rudas, I.J., Fodor, J., Kacprzyk, J. (eds.) Towards Intelligent Engineering and Information Technology. SCI, vol. 243, pp. 365–376. Springer, Heidelberg (2009)
16. Zhang, W., Lin, J., Pen, L., et al.: Application of RS485 for communication and synchronization in distributed electromagnetic exploration system. In: International IEEE Conference on Electric Information and Control Engineering (ICE-ICE), Wuhan, pp. 4815–4818 (2011)
17. Ajay Kumar, V.: Overcoming data corruption in RS485 communication. In: International IEEE Conference on Electromagnetic Interference and Compatibility, Madras, pp. 9–12 (1995)
18. Scanlon, J., Rutgers, K.: Safeguard Your RS-485 communication networks from harmful EMC events. Analog Devices 47 (2013)
19. Novak, J.: Electromagnetic compatibility of fieldbus communication. In: Fieldbus Technology, Industrial Network Standards for Real-Time Distributed Control, pp. 413–433. Springer, Heidelberg (2003)
20. Kryca, M.: Hardware aspects of data transmission in coal mines with explosion hazard. In: Kwiecień, A., Gaj, P., Stera, P. (eds.) CN 2013. CCIS, vol. 370, pp. 517–530. Springer, Heidelberg (2013)
21. Electromagnetic Compatibility (EMC) Part 4-4: Testing and Measurement Techniques-Electrical Fast Transient/Burst Immunity Test (IEC 61000-4-4:2012 (Ed. 3.0))

Speech Recognition Based on Open Source Speech Processing Software

Piotr Kłosowski, Adam Dustor, Jacek Izydorczyk, Jan Kotas, and Jacek Ślimok

Silesian University of Technology, Institute of Electronics
Akademicka Str. 16, 44-100 Gliwice, Poland
{piotr.klosowski,adam.dustor,jacek.izydorczyk}@polsl.pl,
{kotas.janek,jacek.slimok}@gmail.com
http://iele.polsl.pl

Abstract. Creating of speech recognition application requires advanced speech processing techniques realized by specialized speech processing software. It is very possible to improve the speech recognition research by using frameworks based on open source speech processing software. The article presents the possibility of using open source speech processing software to construct own speech recognition application.

Keywords: speech recognition, speech processing, open source software.

1 Introduction

Division of Telecommunication, a part of the Institute of Electronics and Faculty of Automatic Control, Electronics and Computer Science Silesian University of Technology, for many years has been specializing in advanced fields of telecommunication engineering. One of them is speech signal processing [1–4]. Main research areas on this field are: speech synthesis, speech recognition and speaker verification and identification.

Creating of speech recognition application requires advanced speech processing techniques realized by specialized speech processing software [5]. Which software should be used for speech recognition application development? There are many possibilities. To create own speech recognition application can be used:

1. programming languages such as C++, Java, Python, etc.,
2. high level commercial computing environment for implementation of speech recognition algorithms such as: MATLAB with Signal Processing Toolbox,
3. open source speech processing software.

Hypothesis can be formulated as follows: It is possible to improve the speech recognition research by using frameworks based on open source speech processing software. The article focuses on open source applications that can be used to construct their own speech recognition system.

A. Kwiecień, P. Gaj, and P. Stera (Eds.): CN 2014, CCIS 431, pp. 308–317, 2014.

2 Open Source Speech Processing Software for Speech Recognition

This section presents to the most frequently used speech processing open source software tools as a basis for testing and building your own speech recognition application.

2.1 CMU Sphinx

CMU Sphinx is a group of speech recognition systems, which have been developed at Carnegie Mellon University in Pittsburgh, Pennsylvania, United States. Sphinx consists of multiple software systems, not all of which are open source. First version of Sphinx, developed by Kai-Fu Lee, offered a system that demonstrates the feasibility of accurate, large-vocabulary speaker-independent, continuous speech recognition [6]. It has been later replaced by newer, better performing versions. Sphinx 2, the successor of the original, was the first Sphinx system to be released as open source. The biggest improvement was, apart from fewer recognition errors overall, the capability to handle much larger vocabulary size. For 5,000-word speaker-independent speech recognition, the recognition error rate has been reduced to 5 %. It is used in real-time speech recognition systems, however this project is no longer developed. Sphinx 3 is a system designed with accuracy in mind and therefore does not currently allow real-time speech recognition. It is, however, in active development and recent improvements made to its algorithms, as well as hardware performance growth have allowed to achieve near real-time recognition results. It adopts widely used continuous hidden Markov model. Together with SphinxTrain – an acoustic model trainer – it gives access to modeling techniques, such as MLLR (Maximum-Likelihood Linear Regression), LDA/MLLT (Linear Discriminant Analysis/Maximum Likelihood Linear Transform) or VTLN (Vocal Tract Length Normalization). Newest version of Sphinx is Sphinx 4. It is a rewritten Sphinx engine (entirely in Java), providing a more flexible framework for research in speech recognition. In addition to systems mentioned above, a lightweight version of Sphinx, named PockedSphinx, is currently in active development and provides access to speech recognition engine designed specifically for embedded devices, such as mobile phones. More details about the Sphinx project are available at: http://cmusphinx.sourceforge.net.

2.2 SPRACH

SPRACH is an abbreviation for Speech Recognition Algorithms for Connectionist Hybrids. It involves usage of HMM (Hidden Markov Models), ANN (Artificial Neural Networks), statistical inference in said networks, as well as hybrid HMM-ANN technology in order to further improve current research on continuous speech recognition. The project was developed across multiple universities in Europe and therefore one of its main goals was to adapt hybrid speech recognitions to languages other than English, French and Portuguese in particular. More details about the project are available at:
http://www.icsi.berkeley.edu/~dpwe/projects/sprach/sprachcore.html.

2.3   GMTK

GMTK is an acronym for Graphical Models Toolkit and a project that was developed by Prof. Jeff Bilmes, Richard Rogers and a number of other individuals at University of Washington, United States. GMTK is a feature-rich toolkit, allowing rapid prototyping of statistical models, by using DBM (Dynamic Graphical Models) and DBN (Dynamic Bayesian Networks) [7]. Among many of its features, one can specify the following: exact and approximate inference, dense, sparse and deterministic conditional probability tables, native support for ARPA backoff-based factors and factored language models, parameter sharing, gamma and beta distributions etc. [8]. In addition, it includes Markov chains of arbitrary orders, graph viewer in form of Graphical User Interface; supports multiple file formats; allows offline and online mode during parameter learning and prediction. More details about GMTK project are available at: http://ssli.ee.washington.edu/~bilmes/gmtk/.

2.4   SONIC

SONIC is a system developed by Bryan Pellom and Kadri Hacioglu at the University of Colorado, Boulder, United States. It is designed in order to allow development of new algorithms of continuous speech recognition. SONIC's main features include: phonetic aligner, phonetic decision tree acoustic trainer, core recognizer, speaker adaptation, live-mode recognition, voice activity detection, language portability, speech compression interface, as well as application programming interface (API) with examples [9]. It is worth mentioning, that SONIC has been successfully ported to over 15 languages. It is based on CDHMM (Continuous Density Hidden Markov Model). SONIC has been developed for 4 years until May 2005 and at the time of the last update it was still considered unfinished by the authors [10]. More details about SONIC project are available at: http://www.bltek.com/virtual-teacher-side-menu/sonic.html.

2.5   HTK

HTK is a toolkit designed for building HMM (hidden Markov models). This toolkit at its core is multi-purpose and may be used in order to model any time series, however it has been created with speech recognition in mind. HTK consists of two major tool sets. The first, namely HTK training tools, requires both: speech data and its transcription and is done in order to estimate parameters of a set of HMM. Once this stage is complete, HTK recognition tools can be used in order to transcribe unknown speech data [11]. An invaluable advantage of the described toolkit is its extensive documentation providing numerous examples of usage. More details about HTK project are available at: http://htk.eng.cam.ac.uk/.

2.6   ALISE

ALIZE is open source platform for biometrics authentication. It provides single engine for face and voice recognition. Project's goal is to provide access to biometric technologies for industrial and academic usage. It consists of low-level API and high-level executables. Thanks to this attribute, ALIZE allows the user to quickly create speech recognition system, as well as provides tools for industrial voice processing applications. The project has been made open source, because it is believed that allowing broad scientific research on speech recognition algorithms results in quicker improvement of such systems, by making them more accurate and resistant to noise [12]. Project has been evaluated by NIST, SRE, RT and French ESTER and achieved very good results. It is written in C++ and allows multi-platform implementation, including a possible use in embedded devices. More details about the project are available at: http://mistral.univ-avignon.fr/index_en.htm.

2.7 SPRO

SPRO is an open source speech processing toolkit. It was designed for both: speaker and speech recognition. SPRO has been created for variable resolution spectral analysis, but it also supports classic mechanics used in speech processing. It provides runtime commands, as well as standard C library for implementing new algorithms and applications. After compilation, the user is given access to tools used for following purposes: filter-bank based speech analysis, linear predictive speech analysis, comparing streams, extracting speech parameters. The library provides additional signal processing functions, which can be used in custom speech recognition applications, such as FFT, LPC analysis and feature processing, such as lifter, CMS and variance normalization. SPRO is distributed under GNU Public License agreement. More details about the project are available at: https://gforge.inria.fr/frs/index.php?group_id=532.

2.8 Other Open Source Speech Processing Tools

Other open source speech processing tools that can be used to construct own speech recognition application are:

- AT&T FSM LibraryTM – Finite-State Machine Library
  http://www2.research.att.com/~fsmtools/fsm/
- LIBSVM – A Library for Support Vector Machines
  http://www.csie.ntu.edu.tw/~cjlin/libsvm/
- SVMlight – Support Vector Machine
  http://svmlight.joachims.org/
- PVTK – Periodic Vector Toolkit
  http://old-site.clsp.jhu.edu/ws2004/groups/ws04ldmk/PVTK.php
- LAPACK – Linear Algebra PACKage
  http://www.netlib.org/lapack/index.html
- Standard Template Library
  http://www.sgi.com/tech/stl/
- MIT Finite State Toolkit
  http://people.csail.mit.edu/ilh/fst/

3   Examples of Use SPRO to Speech Features Extraction

This section presents to the examples of use open source speech processing tools called SPRO to speech features extraction as a basis for testing and building own speech recognition application. After compilation SPRO provides the following tools for speech recognition:

- scompare – tool dedicated to compare input streams tool,
- scopy – feature manipulation tool,
- sfbank – tool dedicated to filter-bank based speech analysis,
- sfbcep – tool dedicated to filter-bank based speech analysis,
- slpc – tool takes as input a waveform and output linear prediction derived features,
- slpcep – tool dedicated to linear predictive analysis of speech signals,
- splp – tool dedicated to linear predictive analysis of speech signals.

The following examples illustrate the use of SPRO tools to the speech feature extraction.

3.1   Extracting Speech Features Using Filter-Bank Analysis Tools

SPRO provides an implementation of filter-bank analysis – a spectral analysis method based on representing the signal spectrum by the log-energies at the output of a filter-bank, where the filters are overlapping band-pass filters spread along the frequency axis. This representation gives a approximation of the signal spectral shape while smoothing out the harmonic structure if any. When using variable resolution analysis, the central frequencies of the filters are determined so as to be evenly spread on the warped axis and all filters share the same bandwidth on the warped axis. This spectral analysis technique is also applied to Mel frequency warping, a very popular warping in speech analysis which mimics the spectral resolution of the human ear. The Mel warping is approximated by function [13]:

$$\mathrm{mel}(f) = 2595 \cdot \log_{10}\left(1 + \frac{f}{700}\right) . \tag{1}$$

Filter-bank analysis implementation uses triangular filters on the FFT module. The energy at the output of channel i is given by:

$$e_i = \log \sum_{j=1}^{N} h_i(j) \cdot \|X(j)\| \tag{2}$$

where N is the FFT $length^2$ and h_i is the filter's frequency response as depicted above. The filter's response is a triangle centered at frequency f_i with bandwidth $[f_{i-1}, f_{i+1}]$, assuming the f_i's are the central frequencies of the filters determined according to the desired spectral warping.

The SPRO tools named **sfbank** and **sfbcep** are dedicated to filter-bank based speech analysis. The first filter-bank analysis tool **sfbank** takes as input a waveform and output filter-bank magnitude features. For each frame, the FFT is performed on the windowed signal, possibly after zero padding, and the magnitude is computed before being integrated using a triangular filter-bank.

The second SPRO tool **sfbcep** takes as input a waveform and output filter-bank derived cepstral features. The filter-bank processing is similar to what is done in **sfbank**. The cepstral coefficients are computed by DCT'ing the filter-bank log-magnitudes and possibly liftered. Optionally, the log-energy can be added to the feature vector. In **sfbcep**, the frame energy is calculated as the sum of the squared waveform samples after windowing. As for the magnitudes in the filter-bank, the log-energy are thresholded to keep them positive or null. The log-energies may be scaled to avoid differences between recordings. Finally, first and second order derivatives of the cepstral coefficients and of the logenergies can be appended to the feature vectors. When using delta features, the absolute log-energy can be suppressed using the **--no-static-energy** option.

```
$ sfbcep -F PCM16 -f 16000 input.wav output.mfc
```

The file with .mfc extention contains the result of the calculation in raw format. Example Bash script performs speech features extraction from all files on the list saved as timit.1st file is presented below:

```bash
#!/bin/bash

### List of file of speech database
fileslist="timit.1st"
speakerlist="speakers_all.txt"

### Deleting old feature files
rm -r ./timit/features

### Creating empty feature directories
mkdir ./timit/features
for dir in $(cat "$speakerlist")
    do
        mkdir "./timit/features/$dir"
    done

### SPRO features extraction
for file in $(cat $fileslist)
    do
        echo -n "Processing file : $file ... "
        inputfile="./timit/speech/$file.WAV"
        outputfile="./timit/features/$file.raw.mfc"
        ./SPRO/sfbcep -F PCM16 -f 16000 $inputfile $outputfile
        if (( $? )); then echo "ERROR"; exit; else echo "OK"; fi
    done
```

File `timit.lst` contains all processed speech files from TIMIT speech database. The TIMIT corpus of read speech is designed to provide speech data for acoustic-phonetic studies and for the development and evaluation of automatic speech recognition systems. TIMIT contains broadband recordings of 630 speakers of eight major dialects of American English, each reading ten phonetically rich sentences [14]. The TIMIT corpus includes time-aligned orthographic, phonetic and word transcriptions as well as a 16-bit, 16 kHz speech waveform file for each utterance. Corpus design was a joint effort among the Massachusetts Institute of Technology (MIT), SRI International (SRI) and Texas Instruments, Inc. (TI). The speech was recorded at TI, transcribed at MIT and verified and prepared for CD-ROM production by the National Institute of Standards and Technology (NIST). Sample contents of the file `timit.lst` is shown below:

```
speaker001/SA1
speaker001/SA2
speaker001/SI1565
speaker001/SI2195
speaker001/SI935
speaker002/SA1
speaker002/SA2
speaker002/SI1081
speaker002/SI1202
speaker002/SI1711
```

3.2   Extracting Speech Features Using LPC Analysis Tools

LPC (Linear Prediction Coding) is a popular speech coding analysis method which relies on a source/filter model if the speech production process. The vocal tract is modeled by an all-pole filter of order p whose response is given by [5]:

$$H(z) = \frac{1}{1 + \sum_{i=1}^{p} a_i z^{-i}} . \tag{3}$$

The coefficients a_i are the prediction coefficients, obtained by minimizing the mean square prediction error. The minimization is implemented in SPRO using the auto-correlation method.

PLP (Perceptual Linear Prediction) is combination of filter-bank analysis and linear prediction to compute linear prediction coefficients on a perceptual spectrum [15]. The filter-bank power spectrum is filtered using an equal loudness curve and passed through a compression function $f(x) = x^{1/n}$ where usually $n = 3$, thus resulting in an auditory spectrum from which the autocorrelation is computed by inverse discrete Fourier transform. Linear prediction coefficients are then carried out as usual from the autocorrelation.

SPRO provides two different tools `slpc` and `slpcep` for linear predictive analysis of speech signals. The tool `slpc` takes as input a waveform and output linear prediction derived features. For each frame, the signal is windowed after pre-emphasis and the generalized correlation is computed and further used to estimate the reflection and the prediction coefficients which can, in turn, be transformed into log area ratios or line spectrum frequencies. The default is to output the linear prediction coefficients however reflection coefficients can be obtained with the --parcor option, log-area ratios with --lar option and line spectrum pairs with the --lsp one. Optionally, the log-energy can be added to the feature vector. In slpc, the log-energy is taken as the linear prediction filter gain, which is also the variance of prediction error, and thresholded to be positive or null. The log-energies may be scaled to avoid differences between recordings using the --scale-energy option.

The tool slpcep takes as input a waveform and outputs cepstral coefficients derived from the linear prediction filter coefficients. The linear prediction processing steps are as in slpc and cepstral coefficients are computed from the linear prediction coefficients using the recursion previously described. The required number of cepstral coefficients must be less then or equal to the prediction order. As for slpc, the log-energy, taken as the gain of the linear prediction filter, can be added to the feature vectors. Mean and variance normalization of the static cepstral coefficients can be specified with the global --cms and --normalize options but do not apply to log-energies. The normalizations can be global or based on a sliding window whose length is specified with --segment-length. Finally, first and second order derivatives of the cepstral coefficients and of the log-energies can be appended to the feature vectors. When using delta features, the absolute log-energy can be suppressed using the --no-static-energy option.

The tool splp takes as input a waveform and outputs cepstral coefficients derived from a perceptual linear prediction analysis. Note that, although not explicitly mentioned in the program name, splp does output cepstral coefficients, not linear prediction coefficients. The LPC order must be less than or equal to the number of filters in the filter-bank while the number of cepstral coefficients must be less than or equal to the prediction order. The log-energy is taken from the frame waveform as in the filter-bank tools.

Example of extracting speech features using PSRO LPC analysis tools is presented below:

```
$ slpc sa.wav slpc.out
$ scopy -o ascii slpc.out slpc.txt
$ slpcep sa.wav slpcep.out
$ scopy -o ascii slpcep.out slpcep.txt
$ splp sa.wav splp.out
$ scopy -o ascii splp.out splp.txt
```

The files with .txt extentions contains the result of the calculation in ASCII format.

4   Summary

Table 1 presents comparison among presented open source speech processing tools. This paper and comparison can be helpful in choosing the right speech processing tool for a specific speech recognition application.

316 P. Kłosowski et al.

Table 1. The comparison presented open source speech processing tools

Name	Environment	Portability	Flexibility	Features
CMU Sphinx	C++ Java	Unix (Linux) Embedded sys.	high	model training, speech recognition framework
SPRACH	C++ Tcl/Tk Perl	Unix (Linux)	high	speech recognition framework
GMTK	C++ Tcl/Tk	Unix (Linux)	very high	probabilistic modeling framework
SONIC	C++ Tcl/Tk	UNIX (Linux, Solaris) Windows Mac OS	very high	continuous speech recognition
HTK	ANSI C	Unix (Linux) Windows	high	hidden Markov model toolkit
ALIZE	C++	Unix (Linux) Windows Embedded sys.	medium	speech processing and recognition framework
SPRO	C++	Unix (Linux) Windows Embedded sys.	medium	speech features extraction

The most important elements of each speech recognition system are speech features extraction and classification. The paper presents examples of the use open source speech processing software for speech features extraction. Similarly, the open source software can be used to test and create efficient classifiers that are the basis for designing effective speech recognition applications. For testing and construction of the various classifiers can be used e.g. the Hidden Markov Model Toolkit (HTK) [16] or open source platform for biometrics authentication called ALIZE [17]. Use of the open source speech processing software can significantly improve the construction and testing of modern speech recognition application.

Acknowledgements. This work was supported by The National Centre for Research and Development (www.ncbir.gov.pl) under Grant number POIG.01.03.01-24-107/12 (Innovative speaker recognition methodology for communications network safety).

References

1. Kłosowski, P.: Speech Processing Application Based on Phonetics and Phonology of the Polish Language. In: Kwiecień, A., Gaj, P., Stera, P. (eds.) CN 2010. CCIS, vol. 79, pp. 236–244. Springer, Heidelberg (2010)
2. Kłosowski, P., Dustor, A.: Automatic Speech Segmentation for Automatic Speech Translation. In: Kwiecień, A., Gaj, P., Stera, P. (eds.) CN 2013. CCIS, vol. 370, pp. 466–475. Springer, Heidelberg (2013)

3. Dustor, A., Kłosowski, P.: Biometric Voice Identification Based on Fuzzy Kernel Classifier. In: Kwiecień, A., Gaj, P., Stera, P. (eds.) CN 2013. CCIS, vol. 370, pp. 456–465. Springer, Heidelberg (2013)
4. Kłosowski, P.: Improving Speech Processing Based on Phonetics and Phonology of Polish Language. Przeglad Elektrotechniczny R 89(8), 303–307. Sigma-Not (2013)
5. Rabiner, L.R., Schafer, R.W.: Introduction to Digital Speech Processing. Foundations and Trends in Signal Processing 1(1-2), 1–194 (2007)
6. Tsontzos, G., Orglmeister, R.: CMU Sphinx4 speech recognizer in a Service-oriented Computing style. In: IEEE International Conference on Service-Oriented Computing and Applications (SOCA), pp. 1–4 (2011)
7. Bilmes, J., Bartels, C.: Graphical model architectures for speech recognition. IEEE Signal Processing Magazine 22(5), 89–100 (2005)
8. Bilmes, J., Zweig, G.: The graphical models toolkit: An open source software system for speech and time-series processing. In: IEEE International Conference on Acoustics, Speech, and Signal Processing (ICASSP), p. IV-3916–IV-3919 (2002)
9. Pellom, B.: SONIC: The University of Colorado Continuous Speech Recognizer. University of Colorado, Colorado (2001)
10. Pellom, B., Hacioglu, K.: Recent Improvements in the CU SONIC ASR System for Noisy Speech: The SPINE Task. In: Proceedings of IEEE International Conference on Acoustics, Speech, and Signal Processing (ICASSP), Hong Kong (April 2003)
11. Young, S., Evermann, G., Hain, T., Kershaw, D., Moore, G., Odell, J., Ollason, D., Povey, D., Valtchev, V., Woodland, P.: The HTK Book. Cambridge University Engineering Department, Cambridge (2002)
12. Bonastre, J.F., Wils, F., Meignier, S.: ALIZE, a free toolkit for speaker recognition. In: IEEE International Conference on Acoustics, Speech, and Signal Processing (ICASSP 2005), vol. 1, pp. 737–740 (2005)
13. Stevens, S.S., Volkman, J.: The relation of pitch to frequency. American Journal of Psychology 53, 329 (1940)
14. Garofolo, J.S., Lamel, L.F., Fisher, W.M., Fiscus, J.G., Pallett, D.S., Dahlgren, N.L., Zue, V.: TIMIT Acoustic-Phonetic Continuous Speech Corpus. Linguistic Data Consortium, Philadelphia (1993)
15. Hermansky, H.: Perceptual linear predictive (plp) analysis of speech. Journal of the Acoustical Society of America 87(4) (1990)
16. Ziółko, B., Manandhar, S., Wilson, R.C., Ziółko, M., Gałka, J.: Application of HTK to the Polish language. In: International Conference on Audio, Language and Image Processing, ICALIP 2008, pp. 1759–1764 (2008)
17. Fauve, B.G.B., Matrouf, D., Scheffer, N., Bonastre, J.F., Mason, J.S.D.: State-of-the-Art Performance in Text-Independent Speaker Verification Through Open-Source Software. IEEE Transactions on Audio, Speech, and Language Processing 15(7), 1960–1968 (2007)

Automatic RESTful Web Service Identification and Information Extraction

Adam Czyszczoń and Aleksander Zgrzywa

Wrocław University of Technology, Faculty of Computer Science and Management,
Institute of Informatics
Wybrzeże Wyspiańskiego 27, 50370 Wrocław, Poland
{adam.czyszczon,aleksander.zgrzywa}@pwr.wroc.pl
http://www.zsi.ii.pwr.edu.pl

Abstract. Lightweight RESTful Web Services have become more popular and widespread in recent Web application development, thereby dominating the unwieldy SOAP Web Services. However, finding RESTful services on the Web is still unsolved and challenging problem, and it is beyond the scope of current approaches on SOAP Web Service retrieval that rely on formal descriptions. In this paper we present a method for Web scale automatic RESTful Web Service identification and data extraction from HTML pages describing services. This paper also describes first complete and extensive RESTful Web Service test collection that can be used for many benchmarks and make research results comparable. Presented evaluation results prove the effectiveness of presented method.

Keywords: RESTful, Web Services, Web API, identification, information extraction.

1 Introduction

RESTful Web Services have become popular in the last years because of their flexibility, ease of implementation and usage. For this reason, many service providers of publicly accessible Web Services (as for example Google) in many cases has resigned from SOAP Web Services in the favor of RESTful solutions. RESTful Web Services have also become widely adopted in e-business. Lightweight RESTful services increasingly provide support or even the foundation tools for many applications in B2B, B2C or C2C models like for example e-commerce, online invoicing, exchange rates, stock market analysis, data analysis tools or online payments.

The name RESTful Web Service was for the first time officially used in 2007 together with description of its design principles [1]. In the following years, a number of publications on the topic of RESTful Web Service architecture and implementation rules have appeared [2–4]. Furthermore, a number of methods for machine-readable descriptions of RESTful Web Services was proposed: WADL (Web Application Description Language) [5], hRESTS (HTML for RESTful Services) [6], ReLL (Resource Linking Language) [7] or an approach to semi-automatically

A. Kwiecień, P. Gaj, and P. Stera (Eds.): CN 2014, CCIS 431, pp. 318–327, 2014.

build semantic models from example data [8]. However, none of these standards was widely adopted.

Despite the rapidly increasing number of RESTful services on the Internet, their identification still remained a key problem. Descriptions of the services of this class are most frequently given in HTML documents in the form of API (Application Programming Interface) documentation. Moreover, current development of RESTful Web Services is rather self-governed as service providers do not follow any standard practices for service implementation, publication and documentation. In consequence, finding RESTful services requires a lot of manual effort even when using generic search engines or knowing best possible service providers. Currently, finding RESTful services on the Web is still unsolved and challenging problem and it is beyond the scope of current approaches on SOAP Web Service retrieval that rely on formal descriptions like WSDL (Web Service Description Language). Therefore, RESTful services require a different approach of identifying and obtaining descriptions of services, which at the moment is a major obstacle for the development of effective methods of their retrieval.

To solve this problem, we propose in this paper a method for automatic and effective RESTful Web Service identification and information extraction. Presented approach is founded on an algorithm that uses binary classification of RESTful Web services based on their link structure patterns. Extracted data can be later indexed and used for RESTful Web Service retrieval.

2   Related Work

This paper is a continuation to our research on Web Service retrieval method of both SOAP and RESTful services. It is also an extension of our preliminary analysis presented in 2011 [9] that included definition of the RESTful Web service URI (Uniform Resource Identifier) structure and utilized an Artificial Neural Network (ANN) model to classify services based on their link structure patterns. Presented approach allowed to extract service's resources and possible values so generic description of a service could be drawn.

Currently there are only several methods on RESTful Web Service identification and information extraction. The first one was presented in 2008 [10] where authors used faceted classification. Services were classified to manually predefined categories based on terms in the API and available user tags. Presented approach included also indexing and ranking which together allowed for retrieval of Web Services. Despite the fact that proposed method gave better results than *ProgrammableWeb*[1] (popular API directory) or plain Google search, it did not include service identification on the Web. The test collection was composed of services collected from the *ProgrammableWeb* directory which would not work for Web scale crawling of services. Therefore, the method relied on directory and tags that are manually assigned by users. Another drawback is that the number of services grows and some of them change. Such a snapshot of a directory was

---

[1] http://programmableweb.com

not published and therefore we cannot compare the effectiveness on the same dataset.

Another method presented in [11] allows Web scale discovery of publicly available Web Services, both SOAP and RESTful, by performing a focused Web crawl, identifying relevant documents and aggregating available information. In order to detect RESTful services documentation, the authors proposed two experimental approaches. The first one utilized SVM (Support Vector Machine) supervised learning algorithm for text classification of HTML documents. The drawback was similar as to previous method – lack of test collection, especially to use as training data. Another drawback was using Web Services from *Programmable Web* as a positive train data, since not every provided there service conform to RESTful. The second approach used term frequency of certain words like for example "rest", "web", "api", "url" with assigned scores indicating assignment strength (weak, medium, strong). These indicators were used to calculate the score of a page. However, proposed method did not consider extracting information about RESTful Web Services.

3 Methodology

In order to properly identify RESTful Web Service and extract the information, a number of steps must be performed. The first one is to determine whether a webpage actually contains the API documentation about services. If this is true, such pages are analyzed for the presence of URIs that represent RESTful Web Services and for the presence of useful information. In order to handle all the steps concerned with the identification and information extraction process an efficient algorithm need to be used.

3.1 Detecting RESTful Web Service Documentation

In [11] authors described two methods for checking if a website is a documentation of a RESTful Web Service – one using SVM text classification and second one that was keyword-based with term frequency. Because text classification is too complex for web-scale usage and keyword-based approach performed well, we adopted in our approach the second method.

3.2 RESTful Web Service Identification

Every URI that is discovered on a documentation website is checked if it satisfies the conditions of RESTful Web Service structure. In this case, we consider identification as binary classification – that is whether a URI is RESTful or not. The goal is to encode URIs into patterns so they can be applied for supervised learning algorithms but with possibly low number of features. Based on our research conducted in [9] we define RESTful Web Service to follow specific structure:

```
http[s]://authority/[serviceInfo/][resources/][?query]
```

where

- `authority` is composed of `[WebServicePrefix(.-)]hostname`
- `serviceInfo` is composed of `[version/][accessPolicy/][serviceName/]`
- `resources` are composed of `(resourceName/[value/]*)*`
- `queryResources` are composed of `(resourceName=value[&])*`

Afterwards we distinguish four features associated with the hostname: *API prefix, access policy indicator* (usually public or private), *version, WS path name*, and four associated with resources: *path resources, query variables, mixed resources, query resources*.

The next step is to encode URIs for the pattern matching. We use parsing algorithm which divides each incoming URI into parts in order to match RESTful URI structure patterns presented above. Each part is successively analyzed for the presence of particular feature. If a feature is present it is assigned a value of one, if not – zero. In result, a specific binary profile is created for each analyzed URI. The resulting RESTful Web service URI pattern has got following scheme:

`[apiPrefix, accessPolicy, version, pathWSname, pathResources,`
`queryVariables, mixedResources, queryResources, queryWSname],`

for example the URI:

`http://ws.example.com/public/2.0/?WSname.getInfo&type=name`

encoded to the above scheme is as follows. [1 1 1 0 0 0 0 1 1]. This approach allows to keep the number of features low while covering many different pattern configurations. In result, many different classifiers can be applied and work very efficiently in both training and testing.

The pattern matching process also allows to extract information about service's name, version, access policy, resources and resource sample values. Information extracted from the above sample URI is following:

`['accessPolicy': 'public', 'name': 'WSname', 'version': '2.0',`
`'queryResources':['getInfo': 'type=name']].`

3.3   RESTful Web Service Information Extraction

Very often instead of full URI endpoint (e.g. `http://api.example.com/exampleWS/resource`), only the fragments of RESTful service are exposed (e.g. `/exampleWS/resource`). Therefore, it is necessary to extract those fragments from the documentation website. However, for the fragment to be useful to the user, we also need to find its root. Very often the the root URI of a Web Service is different than the root of the documentation website. Therefore, detecting the root may be critical in order to properly identify a service.

Finding root URI and RESTful fragments can be performed using keywords, in a similar manner as in the case of documentation detection. Additionally, fragments are separated by a dot character or slash and can be localized using regular expressions. In our approach combining it with the method described in [11] allows to extract fragments and root URIs easily.

Another very informative elements are example URIs – service fully qualified endpoint URIs that contain sample values. Example URIs can be detected using the pattern matching described above.

3.4 Identification and Information Extraction Algorithm

Based on the above, the following algorithm of RESTful Web Service identification and information extraction is proposed:

Algorithm 1.
Input: Set of webpages on a host (*host*).
Output: Set of RESTful URIs (*restfulURIs*) or RESTful fragments (*fragments*).

*processPages(*host*):*
```
 1:  for webpage ∈ host do
 2:      rootURI ← False
 3:      restfulURIs, fragments ← ∅, ∅
 4:      if isDocumentation(webpage) do
 5:          rootURI ← findRoot(webpage)
 6:          for URI ∈ webpage do
 7:              if filterOut(URI) and isRestful(URI) do
 8:                  restfulURIs ← restfulURIs ∪ URI
 9:              end
10:          end
11:          for fragment ∈ webpage do
12:              if filterOut(fragment) do
13:                  fragments ← fragments ∪ fragment
14:              end
15:          end
16:      end
17:      if rootURI do
18:          return fragments ∪ restfulURIs
19:      end
20:      else do
21:          return restfulURIs
22:      end
23: end
```

The algorithm loops over all pages within a given hostname. The first step is to set initial value to the *rootURI* variable (line 2) and prepare the sets for URIs of RESTful Web Services and fragments (line 3). Lines 4–16 are executed for webpages that contain RESTful documentation (Sect. 3.1). In line 5, using function *findRoot()* (described in Sect. 3.3), the algorithm tries to find the root URI on current webpage.

The loop in lines 6–10 executes for every URI found on the website, rejects URIs to files with common file extensions using function *filterOut()* (here any filter can be applied) and checks if URI has RESTful Web Service structure using function *isRestful()* (this function also extracts data from URI as described in Sect. 3.2). If both conditions are true – that is if the URI is not a file and it is a RESTful Web Service – the URI is added to the set of identified RESTful URIs (line 8). In lines 11–15 the algorithm extracts RESTful fragments in a similar manner.

At this point all the available data is collected. Afterwards, if the *rootURI* was not found, the algorithm returns collected set of RESTful Web Services (line 21). No fragments are returned since we cannot verify them. Otherwise, if the *rootURI* was found, in line 18 the algorithm returns both identified endpoint URIs and fragments. Lines 17–22 are split into separate process because we need to crawl all pages within a host to find the root URI.

4 Test Collection

In many cases authors put only the information about destination URLs of pages containing documentation or crawl service directories. In both cases provided information becomes outdated as any of service providers makes changes to the documentation. Additionally, many services become unavailable after some time. This make it impossible to test elaborated methods of different researchers on the same dataset.

Current lack of test collections for RESTful Web Service data analysis prevents a meaningful effectiveness comparison among the different research in this topic. To solve this problem we prepared the *RESTfulWS_2014*[2] test collection that contains 55 RESTful Web Services from 10 service providers, collected from 217 pages with total 3856 URLs. The collection includes, but is not limited to, many e-business services like online payments, shopping cart interface, mailing service, currency exchange, stock exchange or online invoicing. We consider this test collection to be the first that is complete and accessible for research benchmarks. In Table 1 we present aggregated data of the collection.

Table 1. The aggregated data of *RESTfulWS_2014* test collection

	Total	AVG	MIN	MAX
Web Services	55	6	2	15
Resources	239	24	0	131
Sample values	23	2	0	6
Example URIs	26	3	0	9
Blocks	266	27	3	132
Documentation pages	217	22	1	145
RESTful URIs	125	13	0	29
RESTful fragments	215	22	0	131
URLs	3856	386	6	3514
Local URLs	290	29	0	159

---

[2] The RESTfulWS_2014 test collection is available to download at:
http://www.ii.pwr.wroc.pl/~czyszczon/WebServiceRetrieval

The "sample values" are the template values that certain resources accept. The "example URIs" represent full URIs that of Web Services that also contain sample values, for example `http://www.linkagogo.com/api/rest/private/folder/folder_id`. The "blocks" represent website areas that contain information about services.

The data was collected and evaluated manually. The exception are URLs which were extracted using a regex expression. Introduced in this paper test collection is designed for RESTful Web Service identification and information extraction but it may also be used for any other service classification, Web Service retrieval or for testing any other web-related data mining. The greatest advantage of proposed collection is that all analyzed pages are downloaded so it does not become outdated after service providers change them.

5   Evaluation

Based on presented algorithm we implemented a Web crawler that identifies RESTful Web Services and extracts available information about the services. The crawler allowed to perform an experiment on evaluation of proposed approach. The aim of the experiment was to calculate the effectiveness of the RESTful Web Service identification and information extraction method. Additionally, we used two different supervised learning models for pattern matching – the backpropagation ANN (Artificial Neural Network) and SVM (Support Vector Machine) with linear kernel. Based on our research presented in [9], the training set[3] for the learning process was elaborated manually by selecting appropriate URIs. For the positive dataset we selected 9 RESTful URIs and for the negative set 12 non-RESTful URIs. Our initial experimental results presented in [9], showed that only a small part of all URIs conforms to a RESTful Web Service pattern – it is around 1.3 %. We obtained best training results for the following parameters of backpropagation ANN: learning rate=0.5, momentum=0.1, hidden layers=8 and error threshold=0.0001. In order to evaluate the effectiveness the following classical information retrieval measures were used: precision, recall, F-Measure ($\beta = 1$). The experiment was performed on the *RESTfulWS_2014* test collection.

5.1   Identification

Pattern matching using the backpropagation algorithm and SVM turned out to give the same outcome and therefore results of both of them are shown together on the following graphs and tables. Figure 1 illustrates the identification effectiveness in terms of precision and recall for 10 different hosts. The precision and recall are almost 100 % for every host. The worst results were obtained for 6th host, where low precision was caused by the fact that some of the example URIs were qualified as separate services. Low precision can also be noticed for the 3rd

---

[3] The training set is available to download at:
`http://www.ii.pwr.wroc.pl/~czyszczon/WebServiceRetrieval`

host. This happened because instead of endpoint URLs on the documentation pages, the providers used URI fragments and did not expose neither root URL nor any static example URIs. Examples could be loaded using JavaScript on a click event, however this would require an advanced crawler.

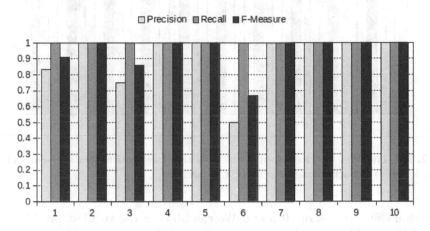

Fig. 1. RESTful Web Service identification – Backpropagation and SVM matching

Table 2 summarizes the results of the identification. Obtained results show that proposed method performed very well. The average precision is equal to 91 % whereas the average recall is equal to 100 %. This results in average F-Measure equal to 94 %. Such a result can be certainly considered as satisfactory.

Table 2. RESTful Web Service identification – Backpropagation and SVM matching

	Pages										Average
	1	2	3	4	5	6	7	8	9	10	
Precision	0.83	1.0	0.75	1.0	1.0	0.5	1.0	1.0	1.0	1.0	**0.908**
Recall	1.0	1.0	1.0	1.0	1.0	1.0	1.0	1.0	1.0	1.0	**1.0**
F-Measure	0.91	1.0	0.86	1.0	1.0	0.67	1.0	1.0	1.0	1.0	**0.943**

5.2   Information Extraction

The second part of the experiment is devoted to the evaluation of proposed information extraction approach. In the experiment the evaluation concerned only three most important elements that were not found during the pattern matching. Those were: (i) service fragments, (ii) resources, (iii) and example URIs. For each host, the precision, recall and F-Measure values were calculated as the mean values of those three elements. However, on certain hosts some of mentioned elements were not present. Therefore, in order not to distort the results, these hosts were not included in the mean. Figure 2 illustrates the information

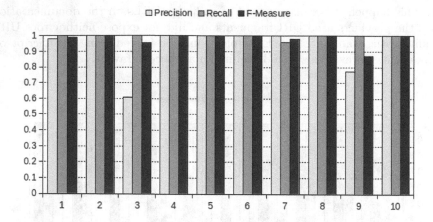

Fig. 2. RESTful Web Service information extraction (service fragments, resources and example URIs) – Backpropagation and SVM matching

extraction effectiveness of 10 hosts. We can observe the smallest precision for the 3rd host and it is for the same reason as in the case of identification.

Table 3 presents summarized results of the information extraction. The average precision is equal to 93.6 %, the average recall is equal to 99.6 %, and average F-Measure equals to 97.9 %. The results show that proposed method succeeded in getting very high precision and recall and can be considered as effective.

Table 3. RESTful Web Service information extraction (service fragments, resources and example URIs) – Backpropagation and SVM matching

	Pages										Average
	1	2	3	4	5	6	7	8	9	10	
Precision	0.98	1.0	0.61	1.0	1.0	1.0	1.0	1.0	0.77	1.0	**0.936**
Recall	1.0	1.0	1.0	1.0	1.0	1.0	0.96	1.0	1.0	1.0	**0.996**
F-Measure	0.99	1.0	0.95	1.0	1.0	1.0	0.98	1.0	0.87	1.0	**0.979**

6   Conclusions and Future Work

In this paper we presented a method for Web-scale automatic RESTful Web Service identification and data extraction which currently is a key problem in RESTful Web Service retrieval. Introduced method included detecting documentation pages that provide descriptions of RESTful Web Services, approach to service identification using URI structure patterns, approach to information extraction of informative service fragments, and finally an algorithm that incorporates presented approaches. Moreover, we described the first complete and extensive RESTful Web Service test collection that can be used for many benchmarks on the RESTful Web Service topic. Furthermore, the collection included also a number of e-business services. Presented evaluation results prove high effectiveness of presented method.

On the other hand proposed method lacked advanced crawling that simulates clicking on elements with assigned JavaScript methods. Presented approach could also include more information extraction elements like for example example output format (XML, JSON) or template resource values. However, both mentioned problems are the subjects of further research. In future work we also plan to use more supervised learning models for pattern matching. Additionally, we want to test other methods of RESTful Web Service identification and information extraction. In the near future we also plan to extend the test collection.

References

1. Richardson, L., Ruby, S.: Restful Web Services, 1st edn. O'Reilly (2007)
2. Alex, R.: Restful web services: The basics (2008),
   www.ibm.com/developerworks/webservices/library/ws-restful/
3. Pautasso, C., Zimmermann, O., Leymann, F.: Restful web services vs. "big"' web services: Making the right architectural decision. In: Proceedings of the 17th International Conference on World Wide Web, WWW 2008, pp. 805–814. ACM, New York (2008)
4. Pautasso, C.: On composing restful services. In: Leymann, F., Shan, T., van den Heuvel, W.J., Zimmermann, O. (eds.) Software Service Engineering. Dagstuhl Seminar Proceedings, vol. 09021. Schloss Dagstuhl – Leibniz-Zentrum fuer Informatik, Germany (2009)
5. Marc, H.: Web application description language. World Wide Web Consortium (2009), www.w3.org/Submission/wadl/
6. Kopecký, J., Gomadam, K., Vitvar, T.: hrests: An html microformat for describing restful web services. In: Proceedings of the 2008 IEEE/WIC/ACM International Conference on Web Intelligence and Intelligent Agent Technology, WI-IAT 2008, vol. 01, pp. 619–625. IEEE Computer Society, Washington, DC (2008)
7. Alarcón, R., Wilde, E.: Restler: crawling restful services. In: Proceedings of the 19th International Conference on World Wide Web, WWW 2010, pp. 1051–1052. ACM, New York (2010)
8. Taheriyan, M., Knoblock, C.A., Szekely, P., Ambite, J.L.: Semi-automatically modeling web apis to create linked apis. In: Proceedings of the First Linked APIs Workshop at the Ninth Extended Semantic Web Conference (May 2012)
9. Czyszczoń, A., Zgrzywa, A.: An artificial neural network approach to RESTful Web services identification, pp. 175–184. Oficyna Wydawnicza Politechniki Wrocławskiej (2011)
10. Ranabahu, A., Nagarajan, M., Sheth, A.P., Verma, K.: A faceted classification based approach to search and rank web apis. In: ICWS, pp. 177–184. IEEE Computer Society (2008)
11. Steinmetz, N., Lausen, H., Brunner, M.: Web service search on large scale. In: Baresi, L., Chi, C.-H., Suzuki, J. (eds.) ICSOC-ServiceWave 2009. LNCS, vol. 5900, pp. 437–444. Springer, Heidelberg (2009)

Generating User Interfaces for XML Schema Documents with a Presentation Language

Dariusz Rumiński, Krzysztof Walczak, and Jacek Chmielewski

Poznań University of Economics,
Niepodległości 10, 61-875 Poznań, Poland
{ruminski,walczak,chmielewski}@kti.ue.poznan.pl
http://www.kti.ue.poznan.pl

Abstract. Paper documents of public administration are being gradually replaced by XML documents with structure formally expressed as XML Schema definitions. However, due to rapid changes in legal and administrative regulations, many IT departments face the problem of fast and effective adaptation of user interfaces (UIs) of applications to changes in versions of public administration documents. This paper presents an approach to automatic transformation of XML Schema documents into useful and effective web form documents using the Form Presentation Language (FPL). The result of transformation is ready to use for end-users, while the style of transformation may be modified with minimal effort without specific technical skills.

Keywords: web forms, XML schema, FPL, Form Presentation Language, ASIS.

1 Introduction

Nowadays, paper documents of public administration are being gradually replaced by electronic documents encoded in XML. Structure of such documents is formally expressed as XML Schema definitions (XSDs). However, due to rapid changes in legal and administrative regulations, many IT departments face the problem of fast and effective adaptation of user interfaces (UIs) of every-day applications to subsequent versions of schema documents (XSDs) [1]. In general, XSD documents are not intended for reading (or processing) by end-users and must be transformed into easy-to-use UIs of office applications to support end-users in their daily work by enabling creation and presentation of correct XML documents. Since the structure of the UIs is highly dependent on the structure of the XML documents that need to be processed by the UI, it is reasonable to use the XSD documents as the basis for UI generation.

The XML Schema is a powerful and widely used language for describing data models, especially structures of XML documents. Having an XML Schema document, it is possible to generate a corresponding UI, however, a "raw" result, without convenient layout and graphical representation, would not be appropriate for end-users. The list of common problems also includes: lack of standardized

A. Kwiecień, P. Gaj, and P. Stera (Eds.): CN 2014, CCIS 431, pp. 328–337, 2014.

solution for expressing additional information (such as textual labels, form field descriptions, explanations or tooltips), limited support for localization, difficult presentation of choices, no support for reordering and filtering of fields. Taking into account today's expectations of end-users the list can be extended with: no support for auto-complete function and no way to express more complex interactions, such as retrieval of values of selected fields based on a value entered in another field (e.g., citizen address data pulled from an external system after providing his/her national ID).

In this paper, a new approach to generation of UIs for XML Schema documents is proposed. In this approach, XML Schema documents are automatically transformed into web forms with the use of additional information expressed in a new language specifically designed for this purpose – the Form Presentation Language (*FPL*). FPL describes aspects not included in the XSD, such as visual representation and layout of document structure. With FPL, it becomes possible to transform XSD documents into good-looking, convenient and effective UIs. The result of transformation is ready to use by end-users, while the process of transformation may be adjusted by modifying the accompanying FPL documents with minimal effort and without specific technical skills required.

The remainder of this paper is structured as follows. Section 2 presents the current state of the art in web forms generation. Section 3 introduces the Form Presentation Language. Section 4 describes an application of the proposed approach. Finally, Section 5 concludes the paper.

2   Related Works

An XML Schema definition (XSD), which defines structure of XML documents, contains detailed information about document particles, types of data in each particle, enumerations of possible values, constraints, etc. Almost all information that is needed to build a web form for editing compatible XML documents is available in XSD. Therefore, having these data, a natural question is *How web forms can be generated based on complex XSD documents, given that the result of this process must be useful, effective, and attractive to end-users?*. This question has been asked by many researchers and the challenges are well summarized by Cagle in his article [2], in which the proposed solution is based on the use of the XForms specification [3].

Kasarda et al. proposed a method of semi-automatic transformation of the XML Schema documents to XForms using XHTML as the host language for the generated web forms [4]. The use of XForms documents in connection with formal reasoning techniques has been presented in [5]. Although the above-proposed solutions build web forms, the result is still visually raw and XForms documents are complex and difficult to modify when the structure of XSD documents changes.

Another approach, presented in [6], uses XSLT – eXtensible Stylesheet Language Transformations stylesheets that allow to automatically generate HTML pages from multidimensional models based on an XML document. Although

XSLT is a powerful tool, it is not convenient for processing large and complex documents, whose structure can often change. Any change in documents structure requires an XSLT expert, in addition to a graphic designer creating the HTML and CSS documents.

A solution based on marked-up ontologies to generate web forms has been presented in [7]. In this approach, an ontology is modelled as a populated conceptual schema and implemented as an XML Schema and a relational database. A form developer marks up this schema and the resulting marked-up ontology is translated into presentation annotations over the underlying XSD. These annotations are used to generate the XHTML/JavaScript code for the form.

A semantic approach to generating personalized UIs was also presented in [8]. Authors of this publication present an architecture where UIs are dynamically generated for web services using the OWL-S ontology [9]. Since the ontology enables mapping of OWL-S input and output only to the WSDL description, this approach does not meet the requirement of generating UIs on the basis of XML Schema documents.

The *xsd-forms* project uses XSD annotation elements to include presentation information tagged using a special grammar. Such annotated schema is then used to automatically generate an HTML form with generic CSS and JavaScript that can be later overridden and customized. However, this solution cannot be used when the web forms are based on external, read-only XSD documents provided by an external source (e.g., governmental catalogue of official documents).

Another project, called *JAXFront* [10], takes a slightly different approach. It uses a raw XSD and an external *XUI* [11] file with presentation information and rules of form behaviour. This set of data combined with an optional source XML document may be used to generate a web form, as well as a Java GUI or a static PDF. The drawback of the JAXFront solution is that the presentation file must include all elements that should appear in the final form. It is not possible to augment only selected elements and generate the rest using only the data available in XSD.

The *DynaForm* solution tackles the problem mentioned for JAXFront by automatically generating a default presentation description from the source XSD [12]. Such a default presentation description covers all form elements and enables augmentation of only selected elements. Unfortunately, the DynaForm solution supports only a subset of XSD features and lacks tools to express rules of dynamic form behaviour.

3   FPL – Form Presentation Language

The Form Presentation Language is an XML-based language which has been developed in the Department of Information Technology at the Poznan University of Economics within the project "ITSOA – New information technologies for electronic economy and information society based on service-oriented architecture". FPL documents provide supporting information enabling transformation of XML Schema documents into usable web forms. FPL can be used for describing graphical view of XSD components, i.e., fragments of XML Schema documents that correspond to particular web form fields. FPL can also express human-readable labels and tooltips of XSD components. In addition, the FPL language enables describing web services that are used for supplying data into the generated web forms. Finally, it supports also the auto-complete function that expresses more complex interactions, such as retrieval of values for selected fields based on a value entered in another field (e.g., citizen address data pulled from an external system after providing his national ID).

3.1   Structure of FPL Documents

In Figure 1, the structure of FPL elements used for describing presentation properties of XSD components is presented. The following subsections describe selected FPL elements and their usage.

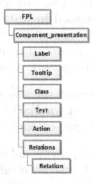

Fig. 1. Structure of FPL presentation elements

***FPL* Element.** The *FPL* element is the root of an FPL document. It is responsible for grouping *Component_presentation* elements that describe presentation of particular components. Information about XSD namespaces is stored in attributes. The namespace prefixes and names correspond to the ones defined in the XSD document.

***Component_presentation* Element.** The *Component_presentation* element is responsible for describing presentation of XSD components. Each *Component_presentation* element corresponds to only one schema component. The *Component_presentation* element has three attributes: *id*, *ref* and *presentationType*. The *id* attribute is responsible for the identification of the web form field. The *ref* attribute points to the XSD component using an XPath location expression. The *presentationType* attribute specifies the cardinality of the component, with the possibility to override the default component cardinality specified in the XSD. The attribute may take one of the following values: *TAKEN_FROM_XSD*, *HIDDEN*, *OPTIONAL*, and *REQUIRED*.

The *presentationType* attribute is needed to modify presentation of external XSDs. Building own schemas based on components from external XSDs is a good practice. However, some of the imported XSD components may be too general and some components may be used only in specific use cases. Therefore, there is a need to override the cardinality of particular components defined in external XSDs during the process of web form generation. Using the *presentationType* attribute, it is possible to declare that some of XSD components should be omitted during the process of web form creation (the "HIDDEN" value). It is also possible to specify that some of XSD components should be interpreted as optional or required during the process of form validation. By default, the setting is inherited from the XSD, which is represented by the *TAKEN_FROM_XSD* value.

Label Element. The *Label* element contains a presentation-friendly label of the XSD component, which will be displayed within the form instead of the raw component name defined in XSD. This enables presentation of form elements with proper names that are not used in XSD schemas, including long names, names with proper capitalization, spaces and diacritical characters. The use of the *Label* element enables also presentation of the interfaces in different languages. *Label* is a child of the *Component_presentation* element.

Class Element. The *Class* element specifies names of CSS styles used to format the graphical representation of the XSD component. The *Class* element is another child of the *Component_presentation* element.

Text Element. The *Text* element provides information about how to present textual elements. The *Text* element is another child of the *Component_presentation* element. It consists of two attributes – the *multiline* attribute holds a Boolean value, which specifies whether the field is single-line or multi-line, the *type* attribute describes the length of the field and must be set to one of the following predefined constants: *VERY_LONG, LONG, SHORT* or *VERY_SHORT*.

3.2   Using SOA Web Services in FPL

The following FPL elements provide description of SOA web services that are used to supply data to web forms generated on the basis of XSD and FPL.

Listing 1.1. Web service declaration in FPL

```
<Action>
  <Endpoint>http://egov.kti.ue.poznan.pl:8080/axis2/services/KnBaseEntityService</Endpoint>
  <Method>getIndividualByPesel</Method>
  <Params>
    <Param>pesel</Param>
  </Params>
</Action>
```

Action Element. The *Action* element groups information about a SOAP web service: *method, parameters* and *endpoint*. *Action* is another child of the *Component_presentation* element. Listing 1.1 provides a description of a sample web service in FPL.

Relations **Element.** To insert the data returned by a web service into an appropriate form field, there is a need to indicate relations between services and fields. The following elements enable specifying such relations. The *Relations* element is a child of the *Component_presentation* element and is used to group particular *Relation* elements.

Relation **Element.** The *Relation* element specifies a relation between a parameter name located in a SOAP response returned by a web service (a parameter called *serviceParamName*) and an identifier of the *Component_presentation* element. This identifier is provided as the element value. The *Relation* element is a child of the *Relations* element.

An example of relations declaration is presented in Listing 1.2, where two relations are shown. In the example, the returned SOAP response message contains information about an object consisting of two values – first name and last name. These values are inserted into form fields whose identifiers are specified in the *Relation* elements (cf. Listing 1.4).

Listing 1.2. *Relations* element declaration

```
<Relations>
   <Relation serviceParamName="firstName">x522</Relation>
   <Relation serviceParamName="lastName">x525</Relation>
</Relations>
```

3.3   Creating FPL Descriptions

An FPL description can be created off-line by a form designer who does not need to have programming skills. Such a description is then used on-line while generating web forms.

To create an FPL description, a form designer uses a graphical tool called *FPL Creator*. The tool uses an XML Schema document as the input and generates an FPL description. For the generation of FPL, a user can choose between two options.

The first option is an automatic process of FPL description generation. It takes place without human interference and default values are taken to describe graphical representation of XSD components. The auto-generated FPL description can be then selectively customized. The second option is an interactive process whereby a form designer can prepare a custom FPL description (Fig. 2). In the interactive mode, for each XSD component, a form designer can change graphical properties such as the label, the tooltip, font attributes, field attributes and the presentation type.

3.4   Example FPL Form Description

Listing 1.3 provides an example of an XSD document, while Listing 1.4 shows an FPL description used for visualization of the XSD document. For the purpose of this publication, the listings have been shortened and contain only the most important fragments of the XSD and FPL code.

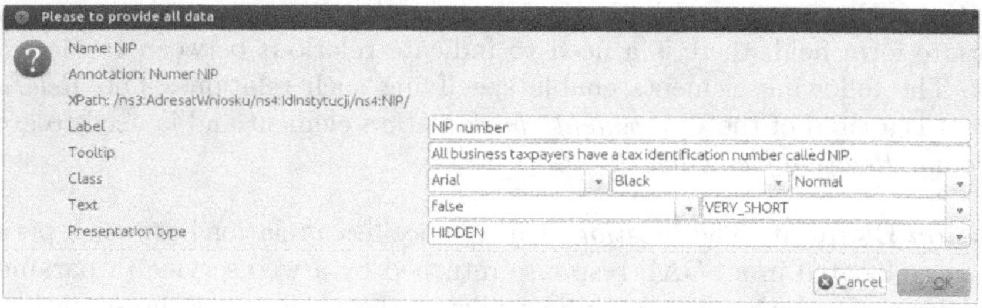

Fig. 2. Setting XSD component properties

Listing 1.3. Example XSD document for visualization

```
<xs:element name="DanePelnomocnika" type="ns6:OsobaTyp"/>
<xsd:complexType name="OsobaTyp">
 <xsd:sequence>
  <xsd:element ref="ns6:IdOsoby" minOccurs="0"/>
  <xsd:group ref="ns6:ImieNazwiskoGrupa"/>
 </xsd:sequence>
</xsd:complexType>
<xsd:group name="ImieNazwiskoGrupa">
 <xsd:sequence>
  <xsd:element ref="ns6:Imie"/>
  <xsd:element ref="ns6:ImieDrugie" minOccurs="0"/>
  <xsd:element ref="ns6:Nazwisko" maxOccurs="3"/>
 </xsd:sequence>
</xsd:group>
<xsd:element name="IdOsoby" type="ns6:IdOsobyTyp"/>
<xsd:complexType name="IdOsobyTyp">
 <xsd:sequence>
  <xsd:element ref="ns6:PESEL" minOccurs="0"/> ...
 </xsd:sequence>
</xsd:complexType>
<xsd:element name="PESEL" type="ns6:PESELTyp"/>
<xsd:simpleType name="PESELTyp"> ...
</xsd:simpleType> ...
```

In the FPL document, every XSD component is described by an *Component_presentation* element. The description provides presentation information for four schema components: a Complex Type "IdOsoby" (*Component_presentation* with id *s121*) and three Simple Types "PESEL", "Imie" and "Nazwisko" (*Component_presentation* with ids *x518*, *x522* and *x525*).

The *Relations* and *Action* elements in FPL have been declared as children of the *Component_presentation* element with the identifier *x518*. The element *x518* corresponds to the XSD element named *PESEL*. This element represents a form field, whose value is required by a web service method. This web service method is called by a 'proxy' servlet that gets data such as web service endpoint, name of method and parameters, from web form's JavaScript code. The JavaScript code sends these data asynchronously and gets a response from a servlet. In turn, the *Relation* elements indicate where the returned response data should be put in the form – in this case form fields with ids *x522* and *x525*.

Listing 1.4. An example of FPL describing XSD from Listing 1.3

```
<Component_presentation id="s121" ref="/ns2:DanePelnomocnikow/ns2:DanePelnomocnika/ns6:IdOsoby/"
      presentationType="TAKEN_FROM_XSD">
   <Label>Id Osoby</Label>
   <Tooltip>Id Osoby</Tooltip>
   <Class>complexType</Class>
</Component_presentation>
<Component_presentation id="x518" ref="/ns2:DanePelnomocnikow/ns2:DanePelnomocnika/ns6:IdOsoby/
      ns6:PESEL/" presentationType="TAKEN_FROM_XSD">
   <Label>Numer PESEL</Label>
   <Tooltip>Numer PESEL</Tooltip>
   <Class>simpleType</Class>
   <Text multiline="false" type="VERY_LONG"/>
   <Relations>
      <Relation serviceParamName="firstName">x522</Relation>
      <Relation serviceParamName="lastName">x525</Relation>
   </Relations>
   <Action>
      <Endpoint>http://egov.kti.ue.poznan.pl:8080/axis2/services/KnBaseEntityService</Endpoint>
      <Method>getIndividualByPesel</Method>
      <Params><Param>pesel</Param></Params>
   </Action>
</Component_presentation>
   <Component_presentation id="x522" ref="/ns2:DanePelnomocnikow/ns2:DanePelnomocnika/ns6:Imie/"
      presentationType="TAKEN_FROM_XSD">
   <Label>Imie</Label>
   <Tooltip>Imie</Tooltip>
   <Class>simpleType</Class>
   <Text multiline="false" type="VERY_LONG"/>
</Component_presentation>
   <Component_presentation id="x525" ref="/ns2:DanePelnomocnikow/ns2:DanePelnomocnika/ns6:
      Nazwisko/" presentationType="TAKEN_FROM_XSD">
   <Label>Nazwisko</Label>
   <Tooltip>Nazwisko</Tooltip>
   <Class>simpleType</Class>
   <Text multiline="false" type="VERY_LONG"/>
</Component_presentation>
```

4   Application

The presented approach to transformation of XSD Schema documents into web forms has been applied within the Adaptable SOA Interface System (ASIS) [13]. ASIS is responsible for dynamic web forms generation within the PEOPA Plat form – a comprehensive e-government solution for public administrative units for modeling and executing administrative procedures [1].

The overall ASIS architecture is presented in Fig. 3. ASIS takes into account three kinds of XML files while generating web forms. The first file is an XML Schema that defines the structure of XML documents. The second file is an FPL description which provides additional presentation information for XSD components and specifies SOA web services that supply data to the web form. The third – optional – file is an XML document that contains saved form data. ASIS communicates with the PEOPA platform using the SOAP protocol.

Figure 4 shows a fragment of a web form generated on the basis of XSD and FPL presented in Listings 1.3 and 1.4. The presented web form facilitates work of an end-user. For instance, when the web form is long and complex, a user can click on a grey bar with an arrow that reflects an XSD complex type, which causes hiding or showing embedded form components. The form can also retrieve data from external SOA web services to automatically fill-in selected fields. If an *Element_presentation* element has been declared with a SOA web service description, ASIS creates additional button next to the field whose value is required while preparing a corresponding SOAP message. In Fig. 4, a button responsible for asynchronous sending of the SOAP message was placed by ASIS

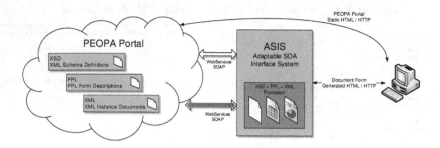

Fig. 3. Web forms generation for the PEOPA platform with ASIS

next to the *Numer PESEL* form field. When a user clicks on this button, the web form sends an asynchronous request to the SOAP web service. Next, the response data are processed and inserted into the form fields specified in the FPL *Relations* element.

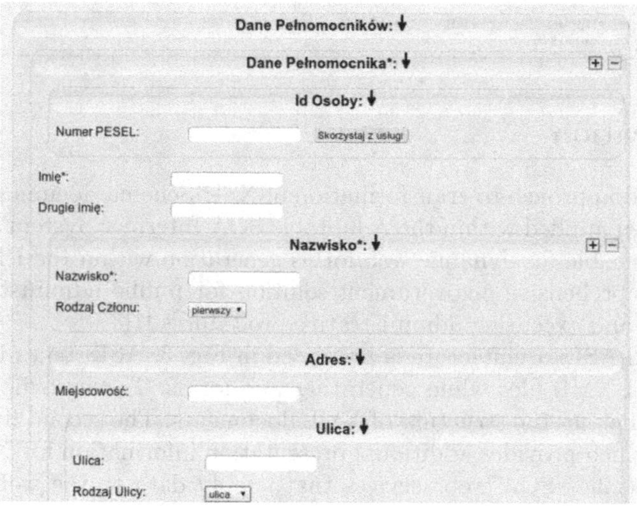

Fig. 4. Web form generated for example XSD (Listing 1.3) with FPL (Listing 1.4)

5   Conclusions

In this paper, an approach to automatic transformation of XML Schema documents into web forms using FPL descriptions has been presented. In this approach, XML Schema provides information about structure of the user interfaces, while FPL descriptions provide additional information enabling generation of useful and effective web forms. Additionally, FPL enables specifying web services used to automatically retrieve form fields data.

The presented approach has been used and validated within the Adaptable SOA Interface System in the PEOPA e-administration platform. Generating web forms on the basis of XSD and FPL descriptions significantly simplifies the work of form designers and therefore can be used for documents with frequently changing XSD schemas. Form designers do not need to have programming skills, which would be needed for creating the web forms manually. Instead, a designer can use the *FPL Creator* application for defining the final view of a web form.

At the same time, with the use of the presented approach, it is possible to provide end-users with convenient graphical interfaces for creating and editing XML documents.

References

1. Strykowski, S., Wojciechowski, R.: Composable Modeling and Execution of Administrative Procedures. In: Kő, A., Leitner, C., Leitold, H., Prosser, A. (eds.) EDEM/EGOVIS 2012. LNCS, vol. 7452, pp. 52–66. Springer, Heidelberg (2012)
2. Cagle,   K.:   XForms,   XML   Schema,   and   ROX, http://www.xml.com/pub/a/2007/08/16/xforms-xml-schema-and-rox.html
3. W3C Consortium: XForms 1.1, http://www.w3.org/TR/xforms11/
4. Kasarda, J., Nečaský, M., Bartoš, T.: Generating XForms from an XML Schema. In: Zavoral, F., Yaghob, J., Pichappan, P., El-Qawasmeh, E. (eds.) NDT 2010, Part II. CCIS, vol. 88, pp. 706–714. Springer, Heidelberg (2010)
5. Cheow, P., Governatori, G.: Representing and Reasoning on XForms Document. In: ADC 2004 Proc. of the 15th Australasian Database Conf., vol. 27, pp. 141–150. ACM (2004)
6. Luján-Mora, S., Medina, E., Trujillo, J.: A Web-Oriented Approach to Manage Multidimensional Models through XML Schemas and XSLT. In: Chaudhri, A.B., Unland, R., Djeraba, C., Lindner, W. (eds.) EDBT 2002 Workshops. LNCS, vol. 2490, pp. 29–44. Springer, Heidelberg (2002)
7. Dumas, M., Aldred, L., Heravizadeh, M., ter Hofstede, A.H.M.: Ontology markup for web forms generation. In: Workshop on Real World RDF and Semantic Web Applications (May 2002)
8. Khushraj, D., Lassila, O.: Ontological Approach to Generating Personalized User Interfaces for Web Services. In: Gil, Y., Motta, E., Benjamins, V.R., Musen, M.A. (eds.) ISWC 2005. LNCS, vol. 3729, pp. 916–927. Springer, Heidelberg (2005)
9. Ankolekar, A., et al.: DAML-S: Web Service Description for the Semantic Web. In: Horrocks, I., Hendler, J. (eds.) ISWC 2002. LNCS, vol. 2342, pp. 348–363. Springer, Heidelberg (2002)
10. JAXFront, http://www.jaxfront.org/pages/home.html
11. XUI, http://www.xoetrope.com/xui
12. Raudjärv, R.: DynaForm, https://code.google.com/p/xsd-web-forms/
13. Walczak, K., Wiza, W., Chmielewski, J.: Adaptation of User Interfaces in SOA Applications. e-Minds: International Journal on Human-Computer Interaction 2(8), 3–17 (2012)

Relational Database Index Selection Algorithm

Radoslaw Boronski and Grzegorz Bocewicz

University of Technology of Koszalin,
ul. Sniadeckich 2 75-453 Koszalin, Poland
radoslaw.boronski@tu.koszalin.pl, bocewicz@ie.tu.koszalin.pl
http://tu.koszalin.pl

Abstract. The Index Selection Problem (ISP) is an important element of research in the field of optimization of relational database systems. Commonly used commercial tools are based on a methodology that enables tables indexing for independent SQL queries. The article presents an original method, based on a genetic algorithm, for indexing tables for groups of queries in a relational database (MDI). Conducted experiments have shown that the use of indices for a group of queries can reduce the group execution time by 15 % as well as can reduce the memory needs by 68–90 %.

Keywords: index, indexing, database, query, optimization, SQL, RDBMS.

1 Introduction

Index Selection Problem (ISP) was broaden discussed in the literature [1–5]. There are different approaches for optimal selection of indices for a single query [6–8] and grouped queries [9, 10]. They include methods based on estimates of the cost-based optimizer, methods that operate on the execution plan or methods of analysing only the semantics of a query. Despite the lapse of years, the problem of selection indices is still valid and current. Although the proposed solutions present interesting attempts to solve the problem and have many benefits (i.e. multiple columns indexing, redundant structures elimination [7, 11]), they are not sufficient to implement them in practice [8]. In previous studies too little attention was paid to the numerous disadvantages, such as: not taking into account the size and index creation time or single query tables indexing. Omission of other queries relating to the same table could result in the creation of too many similar indices.

Bearing the above, it is worth noting that currently available commercial tools (i.e. Oracle Access Advisor, Toad) do not use the full capabilities of tables indexing within group of queries. It can therefore be inferred that there is a need for an automatic mechanism that would allow the noticing of the relationship between group queries and would suggest better indices. If one looked at sample queries as a group and tried to find relations in this group, one could try to create better indices that shorten the group's execution time and at the same

A. Kwiecień, P. Gaj, and P. Stera (Eds.): CN 2014, CCIS 431, pp. 338–347, 2014.

time would take up less disk space and reduce the time needed to create a tree structure [9, 10, 12, 13].

The aim of the paper is to propose a new concept of the tables indexing in the relational database systems. In contrast to known approaches [1, 2, 4, 6], that consider indexing for individual queries, a proposed solution allows to look at the queries as a group and to try to find good indices for a group as a whole. To the best of our knowledge, there is no research paper on grouped database queries, which are a key feature of a production databases cyclic processing. In that context this work can be seen as a continuation to our former research conducted in [9, 10, 12, 13], where the properties of grouped SQL queries have been studied and the original index selection proposition have been presented. In that context, our paper extends the index selection problem and as a solution, proposes an original self-tuning algorithm, which could be used in any relational database system.

Section 2 presents the new problem statement on Index Selection for Grouped Queries in relational database. Section 3 presents an original index selection method (we called it MDI) based on a genetic algorithm. Section 4 shows the results of the experiment. Section 5 summarizes the issue and outlines the direction for further research.

2   Index Selection Problem for Grouped Queries

In classical approach, the ISP problem lies in the search for sets of indices that minimize the response time for the query in the relational database.

In such context, the problem discussed in this paper extends the ISP problem of elements related to the size of memory used to store indices and application of grouped queries (a set of queries). The problem of this type is defined as follows:

Given is a set of tables T stored in the database DB:

$$T = (T_1, \ldots, T_i, \ldots, T_{lt}) , \tag{1}$$

where: T_i is a i-th table, lt – number of DB database tables, described by a set of columns included in the tables:

$$K = K_1 \cup \ldots \cup K_i \cup \ldots \cup K_n , \tag{2}$$

where: $K_i = \{k_{i,1}, \ldots, k_{i,l(i)}\}$ is a set of columns for table T_i, $k_{i,j}$ is a j-th column of table T_i, $l(i)$ is a number of columns for table T_i.

Each column $k_{i,j}$ corresponds to set of values $V_{i,j}$ included in this column.

For the set of tables T various queries Q_m can be formulated (in SQL these are $SELECT$ queries). These queries are put against the specified set of columns $KQ_m \subseteq K$. The result of query Q_m is set:

$$A_m \subseteq \prod_{k_{i,j} \in KQ_m} V_{i,j} \tag{3}$$

where: $\prod_{i=1}^{n} Y_i = Y_1 \times Y_2 \times \ldots \times Y_n$ is a Cartesian product of sets $Y_1, Y_2, \ldots, Y_n$.

For a given database DB it is taken into account that A_m is a result of following function:

$$A_m = q_m\big(KQ_m, Op\,(DB)\big) \tag{4}$$

where: KQ_m is a subset of columns used in query Q_m, $Op(DB)$ is a set of operators available in database DB of which relation describing query Q_m is built.

The time associated with the determination of the A_m set is dependent on the DB database used (search algorithms, indices structures) and set of adopted J_m indices. This set is defined as follows:

$$J_m \subseteq P\left(\bigcup_{i\in\{j_1,\dots,j_b\}} \bigcup_{a=1}^{|KQ_m^i|} VQ_m^i(a)\right) \tag{5}$$

where:

$P(Y)$ is a power set of Y set, $|\,Y\,|$ is a Y set cardinality,

$\bigcup\limits_{i=1}^{n} Y_i = Y_1 \cup Y_2 \cup \dots \cup Y_n$ is a sum of $Y_1, Y_2, \dots, Y_n$ sets,

KQ_m^i is a set of columns for T_i table present in Q_m query, as follows:

$KQ_m = KQ_m^{j_1} \cup KQ_m^{j_2} \cup \dots \cup KQ_m^{j_b}$; $j_1, j_2, \dots, j_b$ are tables numbers of which columns are present in Q_m query,

$VQ_m^i(a)$ is a set of a-element tuples formed from a set of unique elements of KQ_m^i set:

$$VQ_m^i(a) = \begin{cases} \{(x) \mid x \in KQ_m^i\} & \text{for } a = 1 \\ \{(x_1,\dots,x_a) \mid x_1,\dots,x_a \in KQ_m^i; \\ \qquad\qquad x_1 \neq x_2 \neq \dots \neq x_a\} & \text{for } a > 1 \end{cases} \tag{6}$$

where

$$a \in \{1,\dots,|\,KQ_m^i\,|\} \ .$$

The set of J_m contains tuples (records) consisting of columns of a single table. In other words, a single tuple which is part of J_m set, contains columns of a table, for which the index is built.

In general, it is assumed that the execution time of the Q_m query (determination of A_m) in the specified DB database is defined as: $t(Q_m, J_m, DB)$. Because our considerations apply only to the effect on J_m indices (DB database working conditions are considered immutable), execution time of Q_m, will be defined as: $t_m(J_m)$.

Furthermore, J_m indices set, together with tables referenced in query Q_m are adequately characterized by the size of memory used: $S(J_m)$ – indices size, $S(Q_m)$ – indexed tables size (tables used in Q_m queries). The ratio of these volumes is determined by the J_m indices size against the tables size present in the Q_m query: $r(J_m, Q_m) = \frac{S(J_m)}{S(Q_m)}$.

In general, one can consider a $Q = \{Q_1, Q_2, \dots, Q_n\}$ grouped queries characterized by the following values:

– set of indices:

$$J \subseteq P \left(\bigcup_{m=1}^{n} \bigcup_{i_m \in \{j_{m,1},...,j_{m,b}\}} \bigcup_{a=1}^{|KQ_m^i|} VQ_m^i(a) \right) , \qquad (7)$$

– query execution time of the whole Q group: $t(J) = \sum\limits_{m=1}^{n} t_m(J)$,

– disk space usage ratio: $r(J,Q) = \dfrac{S(J)}{\sum\limits_{m=1}^{n} S(Q_m)}$.

In context of so-defined parameters, considered the Index Selection Problem for grouped queries binds to the response to the question: *What set of J indices, created for the immutable group of Q queries, that satisfy the condition: $r(J,Q) < r_h$ (coefficient of utilization of space does not exceed a preset r_h value), minimizes the queries execution time in the DB database: $t(J) \to min$?*

Concept of the proposed Index Selection Problem is illustrated in Fig. 1.

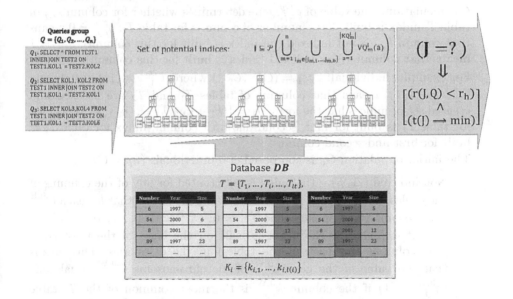

Fig. 1. Index Selection Problem in relational database

3   Index Selection Algorithm – MDI

In order to answer the question a new approach has been developed based on the concept of an evolutionary algorithm. The steps involved in the proposed index selection algorithm for set of J indices for the associated group of Q queries:

1. **Generation of initial population** G_0. It is assumed that every G_x population contains np number of individuals. The approach assumes arbitrarily that $np = 6$. In general, a G_x population is a set of $\Delta_{x,y}$ individuals:

$$G_x = \{\Delta_{x,1}, \Delta_{x,2}, \ldots, \Delta_{x,np}\} \tag{8}$$

where: $\Delta_{x,y}$ – is a y individual of G_x population.
Individual $\Delta_{x,y}$ of the G_x population is a sequence of chromosomes, defining the columns of tables on which indices are built:

$$\Delta_{x,y} = \left(C_{i_1}^{x,y}, C_{i_2}^{x,y}, \ldots, C_{i_r}^{x,y}\right) \tag{9}$$

where: r is a number of indexed tables from Q queries group; $i_1, i_2, \ldots, i_r$ are numbers of tables, of which columns are used in Q queries group; $C_i^{x,y}$ is a chromosome of individual $\Delta_{x,y}$, corresponding to table T_i:

$$C_i^{x,y} = \left(g_{i,1}^{x,y}, g_{i,2}^{x,y}, \ldots, g_{i,l(i)}^{x,y}\right) \tag{10}$$

where: $g_{i,j}^{x,y} \in \{0,1\}$ is a j-th gene of i-th chromosome of individual $\Delta_{x,y}$ of G_x population. The value of $g_{i,j}^{x,y}$ gene determines whether for column $k_{i,j}$ of table T_i index is built. $l(i)$ – number of columns for table T_i. $g_{i,j}^{x,y} = 0$ means that column $k_{i,j}$ is not indexed (no index is built for this column), $g_{i,j}^{x,y} = 1$ means that column $k_{i,j}$ is indexed (index is built for this column).
For example, individual: $\Delta_{x,1} = (C_1^{x,1}, C_2^{x,1})$ where: $C_1^{x,1} = (1,1,0,0,1)$ and $C_2^{x,1} = (1,1)$, means that columns of tables T_1 and T_2 are indexed and created set of indices has the form: $J_1 = \{(k_{1,1}, k_{1,2}, k_{1,5}), (k_{2,1}, k_{2,2})\}$ (index for table T_1 is built for first, second and fifth column, index for table T_2 is built for first and second column).
The initial population G_0 is composed of 6 individuals ($np = 6$):

- Non-indexed ($\Delta_{0,B}$) – the index is not created for any of the columns of the tables present in the group of queries. This means that for each $C_i^{0,B}$ chromosome value of each gene $g_{i,j}^{0,B} = 0$: $C_i^{0,B} = (0,0,\ldots,0)$.
- Weighted ($\Delta_{0,W}$) – the index is built on the basis of the most common columns (for each of the tables) in the Q queries. This means that the value of the gene $g_{i,j}^{0,W}$ of the chromosome $C_i^{0,W}$ equals one ($g_{i,j}^{0,W} = 1$) if the column $k_{i,j}^{0,W}$ is the most common of the T_i table columns within group of Q queries. The values of other $C_i^{0,W}$ chromosome genes equal zero $g_{i,j}^{0,W} = 0$. For example, $C_1^{0,W} = (1,0,0)$ means that for table T_1, first column is indexed and created set of indices has the form: $C_1^{0,W} = \{(k_{1,1})\}$ (index for table T_1 is built for first column).
- Halved ($\Delta_{0,P}$) – the index is built for the first half of the columns of each table found in the group of Q queries. This means that for each $C_i^{0,P}$ chromosome, the value of $g_{i,j}^{0,P} = 1$ only for a gene that meets the condition $j < \frac{l(i)}{2}$. For remaining genes $g_{i,j}^{0,P} = 0$.

- Mutated B ($\Delta_{0,\mathrm{MB}}$) – the index is built as in the case of the Non-indexed ($\Delta_{0,\mathrm{B}}$) individual, except that each gene $g_{i,j}^{0,\mathrm{MB}}$ of $C_i^{0,\mathrm{MB}}$ chromosome is subjected to mutation with probability of P_k.
- Mutated W ($\Delta_{0,\mathrm{MW}}$) – the index is built as in the case of the Weighted ($\Delta_{0,\mathrm{W}}$) individual, except that each gene $g_{i,j}^{0,\mathrm{MW}}$ of $C_i^{0,\mathrm{MW}}$ chromosome is subjected to mutation with probability of P_k.
- Mutated P ($\Delta_{0,\mathrm{MP}}$) – the index is built as in the case of the Halved ($\Delta_{0,\mathrm{P}}$) individual, except that each gene $g_{i,j}^{0,\mathrm{MP}}$ of $C_i^{0,\mathrm{MP}}$ chromosome is subjected to mutation with probability of P_k.

G_0 initial population is in this case a set of 6 initial individuals:

$$G_0 = \{\Delta_{0,\mathrm{B}}, \Delta_{0,\mathrm{W}}, \Delta_{0,\mathrm{P}}, \Delta_{0,\mathrm{MB}}, \Delta_{0,\mathrm{MW}}, \Delta_{0,\mathrm{MP}}\} \ . \tag{11}$$

In general, G_x (for $x > 0$) population is a set of individuals:

$$G_x = \{\Delta_1, \Delta_2, \dots, \Delta_6\} \ . \tag{12}$$

2. **Selecting "the best" individual** $\Delta_{\min}$ from the G_x population, which corresponds to the $J_{\min}$ index that guarantees the shortest execution time of the Q queries group.
3. **Checking the stop conditions** for Ht and p parameters for $\Delta_{\min}$ individual. Ht parameter specifies the execution time of the Q query group, after which the algorithm must stop ($t(J_\Delta) \leq Ht$). Parameter p specifies the number of G_x populations, after which the algorithm must stop ($G_x = p$). Ht and p parameters are independent of each other and the algorithm terminates, if at least one of the conditions is met.
4. **New G_{x+1} population build:**
   - 6 pairs of individuals from G_x population are randomed,
   - an individual is selected from each of the randomed pairs determining indices of lesser execution time of Q grouped queries. Selected individual $\Delta_{x,y}$ becomes new individual $\Delta_{(x+1),y}$ of population G_{x+1},
   - for individual $\Delta_{(x+1),y}$, each gene $g_{i,j}^{(x+1),y}$ of chromosome $C_i^{(x+1),y}$ is subjected to mutation with probability of P_k. This means that chromosome C_i takes form:

$$C_i^{(x+1),y} = \left(P_k\left(g_{i,1}^{(x+1),y)}\right), P_k\left(g_{i,2}^{(x+1),y)}\right), \dots, P_k\left(g_{i,l(i)}^{(x+1),y}\right) \right) \ , \tag{13}$$

   - for individual $\Delta_{(x+1),y}$, each gene $g_{i,j}^{(x+1),y}$ of chromosome $C_i^{(x+1),y}$ is subjected to crossover. This means that within one $C_i^{(x+1),y}$ chromosome a random $g_{i,j}^{(x+1),y}$ gene location exchange takes place. As a result, a new $'g_{i,j}^{(x+1),y}$ genes sequence for chromosome $'C_i^{(x+1),y}$ is built:

$$'C_i^{(x+1),y} = \left('g_{i,1}^{(x+1),y}, 'g_{i,2}^{(x+1),y}, \dots, 'g_{i,l(i)}^{(x+1),y}\right) \ . \tag{14}$$

Each successive G_{x+1} population is therefore a set of 6 new $'\Delta$ individuals:

$$G_{x+1} = \{'\Delta_{(x+1),1}, '\Delta_{(x+1),2}, \dots, '\Delta_{(x+1),6}\} \ . \tag{15}$$

5. **Return to step 2.**

4   Conditional Experiments

The purpose of the experiments was to verify the effectiveness of the MDI algorithm for a group of Q queries.

The research was carried out on a large (1.2 TB) relational Oracle database (11.2.0.3) of the leading car manufacturer in the United States. For the purposes of the test copies of 3 tables were created, containing 107 records of the production data. The need to create a copy of the tables was justified by the possibility of accidental production table locking (i.e. simultaneous reading by two or more database processes). For the purpose of the experiment, a computer program was created using free Pascal-compatible environment Lazarus (Free Pascal IDE). The application interface is shown in Fig. 2.

Of the hundreds of available database SQL queries, three Q^i groups were optionally selected. The objective was to enable the development of results not worse than commercial solutions, during one full group run (24 hours). A characteristic feature of the discussed database was that the individual Q^i query was repeated cyclically. For example, test query Q^i ran in the database once a day. Experience and test measurements were carried out for each Q^i group separately.

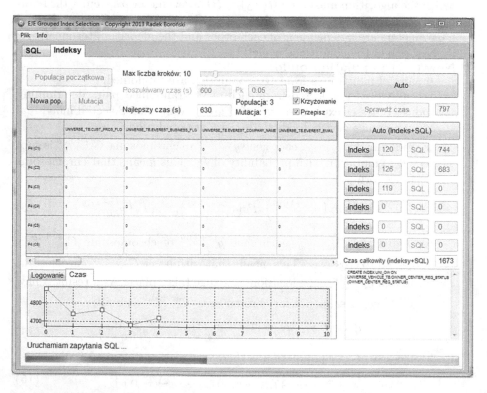

Fig. 2. Example of the computer program runs for a group of queries Q^1

A comparative analysis was carried out for the studied groups of Q^i queries. Indices creation times together with queries execution times for the MDI algorithm were compared with the times obtained by the action of external index advisors (DI). To accomplish this task, Oracle Access Advisor tools was used, which is part of a commercial database management system Oracle 11g. Both methods worked out indices, which were then compared. Generating one step (one complete cycle) corresponded to the generation of one population in the MDI algorithm. The experiment was run for 60 days. Tested were 3 groups of Q^i queries. The results of the experiment are shown in Tables 1, 2, 3.

Table 1. Comparison of the results for the group Q^1

Group Q^1: $S(Q_i) = 4\,075\,\text{MB}$	DI	MDI	Diff.
Group execution time $t(J)$ [s]	701	630	**-71**
Indices number	1	1	0
Indices size $S(J)$ [MB]	728	232	**-496**
$r(J_m, Q_m) = \frac{S(J_m)}{S(Q_m)}$.	0.178	0.056	-0.121
Time required to obtain result [s]	20	12\,020	12\,000

Table 2. Comparison of the results for the group Q^2

Group Q^2: $S(Q_i) = 19\,199\,\text{MB}$	DI	MDI	Diff.
Group execution time $t(J)$ [s]	627	535	**-92**
Indices number	4	1	**-3**
Indices size $S(J)$ [MB]	814	296	**-518**
$r(J_m, Q_m) = \frac{S(J_m)}{S(Q_m)}$	0.042	0.015	-0.027
Time required to obtain result [s]	5	54\,000	53\,995

Table 3. Comparison of the results for the group Q^3

Group Q^3: $S(Q_i) = 23\,274\,\text{MB}$	DI	MDI	Diff.
Group execution time $t(J)$ [s]	1264	1252	**-12**
Indices number	7	1	**-6**
Indices size $S(J)$ [MB]	2434	232	**-2202**
$r(J_m, Q_m) = \frac{S(J_m)}{S(Q_m)}$.	0.104	0.009	-0.095
Time required to obtain result [s]	64	28\,000	27\,936

5   Summary

Theoretical considerations related to the group of indices, verified experimentally on a large production database in real conditions produced a similar effect. It is worth noting that because of the analysis of the queries group as a whole, gains occurred in each test case (Q^1, Q^2, Q^3) in the area of group's execution time and memory usage. Time needed to analyze a group of queries and to propose indices is much smaller in case of an index advisor compared with the proposed method. This is due to the way in which the advisor works and how researches and processes group of queries, analyzing only the cost of queries and not the actual time of their execution (feature of MDI). Often, such an analysis is flawed because it is based only on estimates rather than measurements. In DI case, there are not taken into account the relevant elements of the database processing such as: the physical structure of the processing server or other queries groups executed at the same time. The disadvantage of MDI approach is the number of additional iterations required to stabilize the population and achieve the optimum value. However, the MDI may operate during the cyclic queries execution, which does not cause a delay in obtaining results.

The further work will focus on researching the relationship between the queries group columns occurrence and queries execution time and memory gains.

References

1. Barcucci, E., Pinzani, R., Sprugnoli, R.: Optimal selection of secondary indices. IEEE Transactions on Software Engineering 16(1), 32–38 (1990)
2. Bruno, N., Chaudhuri, S.: An online approach to physical design tuning. In: International Conference on Data Engineering, pp. 826–835 (2007)
3. Bruno, N., Chaudhuri, S.: Automatic physical database tuning: a relaxation-based approach. In: ACM SIGMOD International Conference on Management of Data, pp. 227–238 (2005)
4. Caprara, A., Fischetti, M., Maio, D.: Exact and approximate algorithms for the index selection problem in physical database design. IEEE Transactions on Knowledge and Data Engineering 7(6), 955–967 (1995)
5. Kratica, J., Ljubić, I., Tosic, D.: A Genetic Algorithm for the Index Selection Problem. In: Raidl, G.R., et al. (eds.) EvoWorkshops 2003. LNCS, vol. 2611, pp. 280–290. Springer, Heidelberg (2003)
6. Chaudhuri, S., Narasayya, V.: An efficient Cost-Driven Index Selection Tool for MS SQL Server. Very Large Data Bases Endowment Inc. (1997)
7. Sattler, K.-U., Schallehn, E., Geist, I.: Autonomous query-driven index tuning. In: International Database Engineering and Applications Symposium, pp. 439–448 (2004)
8. Schnaitter, K., Abiteboul, S., Milo, T., Polyzotis, N.: On-line index selection for shifting workloads. In: International Workshop
9. Boronski, R., Bocewicz, G., Wójcik, R.: Grouped queries indexing for relational database. In: eKNOW 2013: The Fifth International Conference on Information, Process, and Knowledge Management, Iaria Journals, pp. 123–129 (2013)

10. Boroński, R., Bocewicz, G.: Multi-criteria index selection for grouped SQL queries. In: Kwiecień, A., Gaj, P., Stera, P. (eds.) CN 2013. CCIS, vol. 370, pp. 573–581. Springer, Heidelberg (2013)
11. Kołaczkowski, P., Rybiński, H.: Automatic Index Selection in RDBMS by Exploring Query Execution Plan Space. In: Ras, Z.W., Dardzinska, A. (eds.) Advances in Data Management. SCI, vol. 223, pp. 3–24. Springer, Heidelberg (2009)
12. Boronski, R., Bocewicz, G.: Indices driven mechanism for grouped SQL queries. Pomiary, Automatyka, Robotyka 2, 135–142 (2013)
13. Boronski, R.: Indices Selection for Blocks of Related SQL Queries. Applied Computer Science 8(2), 3–22 (2012)

Author Index